CW00739289

FREEDOM OF TRANSIT AND ACCESS TO GAS PIPELINE NETWORKS UNDER WTO LAW

Gas transit is network-dependent and it cannot be established without the existence of pipeline infrastructure in the territory of a transit state or the ability to access this infrastructure. Nevertheless, at an inter-regional level, there are no sufficient pipeline networks allowing gas to travel freely from a supplier to the most lucrative markets. The existing networks are often operated by either private or state-controlled vertically integrated monopolies who are often reluctant to release unused pipeline capacity to their potential competitors. These obstacles to gas transit can diminish the gains from trade for states endowed with natural gas resources, including developing land-locked countries, as well as undermine WTO Members' energy security and their attempts at sustainable development.

This book explains how the WTO could play a more prominent role in the international regulation of gas transit and promote the development of an international gas market.

Vitaliy Pogoretskyy works as Counsel at the Advisory Centre on WTO Law (ACWL), where he assists developing-country Members of the ACWL and the least-developed countries in the WTO dispute settlement proceedings, and provides to these countries legal advice and training on WTO law. He has also worked at the Rules Division of the WTO Secretariat, FratiniVergano – European Lawyers, and as a civil servant of the Government of Ukraine.

CAMBRIDGE INTERNATIONAL TRADE AND ECONOMIC LAW

Series Editors

Dr Lorand Bartels, *University of Cambridge*
Professor Thomas Cottier, *University of Berne*
Professor William Davey, *University of Illinois*

As the processes of regionalisation and globalisation have intensified, there have been accompanying increases in the regulations of international trade and economic law at the levels of international, regional and national laws.

The subject matter of this series is international economic law. Its core is the regulation of international trade, investment and cognate areas such as intellectual property and competition policy. The series publishes books on related regulatory areas, in particular human rights, labour, environment and culture, as well as sustainable development. These areas are vertically linked at the international, regional and national levels, and the series extends to the implementation of these rules at these different levels. The series also includes works on governance, dealing with the structure and operation of related international organisations in the field of international economic law, and the way they interact with other subjects of international and national law.

Books in the series

Freedom of Transit and Access to Gas Pipeline Networks under WTO Law
Vitaliy Pogoretskyy

Reclaiming Development in the World Trading System
Yong-Shik Lee

Developing Countries and Preferential Services Trade
Charlotte Sieber-Gasser

WTO Dispute Settlement and the TRIPS Agreement: Applying Intellectual Property Standards in a Trade Law Framework
Matthew Kennedy

Establishing Judicial Authority in International Economic Law Edited by Joanna Jemielniak, Laura Nielsen and Henrik Palmer Olsen

Trade, Investment, Innovation and Their Impact on Access to Medicines: An Asian Perspective
Locknie Hsu

The Law, Economics and Politics of International Standardisation
Panagiotis Delimatsis

The WTO and International Investment Law: Converging Systems
Jürgen Kurtz

Export Restrictions on Critical Minerals and Metals: Testing the Adequacy of WTO Disciplines
Ilaria Espa

Optimal Regulation and the Law of International Trade: The Interface between Societal Values and WTO Law
Boris Rigod

The Social Foundations of World Trade: Norms, Community and Constitution
Sungjoon Cho

Public Participation and Legitimacy in the WTO
Yves Bonzon

The Challenge of Safeguards in the WTO
Fernando Piérola

General Interests of Host States in International Investment Law
Edited by
Giorgio Sacerdoti, Pia Acconci, Mara Valenti and Anna De Luca

The Law of Development Cooperation: A Comparative Analysis of the World Bank, the EU and Germany
Philipp Dann

WTO Disciplines on Subsidies and Countervailing Measures: Balancing Policy Space and Legal Constraints
Dominic Coppens

Domestic Judicial Review of Trade Remedies: Experiences of the Most Active WTO Members
Müslüm Yilmaz

International Organizations in WTO Dispute Settlement: How Much Institutional Sensitivity?
Marina Foltea

Public Services and International Trade Liberalization: Human Rights and Gender Implications
Barnali Choudhury

The Law and Politics of WTO Waivers: Stability and Flexibility in Public International Law
Isabel Feichtner

African Regional Trade Agreements as Legal Regimes
James Thuo Gathii

Liberalizing International Trade after Doha
David Gantz

Processes and Production Methods (PPMs) in WTO Law: Interfacing Trade and Social Goals
Christiane R. Conrad

Non-discrimination in International Trade in Services: 'Likeness' in WTO/ GATS
Nicolas Diebold

The Law, Economics and Politics of Retaliation in WTO Dispute Settlement
Edited by
Chad P. Bown and Joost Pauwelyn

The Multilateralization of International Investment Law
Stephan W. Schill

Trade Policy Flexibility and Enforcement in the WTO: A Law and Economics Analysis
Simon A. B. Schropp

FREEDOM OF TRANSIT AND ACCESS TO GAS PIPELINE NETWORKS UNDER WTO LAW

VITALIY POGORETSKYY
Advisory Centre on WTO Law

CAMBRIDGE
UNIVERSITY PRESS

University Printing House, Cambridge CB2 8BS, United Kingdom

One Liberty Plaza, 20th Floor, New York, NY 10006, USA

477 Williamstown Road, Port Melbourne, VIC 3207, Australia

4843/24, 2nd Floor, Ansari Road, Daryaganj, Delhi – 110002, India

79 Anson Road, #06–04/06, Singapore 079906

Cambridge University Press is part of the University of Cambridge.

It furthers the University's mission by disseminating knowledge in the pursuit of education, learning, and research at the highest international levels of excellence.

www.cambridge.org
Information on this title: www.cambridge.org/9781107163645
10.1017/9781316681497

© Vitaliy Pogoretskyy 2017

This publication is in copyright. Subject to statutory exception and to the provisions of relevant collective licensing agreements, no reproduction of any part may take place without the written permission of Cambridge University Press.

First published 2017

A catalogue record for this publication is available from the British Library.

Library of Congress Cataloging-in-Publication Data
Names: Pogoretskiĭ, V. N. (Vitaliĭ Nikolaevich) author.
Title: Freedom of transit and access to gas pipeline networks under WTO law / Vitaliy Pogoretskyy.
Description: Cambridge [UK] ; New York : Cambridge University Press, 2017. |
Series: Cambridge international trade and economic law ; 35
Identifiers: LCCN 2017000017 | ISBN 9781107163645 (hardback)
Subjects: LCSH: Petroleum law and legislation. | Energy industries – Law and legislation. | Petroleum industry and trade. | World Trade Organization. | Foreign trade regulation. | Natural gas pipelines – Security measures.
Classification: LCC K3915 .P64 2017 | DDC 343.09/396–dc23
LC record available at https://lccn.loc.gov/2017000017

ISBN 978-1-107-16364-5 Hardback

Cambridge University Press has no responsibility for the persistence or accuracy of URLs for external or third-party Internet Web sites referred to in this publication and does not guarantee that any content on such Web sites is, or will remain, accurate or appropriate.

To my wife and best friend, Charlene, my parents
and my grandfather, Victor

CONTENTS

FIGURES AND TABLES

FOREWORD

I gladly comply with Vitaliy Pogoretskyy's request to write a foreword by way of introduction to his book, which I previously read in the form of a PhD thesis as an external examiner. This book addresses the very contentious subject of the freedom of gas transit and access to gas pipeline networks, which is not sufficiently explored, and, at first glance, may seem exceedingly technical and complex. Dr Pogoretskyy's presentation of this subject, however, is both interesting and highly thought-provoking, addressing many issues relevant to regulatory authorities, WTO lawyers, energy analysts and generalists in the field of public international law. Energy security, sustainable development, the right of access to the sea for land-locked countries and the risk of the fragmentation of international law are among the topical issues discussed.

Only a handful of WTO rules directly regulates the issue of access to energy networks, including for purposes of gas transit. These rules are not located in the multilateral agreements on trade in goods. They can be found mainly in the positive commitments on trade in services, or in the accession commitments of a few individual WTO Members. Does this mean that the issue of gas transit, which is highly network-dependent, is of little concern to the WTO? Does this issue fall within the realm of WTO provisions regulating trade in goods, services or both? Similar questions can be asked with regard to other areas of network-dependent trade. Dr Pogoretskyy's book examines these questions by providing a comprehensive and original analysis of all major WTO provisions directly and indirectly relevant to the subject of gas transit.

This book makes a particularly useful contribution to the existing literature by placing the discussion of WTO provisions regulating gas transit within the broader context of public international law. As is well known, the interaction between WTO law and public international law has been an important and challenging topic in the academic literature. This book presents a thorough review of many non-WTO legal sources relevant to WTO transit obligations. These sources are then used to interpret WTO law by applying, among others, the principle of systemic integration. In his book, Dr Pogoretskyy demonstrates his in-depth understanding of the methods of treaty interpretation, WTO jurisprudence, public international law and the relevant regulatory context.

Offering an interesting perspective on the relationship between WTO law and public international law, this book in a timely manner fills a very important gap in the literature. I have no doubts it will be a valuable reference source for the on-going discussions on the place of energy in the WTO, as well as for a broader debate on the role of public international law in the WTO legal system.

Prof Dr Peter Van den Bossche
Member, Appellate Body, World Trade Organization, Geneva
Director of Studies, World Trade Institute, Bern
Geneva, July 2016

PREFACE, ACKNOWLEDGEMENTS AND DISCLAIMER

This monograph is based on more than five years of PhD research, inspired by my general interest in energy and the theme of the fragmentation of international law, which, after the landmark rulings in *US – Gasoline, US – Shrimp, Korea – Procurement* and *EC – Approval and Marketing of Biotech Products*, has got its second wind in the WTO case law and is vividly discussed in the academic literature. This monograph explores whether the problem of the fragmentation of international law is also pertinent in the context of the WTO's regulation of energy trade, in particular gas transit. I set out my preliminary ideas on this subject in an article: Vitaliy Pogoretskyy, 'Freedom of Transit and the Principles of Effective Right and Economic Cooperation: Can Systemic Interpretation of GATT Article V Promote Energy Security and the Development of an International Gas Market?', 16(2) *Journal of International Economic Law* (2013) 313, which were further developed into a PhD thesis, defended at the Centre for Energy, Petroleum and Mineral Law and Policy (CEPMLP, the University of Dundee), and, subsequently, into this monograph.

During the period of my research, I spent two and a half years as a full-time PhD researcher at the CEPMLP, and three years working in the WTO Secretariat, Rules Division and the Advisory Centre on WTO Law (ACWL). In all of these places, I found an intellectually stimulating environment for completing this monograph. I also met many great personalities and competent experts in various fields: international lawyers, energy economists, political scientists, energy industry professionals, who, in one way or another, shaped the outcome of this study. Their contributions took different forms, including a thorough and critical review of particular chapters of this monograph, offering valuable and creative advice or sharing professional experience and knowledge while supervising my professional activities, which I subsequently applied to my research. This monograph has greatly benefited from all of these contributions and I would like to thank all of my colleagues and friends for their help and support. I am also profoundly grateful to the

CEPMLP for its PhD Research Scholarship, without which this research would not be feasible.

Among the individuals whom I would like to thank are my PhD supervisors, Professor Melaku Geboye Desta and Professor Joost Pauwelyn. They were excellent mentors and their comprehensive reviews of my PhD chapters were truly priceless. I would also like to thank Professor Peter Van den Bossche, Dr Jacques Hartmann and Dr Sarah Hendry for challenging my ideas during the examination of my PhD thesis and providing critical comments. I am deeply grateful to Professor Van den Bossche for agreeing to write the foreword for this monograph.

At the CEPMLP, I greatly benefited from intellectual discussions with Dr Abba Kolo, Professor Kaj Hobér, Professor Jonathan Stern, Stephen Dow and Dr Sergei Vinogradov, as well as many of my research fellows, including my good friends Dr Daniel Behn and Dr Ana Maria Daza-Clark. Among my former colleagues at the WTO Secretariat, I would like to thank Mark Koulen and Professor Gabrielle Marceau for providing valuable comments on the draft of my article mentioned earlier and Mireille Cossy, Dr Nora Neufeld, Graham Cook and Pierre Latrille for insightful comments on various aspects of my research. I would like to thank my colleagues at the ACWL for their understanding and support throughout the first year and a half of my tenure, during which I had to combine my professional responsibilities with completing my research, and for sharing their profound knowledge and expertise in WTO law that I applied to this study.

Many other friends and colleagues contributed to this research. I am grateful to Dr Yulia Selivanova and Graham Coop for facilitating my access to the negotiating history of the Energy Charter Treaty, and Dr Daniel Behn, Sergii Melnyk, Tetyana Payosova, and Alexander Volkov for reviewing some parts of my study. I would like to thank Dr Carol Ní Ghiollarnáth for her help in editing different versions of this monograph.

Last but not least, my warmest thanks go to my wife, Charlene, my parents, my brother Nikolay, and my grandparents. Without their love and support throughout more than five years of my research, this monograph would not have seen the light of day.

By way of a disclaimer, all errors in this monograph are mine alone and the views expressed in this study should not be attributed to the organisations with which I have been affiliated.

TABLE OF CASES

Permanent Court of International Justice

Short Form (when applicable)	Full Citation
Jurisdiction of the European Commission of the Danube	*Jurisdiction of the European Commission of the Danube between Galatz and Braila*, Collection of Advisory Opinions, Series B. – No. 14, 8 December 1927
	'Lotus', Collection of Judgments, Series A. – No. 10, 7 September 1927
	Oscar Chinn, Collection of Judgments, Orders and Advisory Opinions, Judgment, Series A/B. – No. 63, 12 December 1934
	Polish Postal Service in Danzig, Collection of Advisory Opinions, Series B. – No. 11, 16 May 1925
	Railway Traffic between Lithuania and Poland, Collection of Judgments, Orders and Advisory Opinions, Advisory Opinion, Series A/B. – No. 42, 15 October 1931
	S.S. 'Wimbledon', Collection of Judgments, Series A. – No. 1, 17 August 1923
	S.S. 'Wimbledon', Dissenting Opinion of Judge Schücking, Collection of Judgments, Series A. – No. 1, 17 August 1923, p. 43
River Oder	*Territorial Jurisdiction of the International Commission of the River Oder*, Collection of Judgments, Series A. – No. 23, 10 September 1929

International Court of Justice

Short Form (when applicable)	Full Citation
Elimination of All Forms of Racial Discrimination	*Application of the International Convention on the Elimination of All Forms of Racial Discrimination (Georgia v. Russian Federation)*, Judgment of 1 April 2011: I.C.J. Reports 2011, p. 70
Oil Platforms	*Case Concerning Oil Platforms (Islamic Republic of Iran v. United States of America)*, Judgment of 6 November 2003: I.C.J. Reports 2003, p. 161
	Case Concerning Oil Platforms (Islamic Republic of Iran v. United States of America), Separate Opinion of Judge Higgins, Judgment of 6 November 2003: I.C.J. Reports 2003, p. 161, p. 225
	Case Concerning Oil Platforms (Islamic Republic of Iran v. United States of America), Separate Opinion of Judge Simma, Judgment of 6 November 2003: I.C.J. Reports 2003, p. 161, p. 324
Right of Passage over Indian Territory	*Case Concerning Right of Passage over Indian Territory*, Judgment of 12 April 1960: I.C.J. Reports 1960, p. 6
	Case Concerning Right of Passage over Indian Territory, Dissenting Opinion of Judge Moreno Quintana (translation), Judgment of 12 April 1960: I.C.J. Reports 1960, p. 6, p. 88
	Case Concerning Right of Passage over Indian Territory, Dissenting Opinion of Judge Chagla, Judgment of 12 April 1960: I.C.J. Reports 1960, p. 6, p. 116

(*cont.*)

Short Form (when applicable)	Full Citation
	Case Concerning Right of Passage over Indian Territory, Dissenting Opinion of Judge Fernandes (translation), Judgment of 12 April 1960: I.C.J. Reports 1960, p. 6, p. 123
	Case Concerning Right of Passage over Indian Territory, Separate Opinion of Judge V. K. Wellington Koo, Judgment of 12 April 1960: I.C.J. Reports 1960, p. 6, p. 54
	Continental Shelf (Tunisia/Libyan Arab Jamahiriya), Judgment of 24 February 1982, I.C.J. Reports 1982, p. 18
	Continental Shelf (Tunisia/Libyan Arab Jamahiriya), Separate Opinion of Judge Jiménez de Aréchaga, Judgment of 24 February 1982: I.C.J. Reports 1982, p. 18, p. 100
Navigational and Related Rights	*Dispute Regarding Navigational and Related Rights (Costa Rica v. Nicaragua)*, Judgment of 13 July 2009: I.C.J. Reports 2009, p. 213
	Dispute regarding Navigational and Related Rights (Costa Rica v. Nicaragua), Declaration of Judge Gilbert Guillaume, Judgment of 13 July 2009: I.C.J. Reports 2009, p. 213, p. 290
	Gabčíkovo-Nagymaros Project (Hungary v. Slovakia), Separate Opinion of Vice-President Weeramantry, Judgment of 25 September 1997: I.C.J. Reports 1997, p. 7, p. 88

(*cont.*)

Short Form (when applicable)	Full Citation
	International Status of South-West Africa, Separate Opinion by Sir Arnold McNair, Advisory Opinion of 11 July 1950: I.C.J Reports 1950, p. 128, p. 146
	Legality of the Threat or Use of Nuclear Weapons, Advisory Opinion of 8 July 1996, I.C.J. Reports 1996, p. 226
Maritime Delimitation and Territorial Questions	*Maritime Delimitation and Territorial Questions between Qatar and Bahrain*, Merits, Judgment of 16 March 2001, I.C.J. Reports 2001, p. 40
	Military and Paramilitary Activities in and against Nicaragua (Nicaragua v. United States of America), Judgment of 27 June 1986, I.C.J. Reports 1986, p. 14
	North Sea Continental Shelf, Judgment of 20 February 1969: I.C.J. Reports 1969, p. 3
	Nuclear Tests (New Zealand v. France), Judgment 20 December 1974: I.C.J. Reports 1974, p. 457
Pulp Mills	*Pulp Mills on the River Uruguay (Argentina v. Uruguay)*, Judgment of 20 April 2010: I.C.J. Reports 2010, p. 14
Corfu Channel	*The Corfu Channel Case*, Judgment of 9 April 1949: I.C.J. Reports 1949, p. 4
	The Corfu Channel Case, Separate Opinion of Judge Alejandro Alvarez, Judgment of 9 April 1949: I.C.J. Reports 1949, p. 4, p. 39

GATT Panel Reports

Short Form (when applicable)	Full Citation
Australia – Ammonium Sulphate	Working Party Report, *The Australian Subsidy on Ammonium Sulphate*, GATT/CP.4/39, adopted 3 April 1950, BISD II/188
EEC – Oilseeds	GATT Panel Report, *European Economic Community – Payments and Subsidies Paid to Processors and Producers of Oilseeds and Related Animal-Feed Proteins*, L/6627, adopted 25 January 1990, BISD 37S/86
Germany – Sardines	GATT Panel Report, *Treatment by Germany of Imports of Sardines*, G/26, adopted 31 October 1952, BISD 1S/53
Germany – Starch Duties	GATT Panel Report, *German Import Duties on Starch and Potato Flour*, W.9/178, 16 February 1955, unadopted, BISD 3S/77
Japan – Semi-conductors	GATT Panel Report, *Japan – Trade in Semi-conductors*, L/6309, adopted 4 May 1988, BISD 35S/116
US – Nicaraguan Trade	GATT Panel Report, *United States – Trade Measures Affecting Nicaragua*, L/6053, 13 October 1986, unadopted
US – Sugar Waiver	GATT Panel Report, *United States – Restrictions on the Importation of Sugar and Sugar-Containing Products Applied under the 1955 Waiver and under the Headnote to the Schedule of Tariff Concessions*, L/6631, adopted 7 November 1990, BISD 37S/228
US – Superfund	GATT Panel Report, *United States – Taxes on Petroleum and Certain Imported Substances*, L/6175, adopted 17 June 1987, BISD 34S/136

WTO Panel and Appellate Body Reports

Short Form (when applicable)	Full Citation
Argentina – Footwear (EC)	Appellate Body Report, *Argentina – Safeguard Measures on Imports of Footwear*, WT/DS121/AB/R, adopted 12 January 2000, DSR 2000:I, 515
Argentina – Import Measures	Appellate Body Reports, *Argentina – Measures Affecting the Importation of Goods*, WT/DS438/AB/R/WT/DS444/AB/R/WT/DS445/AB/R, adopted 26 January 2015
Brazil – Retreaded Tyres	Appellate Body Report, *Brazil – Measures Affecting Imports of Retreaded Tyres*, WT/DS332/AB/R, adopted 17 December 2007, DSR 2007:IV, 1527
Canada – Autos	Appellate Body Report, *Canada – Certain Measures Affecting the Automotive Industry*, WT/DS139/AB/R, WT/DS142/AB/R, adopted 19 June 2000, DSR 2000:VI, 2985
Canada – Renewable Energy /Canada – Feed-in Tariff Program	Appellate Body Reports, *Canada – Certain Measures Affecting the Renewable Energy Generation Sector/ Canada – Measures Relating to the Feed-in Tariff Program*, WT/DS412/AB/R/WT/DS426/AB/R, adopted 24 May 2013
China – Auto Parts	Appellate Body Reports, *China – Measures Affecting Imports of Automobile Parts*, WT/DS339/AB/R/WT/DS340/AB/R/WT/DS342/AB/R, adopted 12 January 2009, DSR 2009: I, p. 3
China – Electronic Payment Services	Panel Report, *China – Measures Affecting Electronic Payment Services*, WT/DS413/R, adopted 31 August 2012

(*cont.*)

Short Form (when applicable)	Full Citation
China – Publications and Audiovisual Products	Appellate Body Report, *China – Measures Affecting Trading Rights and Distribution Services for Certain Publications and Audiovisual Entertainment Products*, WT/DS363/AB/R, adopted 19 January 2010, DSR 2010:I, 3
China – Publications and Audiovisual Products	Panel Report, *China – Measures Affecting Trading Rights and Distribution Services for Certain Publications and Audiovisual Entertainment Products*, WT/DS363/R and Corr.1, adopted 19 January 2010, as modified by Appellate Body Report WT/DS363/AB/R, DSR 2010:II, 261
China – Raw Materials	Appellate Body Reports, *China – Measures Related to the Exportation of Various Raw Materials*, WT/DS394/AB/R/WT/DS395/AB/R/WT/DS398/AB/R, adopted 22 February 2012
China – Raw Materials	Panel Reports, *China – Measures Related to the Exportation of Various Raw Materials*, WT/DS394/R/WT/DS395/R/WT/DS398/R/ and Corr.1, adopted 22 February 2012, as modified by Appellate Body Reports WT/DS394/AB/R/WT/DS395/AB/R/WT/DS398/AB/R
China – Rare Earths	Panel Reports, *China – Measures Related to the Exportation of Rare Earths, Tungsten, and Molybdenum*, WT/DS431/R and Add.1/WT/DS432/R and Add.1/WT/DS433/R and Add.1, adopted 29 August 2014, upheld by Appellate Body Reports WT/DS431/AB/R/WT/DS432/AB/R/WT/DS433/AB/R

(cont.)

Short Form (when applicable)	Full Citation
Colombia – Ports of Entry	Panel Report, *Colombia – Indicative Prices and Restrictions on Ports of Entry*, WT/DS366/R and Corr.1, adopted 20 May 2009, DSR 2009:VI, 2535
Dominican Republic – Import and Sale of Cigarettes	Panel Report, *Dominican Republic – Measures Affecting the Importation and Internal Sale of Cigarettes*, WT/DS302/R, adopted 19 May 2005, as modified by Appellate Body Report WT/DS302/AB/R, DSR 2005:XV, 7425
EC – Approval and Marketing of Biotech Products	Panel Reports, *European Communities – Measures Affecting the Approval and Marketing of Biotech Products*, WT/DS291/R/ WT/DS292/R/ WT/DS293/R, Add.1 to Add.9, and Corr.1, adopted 21 November 2006, DSR 2006: III-VIII, 847
EC – Asbestos	Panel Report, *European Communities – Measures Affecting Asbestos and Asbestos-Containing Products*, WT/DS135/R and Add.1, adopted 5 April 2001, as modified by Appellate Body Report WT/DS135/AB/R, DSR 2001:VIII, 3305
EC – Bananas III	Appellate Body Report, *European Communities – Regime for the Importation, Sale and Distribution of Bananas*, WT/DS27/AB/R, adopted 25 September 1997, DSR 1997:II, 591
EC – Bananas III (US) (Article 22.6 – EC)	Decision by the Arbitrators, *European Communities – Regime for the Importation, Sale and Distribution of Bananas – Recourse to Arbitration by the European Communities under Article 22.6 of the DSU*, WT/DS27/ARB, 9 April 1999, DSR 1999:II, 725

(cont.)

Short Form (when applicable)	Full Citation
EC – Bananas III (Article 21.5 – Ecuador)	Panel Report, *European Communities – Regime for the Importation, Sale and Distribution of Bananas – Recourse to Article 21.5 of the DSU by Ecuador*, WT/DS27/RW/ECU, adopted 6 May 1999, DSR 1999:II, 803
EC – Chicken Cuts	Appellate Body Report, *European Communities – Customs Classification of Frozen Boneless Chicken Cuts*, WT/DS269/AB/R, WT/DS286/AB/R, adopted 27 September 2005, and Corr.1, DSR 2005:XIX, 9157
EC – Hormones	Appellate Body Report, *EC Measures Concerning Meat and Meat Products (Hormones)*, WT/DS26/AB/R, WT/DS48/AB/R, adopted 13 February 1998, DSR 1998:I, 135
EC – Hormones (Canada) (Article 22.6 – EC)	Decision by the Arbitrators, *European Communities – Measures Concerning Meat and Meat Products (Hormones), Original Complaint by Canada – Recourse to Arbitration by the European Communities under Article 22.6 of the DSU*, WT/DS48/ARB, 12 July 1999, DSR 1999:III, 1135
EC – IT Products	Panel Reports, *European Communities and its member States – Tariff Treatment of Certain Information Technology Products*, WT/DS375/R/WT/DS376/R/WT/DS377/R, adopted 21 September 2010, DSR 2010:III, 933-DSR 2010:IV, 1567

(*cont.*)

Short Form (when applicable)	Full Citation
EC – Poultry	Appellate Body Report, *European Communities – Measures Affecting the Importation of Certain Poultry Products*, WT/DS69/AB/R, adopted 23 July 1998, DSR 1998:V, 2031
EC – Sardines	Appellate Body Report, *European Communities – Trade Description of Sardines*, WT/DS231/AB/R, adopted 23 October 2002, DSR 2002:VIII, 3359
EC – Seal Products	Panel Reports, *European Communities – Measures Prohibiting the Importation and Marketing of Seal Products*, WT/DS400/R and Add.1 / WT/DS401/R and Add.1, adopted 18 June 2014, as modified by Appellate Body Reports WT/DS400/AB/R/WT/DS401/AB/R
EC – Tariff Preferences	Appellate Body Report, *European Communities – Conditions for the Granting of Tariff Preferences to Developing Countries*, WT/DS246/AB/R, adopted 20 April 2004, DSR 2004:III, 925
EC and Certain Member States – Large Civil Aircraft	Appellate Body Report, *European Communities and Certain Member States – Measures Affecting Trade in Large Civil Aircraft*, WT/DS316/AB/R, adopted 1 June 2011
EC and Certain Member States – Large Civil Aircraft	Panel Report, *European Communities and Certain Member States – Measures Affecting Trade in Large Civil Aircraft*, WT/DS316/R, adopted 1 June 2011, as modified by Appellate Body Report, WT/DS316/AB/R

(*cont.*)

Short Form (when applicable)	Full Citation
Guatemala – Cement I	Appellate Body Report, *Guatemala – Anti-dumping Investigation Regarding Portland Cement from Mexico*, WT/DS60/AB/R, adopted 25 November 1998, DSR 1998:IX, 3767
India – Patents (US)	Appellate Body Report, *India – Patent Protection for Pharmaceutical and Agricultural Chemical Products*, WT/DS50/AB/R, adopted 16 January 1998, DSR 1998:I, 9
Japan – Alcoholic Beverages II	Appellate Body Report, *Japan – Taxes on Alcoholic Beverages*, WT/DS8/AB/R, WT/DS10/AB/R, WT/DS11/AB/R, adopted 1 November 1996, DSR 1996:I, 97
Japan – Film	Panel Report, *Japan – Measures Affecting Consumer Photographic Film and Paper*, WT/DS44/R, adopted 22 April 1998, DSR 1998:IV, 1179
Korea – Procurement	Panel Report, *Korea – Measures Affecting Government Procurement*, WT/DS163/R, adopted 19 June 2000, DSR 2000:VIII, 3541
Mexico – Telecoms	Panel Report, *Mexico – Measures Affecting Telecommunications Services*, WT/DS204/R, adopted 1 June 2004, DSR 2004:IV, 1537
Peru – Agricultural Products	Appellate Body Report, *Peru – Additional Duty on Imports of Certain Agricultural Products*, WT/DS457/AB/R and Add.1, adopted 31 July 2015
US – Anti-dumping and Countervailing Duties (China)	Appellate Body Report, *United States – Definitive Anti-dumping and Countervailing Duties on Certain Products from China*, WT/DS379/AB/R, adopted 25 March 2011

(*cont.*)

Short Form (when applicable)	Full Citation
US – Carbon Steel	Appellate Body Report, *United States – Countervailing Duties on Certain Corrosion Resistant Carbon Steel Flat Products from Germany*, WT/DS213/AB/R and Corr.1, adopted 19 December 2002, DSR 2002:IX, p. 3779
US – Carbon Steel (India)	Appellate Body Report, *United States – Countervailing Measures on Certain Hot-Rolled Carbon Steel Flat Products from India*, WT/DS436/AB/R, adopted 19 December 2014
US – Continued Zeroing	Appellate Body Report, *United States – Continued Existence and Application of Zeroing Methodology*, WT/DS350/AB/R, adopted 19 February 2009, DSR 2009:III, 1291
US – COOL	Appellate Body Reports, *United States – Certain Country of Origin Labelling (COOL) Requirements*, WT/DS384/AB/R / WT/DS386/AB/R, adopted 23 July 2012
US – Corrosion-Resistant Steel Sunset Review	Appellate Body Report, *United States – Sunset Review of Anti-dumping Duties on Corrosion-Resistant Carbon Steel Flat Products from Japan*, WT/DS244/AB/R, adopted 9 January 2004, DSR 2004:I, 3
US – Cotton Yarn	Appellate Body Report, *United States – Transitional Safeguard Measure on Combed Cotton Yarn from Pakistan*, WT/DS192/AB/R, adopted 5 November 2001, DSR 2001:XII, 6027

(*cont.*)

Short Form (when applicable)	Full Citation
US – Countervailing Measures on Certain EC Products	Panel Report, *United States – Countervailing Measures Concerning Certain Products from the European Communities*, WT/DS212/R, adopted 8 January 2003, as modified by Appellate Body Report WT/DS212/AB/R, DSR 2003:I, 73
US – Export Restraints	Panel Report, *United States – Measures Treating Exports Restraints as Subsidies*, WT/DS194/R and Corr.2, adopted 23 August 2001, DSR 2001: XI, 5767
US – FSC (Article 21.5 – EC)	Appellate Body Report, *United States – Tax Treatment for 'Foreign Sales Corporations' – Recourse to Article 21.5 of the DSU by the European Communities*, WT/DS108/AB/RW, adopted 29 January 2002, DSR 2002: I, 55
US – Gambling	Appellate Body Report, *United States – Measures Affecting the Cross-border Supply of Gambling and Betting Services*, WT/DS285/AB/R, adopted 20 April 2005, DSR 2005:XII, 5663 (Corr.1, DSR 2006:XII, 5475)
US – Gambling	Panel Report, *United States – Measures Affecting the Cross-border Supply of Gambling and Betting Services*, WT/DS285/R, adopted 20 April 2005, as modified by Appellate Body Report WT/DS285/AB/R, DSR 2005:XII, 5797
US – Gambling (Article 22.6 – US)	Decision by the Arbitrator, *United States – Measures Affecting the Cross-border Supply of Gambling and Betting Services – Recourse to Arbitration by the United States*

(*cont.*)

Short Form (when applicable)	Full Citation
	under Article 22.6 of the DSU, WT/DS285/ARB, 21 December 2007, DSR 2007:X, 4163
US – Gasoline	Appellate Body Report, *United States – Standards for Reformulated and Conventional Gasoline*, WT/DS2/AB/R, adopted 20 May 1996, DSR 1996:I, 3
US – Hot-Rolled Steel	Appellate Body Report, *United States – Anti-dumping Measures on Certain Hot-Rolled Steel Products from Japan*, WT/DS184/AB/R, adopted 23 August 2001, DSR 2001:X, 4697
US – Line Pipe	Appellate Body Report, *United States – Definitive Safeguard Measures on Imports of Circular Welded Carbon Quality Line Pipe from Korea*, WT/DS202/AB/R, adopted 8 March 2002, DSR 2002:IV, 1403
US – Offset Act (Byrd Amendment)	Appellate Body Report, *United States – Continued Dumping and Subsidy Offset Act of 2000*, WT/DS217/AB/R, WT/DS234/AB/R, adopted 27 January 2003, DSR 2003:I, 375
US – Section 211 Appropriations Act	Panel Report, *United States – Section 211 Omnibus Appropriations Act of 1998*, WT/DS176/R, adopted 1 February 2002, as modified by Appellate Body Report WT/DS176/AB/R, DSR 2002:II, 683
US – Shrimp	Appellate Body Report, *United States – Import Prohibition of Certain Shrimp and Shrimp Products*, WT/DS58/AB/R, adopted 6 November 1998, DSR 1998:VII, 2755
US – Shrimp (Article 21.5 – Malaysia)	Appellate Body Report, *United States – Import Prohibition of Certain Shrimp and Shrimp Products – Recourse to*

(cont.)

Short Form (when applicable)	Full Citation
	Article 21.5 of the DSU by Malaysia, WT/DS58/AB/RW, adopted 21 November 2001, DSR 2001:XIII, 6481
US – Tuna II (Mexico)	Appellate Body Report, *United States – Measures Concerning the Importation, Marketing and Sale of Tuna and Tuna Products*, WT/DS381/AB/R, adopted 13 June 2012
US – Upland Cotton (Article 22.6 – US I)	Decision by the Arbitrator, *United States – Subsidies on Upland Cotton – Recourse to Arbitration by the United States under Article 22.6 of the DSU and Article 4.11 of the SCM Agreement*, WT/DS267/ARB/1, 31 August 2009, DSR 2009:IX, 3871
US – Wool Shirts and Blouses	Appellate Body Report, *United States – Measure Affecting Imports of Woven Wool Shirts and Blouses from India*, WT/DS33/AB/R, adopted 23 May 1997, and Corr.1, DSR 1997: I, 323
US – Zeroing (EC) (Article 21.5 – EC)	Appellate Body Report, *United States – Laws, Regulations and Methodology for Calculating Dumping Margins ('Zeroing') – Recourse to Article 21.5 of the DSU by the European Communities*, WT/DS294/AB/RW and Corr.1, adopted 11 June 2009, DSR 2009:VII, 2911

Other Tribunals

Short Form (when applicable)	Full Citation
	AES Summit Generation Ltd et al. vs. the Republic of Hungary, ICSID, 23 September 2010, http://www.italaw.com/sites/default/files/casedocuments/ita0014_0.pdf, accessed 3 July 2013
Iron Rhine Arbitration	*Belgium v Netherlands (Iron Rhine Arbitration)*, Permanent Court of Arbitration, 24 May 2005, www.pca-cpa.org, accessed 30 March 2012
	Eritrea – Yemen Arbitration, Permanent Court of Arbitration, 3 October 1996, http://www.pca-cpa.org, accessed 30 September 2011
	Faber Case, in Venezuelan Arbitrations of 1903, prepared by Jackson H. Ralston (assisted by W.T.H. Doyle) (Washington, DC: Government Printing Office, 1904) 600
	Lake Lanoux Arbitration, in E. Lauterpacht (ed.), 27 *International Law Reports* (1957) 100
	North Atlantic Coast Fisheries (United States v. Great Britain), Permanent Court of Arbitration, 7 September 1910, http://www.pca-cpa.org, accessed 30 September 2011
	Saluka Investments BV (the Netherlands) v. the Czech Republic, Partial Award, UNCITRAL, 17 March 2006, http://www.italaw.com/sites/default/files/case-documents/ita0740.pdf, accessed 3 July 2013
Aramco Arbitration	*Saudi Arabia v Arabian American Oil Company*, in E. Lauterpacht (ed.), 27 *International Law Reports* (1963) 117
SAIGA	*The M/V 'SAIGA' (No. 2) Case (Saint Vincent and the Grenadines v. Guinea)*, ITLOS, Separate Opinion of Judge Laing, 1 July 1999, http://www.itlos.org/index.php?id=64, accessed 15 May 2014
	Wintershall A.G. et al vs. the Government of Qatar (Partial Award on Liability), 28 *International Legal Materials* (1989) 798

Decisions of National Courts

Full Citation
Case 115/02 Rioglass & Transremar [2003] ECR I-12719
Case 159/94 Commission v France [1997] ECR I-5819
Case 266/81 Siot [1983] ECR 731
Union of India and Others v. Sukumar Sengupta and Others, Supreme Court of India (3 May 1990), in E. Lauterpacht (ed.), 92 *International Law Reports* (1993), 554

ACRONYMS AND ABBREVIATIONS

ASEAN	Association of Southeast Asian Nations
BIT	Bilateral investment treaty
CIS	Commonwealth of Independent States
COMESA	Common Market for Eastern and Southern Africa
CPC	United Nations Provisional Central Product Classification
DSB	Dispute Settlement Body
DSU	Dispute Settlement Understanding
EAC	East African Community
ECOWAS	Economic Community of West African States
ECT	Energy Charter Treaty
EEZ	Exclusive economic zone
EPS	Electronic payment services
EU	European Union
GA	United Nations General Assembly
GATT	General Agreement on Tariffs and Trade
GATS	General Agreement on Trade in Services
ICJ	International Court of Justice
IEA	International Energy Agency
ILC	International Law Commission
ITA	Information Technology Agreement
ITO	International Trade Organization
LNG	Liquefied natural gas
LOSC	The Law of the Sea Convention
MERCOSUR	Mercado Común del Sur (Southern Common Market)
MFN	Most-favoured-nation obligation
NAFTA	North American Free Trade Agreement
OECD	Organisation for Economic Co-operation and Development
PCIJ	Permanent Court of International Justice
SADC	Southern African Development Community
SCM Agreement	Agreement on Subsidies and Countervailing Measures
SCP	South Caucasus Pipeline
STE	State trading enterprise
TBT	Technical barriers to trade

TELECOM	Telecommunications
TRIMS Agreement	Agreement on Trade-Related Investment Measures
TRIPS Agreement	Agreement on Trade-Related Aspects of Intellectual Property Rights
UN	United Nations
UNCLOS	United Nations Conference on the Law of the Sea
UNCTAD	United Nations Conference on Trade and Development
USA	United States of America
VCLT	Vienna Convention on the Law of Treaties
WTO	World Trade Organization
WWI	World War I
WWII	World War II

I

The Topic and Its Importance, the Scope and Structure of This Study, Overview of Relevant Theoretical Issues

> [T]he major obstacle to the development of new supplies [of energy] is not geology but what happens above ground: namely, international affairs, politics, decision-making by governments.
>
> Daniel Yergin[1]

I. Setting the Context

This monograph discusses the regulation of the transit of natural gas via pipelines[2] under the rules of the World Trade Organization (WTO).[3] This issue is becoming increasingly topical in international trade law, which is evident from the growing number of WTO disputes involving trade in energy[4] and the on-going discussions within the WTO

[1] Yergin, Daniel, 'Ensuring Energy Security', 85 *Foreign Affairs* (2006) 69.

[2] The term 'gas' is shorthand for hydrocarbon deposits existing naturally in a gaseous or a mixed gaseous and liquid state, such as methane. Roberts, Peter, *Gas Sales and Gas Transportation Agreements: Principles and Practice*, 2nd ed. (London: Sweet & Maxwell, 2008) at 5. The term 'pipeline' throughout this monograph does not refer exclusively to a line of pipe but also, where applicable, to pumping machinery and apparatus, including valves, compressors, metering and regulator stations. See Vinogradov, Sergei, 'Challenges of Nord Stream: Streamlining International Legal Frameworks and Regimes for Submarine Pipelines', 9 *Oil, Gas & Energy Law Intelligence* (2011), www.ogel.org, accessed 15 February 2014 at 10.

[3] At the time of writing, the WTO had 162 Members, including all major economies, apart from a few states in the Middle East and Central Asia. See World Trade Organization, 'Members and Observers', www.wto.org/english/thewto_e/whatis_e/tif_e/org6_e.htm, accessed 30 November 2015.

[4] Among the most recent disputes only, see Appellate Body Reports, *Canada – Renewable Energy/Canada – Feed-in Tariff Program*; and the request for consultations submitted to the WTO by the Russian Federation in *European Union and Its Member States – Certain Measures Relating to the Energy Sector* (DS476).

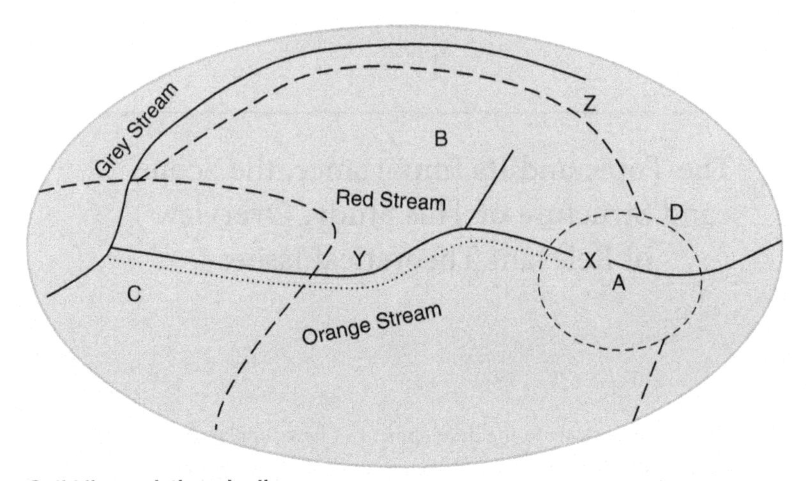

Solid line: existing pipeline
Round dot line: proposed pipeline

Figure 1: Example of Pertinent Problems Arising in Cross-border Gas Transit

community on the need to negotiate specific rules on energy trade.[5] Moreover, this issue raises a number of important legal and technical questions that have not been clarified by WTO panels and examined sufficiently by scholars. This study analyses some of these questions.

The practical challenges to the effective regulation of the pipeline gas transit in the WTO can be illustrated by considering the following hypothetical scenario involving WTO Members[6] A, B, C and D (see Figure 1).

[5] See Lamy, Pascal, *The Geneva Consensus: Making Trade Work for All* (New York: Cambridge University Press) at 117–118; Marceau, Gabrielle, 'The WTO in the Emerging Energy Governance Debate', in J. Pauwelyn (ed.), *Global Challenges at the Intersection of Trade, Energy and the Environment* (Geneva: The Graduate Institute, Centre for Trade and Economic Integration, 2010) 25 at 38–40; Cottier, Thomas et al., 'Energy in WTO Law and Policy', NCCR, Working Paper No 2009/25 (May 2009), http://phase1.nccr-trade.org/images/stories/projects/ip6/IP6%20Working%20paper.pdf, accessed 30 November 2013; Poretti, Pietro and Rios-Herran, Roberto, 'A Reference Paper on Energy Services: The Best Way Forward?', 4 *Oil, Gas & Energy Law Intelligence* (2006), www.ogel.org, accessed 15 February 2014; Evans, Peter, C., 'Strengthening WTO Member Commitments in Energy Services: Problems and Prospects', in P. Sauvé and A. Mattoo (eds.), *Domestic Regulation and Services Trade Liberalization* (Washington, DC: World Bank Publications 2003) 167 at 177–185; Nartova, Olga, *Energy Services and Competition Policies under WTO Law* (Moscow: Infra-M, 2010) at 234–257; and Wälde, Thomas, W. and Gunst, Andreas, J., 'International Energy Trade and Access to Energy Networks', 36 *Journal of World Trade* (2002) 191 at 217.

[6] While some agreements, such as most WTO-covered agreements, use the term 'Member' to denote their member states, others employ the term 'Contracting Parties'. To avoid any confusion, throughout this study, these terms are used interchangeably.

A is an energy-endowed developing country, for which the revenue budget largely depends on gas sales by its domestic producer and exporter company X. Because it is also a land-locked country, for an extensive time period, A has had no choice but to export its gas to transit states B and D. Both B and D also export gas through their domestic companies Y and Z, respectively, to C. The market of C offers a more profitable price for gas than the markets of B and D.

At various inter-governmental conferences and meetings conducted under the auspices of, inter alia, the WTO and the United Nations (UN), A, B, C and D publicly acknowledged the importance of gas trade for increasing the welfare of developing countries and promoting energy security and the sustainable development of states by virtue of access to clean energy sources. They confirmed that, as Members of the WTO, they would comply with WTO rules establishing freedom of transit, including energy transit, namely Article V of the General Agreement on Tariffs and Trade (GATT) 1994. They also vowed to engage in international cooperation on promoting transit via fixed facilities (such as pipelines and grids) and the development of cross-border pipeline infrastructure.

After one such event, private energy-supplying companies incorporated in Member C offered company X and the government of A to buy a significant portion of A's annual gas production for a lucrative price, much higher than the price that A currently receives from B and D. If this situation were to occur in reality, a gas-sales contract could be implemented between X and C's private companies in two ways.

Due to A's land-locked position, gas produced by X would have to travel to the market of C through the territory of B and would, therefore, have to be shipped through the pipeline network operated by company Y, called 'Red Stream'. This would require an arrangement between Y and X or, at an inter-governmental level, between B on one hand and A and/or C on the other. This arrangement would have to stipulate the rules on what will be called in this study a **'third-party access'** right to a given pipeline (*in casu* the Red Stream). If there was no physical capacity in the Red Stream enabling Y to provide third-party access to X or, alternatively, if there was not any pipeline infrastructure in the territory of B, B and A and/or C would have to agree on the ways to expand or construct additional pipeline facilities in the territory of B – that is, **'capacity establishment'** rights. The nature and content of third-party access and capacity establishment rights are discussed in detail in other chapters. However, for the time being, it would suffice to mention that, in the context of gas transit, these rights are essential to the practical implementation of the principle of freedom of transit recognised by WTO law.

In light of the foregoing, if A or its company X were to accept the offer of C's private companies and agree to conclude a gas-sales contract with these companies, A/X would have to seek a separate gas transit agreement with B or Y, setting out conditions for X's third-party access to the Red Stream. In normal circumstances, these conditions would include transit fees covering the cost of transportation and a reasonable profit margin. As explained earlier, if the pipeline capacity in the Red Stream were not sufficient to carry all of A's gas supplies to C, envisaged by the proposed gas-sales agreement, A could offer B to expand its pipeline network by, inter alia, building a new pipeline (e.g. 'Orange Stream'), along the route of the Red Stream at A's expense, to be recouped from the latter's gas sales to C.[7]

However, in a trilateral consultation between A, B and C, B declined to allow X's gas transit via its territory to the market of C. It stated that, first, there is no available pipeline capacity in the Red Stream to allow any gas transit via its territory, because this pipeline is fully utilised by its company Y, which supplies gas to the domestic market as well as to the market of C. Second, B stated that if A wants to build a new pipeline in the territory of B, A must, in exchange, offer reciprocal concessions to B, such as granting Y the right to produce and export gas from the territory of A. B also noted that, while it shares the common objective of the international community[8] to promote energy transit and the development of cross-border pipelines, as expressed in a number of international instruments in which B participates (such as various UN declarations and resolutions), these instruments do not have any legally binding effect on B. Finally, in B's opinion, the lack of a transit route via its territory does not preclude A and C from finding an alternative route, such as a route through the territory of D (by connecting to the 'Grey Stream' pipeline), albeit the latter would be a lengthier, circuitous route.

[7] Assume that, according to A's proposal, the Orange Stream will be operated by Y in the territory of B. The proposal does not in any way affect the property titles of B or Y over pipelines crossing the territory of B. What A merely needs is to lease a certain percentage of the pipeline capacity of the Red Stream and/or the Orange Stream.

[8] In this study, the notion of 'international community' denotes states united by common principles and values as expressed in the 1945 Charter of the United Nations (UN Charter), as well as other commonly recognised principles and rules of public international law. On this notion, see Simma, Bruno and Paulus, Andreas, L., 'The "International Community": Facing the Challenge of Globalization', 9 *European Journal of International Law* (1998) 266.

As explained further in Chapter II, this hypothetical scenario illustrates typical barriers to trade and transit of gas that WTO Members or their private companies face. It demonstrates how, in a technically complex area of gas transit, political considerations can hamper the establishment of a trade connection between suppliers and consumers located in different WTO Members. In turn, such impediments constrain the development of a competitive international gas market, in which the equilibrium between gas supply and demand would be struck by market forces.

The unique feature of gas is that, because of its physical qualities, it can only be shipped via specific modes of transportation, such as pipelines or LNG tankers.[9] However, in any event, to be shipped as an LNG, gas must first be delivered to a vessel via pipelines. Access to the pipeline infrastructure is, therefore, the key prerequisite for the development of an international gas market. Put differently, gas trade and transit are network-dependent – inherently reliant on the existence of adequate infrastructure.

The transit barriers discussed in this chapter would appear to go against the basic logic of free trade promoted by the WTO. As the preamble to the Marrakesh Agreement Establishing the World Trade Organization of 1994 (WTO Agreement) provides, the WTO seeks to establish a progressively liberalised trade system, which aims, inter alia, at expanding the production of and trade in goods and services, while allowing for the optimal use of the world's resources in accordance with the objective of sustainable development.[10] From an economic standpoint, this system is premised on a number of economic theories explaining the gains from trade for the world economy as well as the economies of individual states. These theories include the classical

[9] The particular physical qualities of gas include the relationship between the interdependent variables, such as pressure, volume and temperature of gas, which need to be controlled during its shipment. The term LNG stands for 'liquefied natural gas', which is liquefied through the process of refrigeration to a temperature of −160 °C. Roberts, above n 2 at 6–8. While some other natural resources, such as water and oil, can also be shipped via pipelines, their physical qualities, such as density, are entirely different. For example, even the density of LNG is about half that of oil. See Energy Charter Secretariat, 'Putting a Price on Energy: International Pricing Mechanisms for Oil and Gas' (Brussels: Energy Charter Secretariat, 2007) at 35. On important technical differences between oil pipelines and gas pipelines, see Dow, Stephen, Siddiky, Ishrak A. and Ahmmad, Yadgar K., 'Cross-border Oil and Gas Pipelines and Cross-border Waterways: A Comparison between the Two Legal Regimes', 6 *Journal of World Energy Law and Business* (2013) 107 at 108–109.

[10] WTO Agreement, *WTO Legal Texts*, http://www.wto.org/english/docs_e/legal_e/legal_e.htm, accessed 15 February 2014.

Ricardian theory of 'comparative advantage'[11] and more recent theories of the 'economies of scale in production' and 'enhanced competition'.[12] Based on these theories, the WTO aims to protect the competitive opportunities of its Members in the global market by eliminating discrimination, lowering trade barriers and fostering transparency of trade regulation.[13] Notably, some of these economic theories expressly recognise that the gains from trade for countries endowed with relatively immobile and scarce natural resources, such as natural gas, will, to a large extent, depend on the existence of infrastructure and the distance from world markets.[14] In other words, absent adequate infrastructure, the gains from trade for such countries would be negligent and their participation in the world economy insignificant.

As explained in greater detail in Chapter II, transit barriers to trade in gas may also have other negative consequences, such as the impact on energy security and the sustainable development of states. By hampering access to clean energy sources that are intensively used in industrial processes, not only can a transit state tilt the balance between the competitive opportunities of WTO Members in its own favour, but it can also prevent other Members from shifting their economies away from dirty energy fuels, such as coal.

Another obstacle to trade in gas that can be drawn from the hypothetical scenario is that pipeline facilities in the territory of transit states can be operated by state, private or quasi-private monopolies. These companies, for various reasons, including anti-competitive conduct, may be reluctant to release unused pipeline capacity for access by their potential competitors.[15] However, WTO law concerns only governmental

[11] According to this theory, the best gains from trade could be achieved if WTO Members were to specialise in the production of goods (or services) that they can produce relatively more efficiently than other Members. See Mankiw, Gregory, N. and Taylor, Mark, P., *Economics* (USA: Thomson, 2006) at 51–57.

[12] See the overview of major economic theories explaining the gains from trade in WTO Secretariat, 'World Trade Report 2008: Trade in a Globalizing World' (Geneva, 2008) at xiv–xviii.

[13] World Trade Organization, Secretariat, 'What Is the World Trade Organization?', http://www.wto.org/english/thewto_e/whatis_e/tif_e/fact1_e.htm, accessed 30 September 2011.

[14] WTO Secretariat, 'World Trade Report 2010: Trade in Natural Resources' (Geneva, 2010) at 74.

[15] See ibid. at 167; Wälde, Thomas, W. and Gunst, Andreas, J., 'International Energy Trade and Access to Energy Networks', above n 5 at 210; Slotboom, Marco, 'Recent Developments of Competition Law and the Impact of the Sector Inquiry', in M. M. Roggenkamp and U. Hammer (eds.), *European Energy Law Report VII* (Antwerp: Intersentia, 2010) 97 at 106–108.

measures, as the WTO creates obligations only for its Members.[16] Logically, a purely private measure, which cannot be attributed to a WTO Member, falls outside WTO law.[17] Furthermore, it will be explained in other chapters that international law (including the WTO legal framework) has scant competition rules, which are not sufficient to enforce third-party access against a private pipeline operator abusing its dominant position in the market.[18] The relationship between state and non-state pipeline operators must, therefore, be addressed in this study.

Having said this, it is important to note that private restrictions on the establishment of a gas transit flow are only relevant to the issue of third-party access to pipelines and do not necessarily determine the feasibility of gas transit via a particular state as such. This is because that state can always resort to capacity establishment rights, should it wish to operationalise its transit obligations, but cannot enforce third-party access rights against a domestic private company.

Against the backdrop of our hypothetical scenario, a question can be asked whether the WTO, as a forum governing global trade, has sufficient legal instruments to facilitate the development of an international gas market and ensure energy security and the sustainable development of its Members by requiring WTO Members – transit states – to guarantee freedom of gas transit via their territories.[19] In so doing, WTO law must regulate the essential feature of gas trade, namely its network dependence.

In the following sections, this chapter explains the scope and structure of this study, including specific questions that this study will aim to answer. In Section III, this chapter provides an overview of relevant theoretical issues that permeate the discussion of pipeline gas transit in this monograph.

[16] See GATT Article XXIII (Nullification or Impairment), referring to violation or non-violation complaints resulting from a conduct of *Contracting Parties* (WTO Members). GATT 1994, *WTO Legal Texts*, above n 10. Article 3.1 of the Dispute Settlement Understanding confirms this principle. Dispute Settlement Understanding, ibid.

[17] In WTO law, trade-related measures are broadly divided into governmental and private measures, which are 'counterpoints' excluding each other. See Panel Report, *US – Export Restraints*, para. 8.49. On the rules of attribution and the responsibility of WTO Members for private measures restricting third-party access, see Chapter V, section III.B.

[18] See Wälde and Gunst, above n 5 at 209; and Chapter II, Section III.C.1.

[19] WTO rules relevant to gas transit are discussed in Chapter IV.

II. The Scope and Structure of This Study

A. *The Key Focus of This Study*

1. Regulation of third-party access and capacity establishment rights under WTO law

This monograph aims to answer two main questions. The **first question** (analysed in Chapters III–VI) is how – if at all – WTO law regulates particular aspects of pipeline gas transit: third-party access and capacity establishment rights. It was explained earlier that these rights are essential to effective freedom of gas transit. The nature and precise content of these rights will be further explained in Chapter II. This question can be extended by asking more specifically: (i) whether third-party access rights imply compulsory or negotiated third-party access; (ii) what is the scope of a capacity establishment right (that is, does it imply an obligation for a transit state to construct or expand a pipeline or to allow investment in its territory); and (iii) what is the relationship between the claims to third-party access and capacity establishment? All these questions are answered in the course of this study.

Although some of these questions have previously been addressed by a number of scholars,[20] they are far from resolved. The most obvious gap in the literature on this subject is that previous researchers have always discussed third-party access and capacity establishment rights from a limited perspective of WTO law. However, WTO transit rules are relatively vague and do not expressly regulate these rights. Consequently, it will be seen throughout this study how and why the previous discussions have led to contradicting conclusions and have fallen short of explaining the nature of these rights.

In this study, the boundaries of the previous research will be transcended by analysing the WTO's regulation of third-party access and capacity establishment rights from the perspective of the systemic integration of WTO law with other relevant sources of public international law, [21] in particular, principles of general international law. These principles are freedom of transit, the principle of effective or integrated rights, the principle of economic cooperation, the obligation to negotiate in good faith with a view to achieving a non-discriminatory freedom of transit (derived from the *pactum de contrahendo* nature of GATT Article V:2) and the prohibition of abuse of rights. The WTO provisions

[20] See Chapter IV, Section III.A.1.

[21] See the principle of systemic integration explained in this chapter, Section III.B.

pertaining to the rights that are discussed here are: (i) the obligation to provide freedom of transit under GATT Article V:2 (first sentence); (ii) non-discrimination obligations under GATT Articles V:2 (second sentence) and V:5; and (iii) the non-violation complaint under GATT Article XXIII:1(b). In addition, the relevance of the General Agreement on Trade in Services (GATS) to gas transit and third-party access and capacity establishment rights will be explored.

The assessment of WTO rules governing pipeline gas transit through the prism of systemic integration of public international law sources is the original contribution this study makes to the existing literature on trade in energy. While it has direct implications for trade in gas, the conclusions reached here can be transposed *mutatis mutandis* to other areas of network-bound trade, such as trade in electric power.

The **second question** that will be discussed is how WTO transit rules could be improved through a legislative reform to regulate particular aspects of pipeline gas transit better – namely, third-party access and capacity establishment rights. This question will be analysed in Chapter VI from two perspectives. First, it will be discussed how WTO Members could 'codify' commonly recognised principles of general international law identified for the purposes of answering the first question within the WTO legal system. The term 'codification' is used here as it is generally defined in international law, in particular in the UN Charter, which, in Article 13(1), entrusts the UN General Assembly with encouraging the progressive development of international law and its codification.[22]

[22] Charter of the United Nations of 1945, http://www.un.org/en/documents/charter/, accessed 10 January 2014. The actual process of 'codification' of international law is formally carried out by the International Law Commission (ILC), which prepares a draft report in the form of articles on a particular subject (such as the law of treaties, the law of the sea, most-favoured-nation clause, or state responsibility) and submits this report to the UN General Assembly; the ILC may recommend, inter alia, that the latter body take no action, take note of or adopt the report by resolution, or use it as a basis for an international convention. See Articles 18–24 of the Statute of the International Law Commission, http://legal.un.org/ilc/texts/instruments/english/statute/statute_e.pdf, accessed 15 September 2013. See the analysis of the ILC's methods and achievements in the area of codification in: Villiger, Mark, E., *Customary International Law and Treaties: A Manual on the Theory and Practice of the Interrelation of Sources*, 2nd ed. (The Hague: Kluwer Law International, 1997) at 119–148; Rosenne, Shabtai, *Developments in the Law of Treaties 1945–1986* (New York: Cambridge University Press, 1989) at 1–80; and Roessler, Frieder, *The Legal Structure, Functions and Limits of the World Trade Order: A Collection of Essays* (London: Cameron May, 2000) at 49–67. However, given that the very purpose of codification is to articulate and document the existing principles or rules of general international law, there is no compelling reason why other organisations or institutions specialised in particular areas of international law and consisting of a great

The expression 'codification of international law' means 'the more precise formulation and systematization of rules of international law in fields where there already has been extensive State practice, precedent and doctrine'.[23] In contrast, the term 'progressive development' in Article 13(1) refers to 'the preparation of draft conventions on subjects which have not yet been regulated by international law or in regard to which the law has not yet been sufficiently developed in the practice of States'.[24]

In this connection, relevant rules of the Energy Charter Treaty (ECT) and the 1982 United Nations Convention on the Law of the Sea (LOSC) will be analysed as illustrative examples of possible ways to codify the aforementioned principles of general international law. These legal frameworks address the issues of access to and use of transit facilities in a manner more specific than WTO law, by establishing rights similar to third-party access and capacity establishment. They also appear to reflect the consensus among a large number of the international community members on how these rights must be regulated at a multilateral level.[25]

Second, the possibility of the 'progressive development' of the existing GATS obligations regulating trade in energy services relevant to gas transit will also be explored. This development could be modelled on additional commitments undertaken by Ukraine on pipeline transport, discussed in Chapters IV and VI. Alternatively, as some Members and scholars have proposed, the GATS could provide a legal framework for regulating gas trade in the form of a 'reference paper on energy services', drawing on the relatively successful experience of a similar Reference

number of states, such as the WTO, cannot make their own, albeit informal, contribution to this process. In the end, the ILC has recognised that the texts of international instruments (including multilateral treaties), decisions of international courts and the practice of international organisations constitute evidence of customary international law. See International Law Commission, 'Ways and Means for Making the Evidence of Customary International Law More Readily Available', *Yearbook of the International Law Commission*, vol. 2 (1950) at 368–370, 372. Rosenne also states: 'There is no doubt that to limit the concept of codification convention only to conventions adopted as the consummation of the study of a topic by the International Law Commission would rapidly lead to serious distortions and misconceptions about the substantive content of international law today.' Rosenne, ibid. at 6.

[23] See this term defined in Article 15 of the ILC Statute, above n 22. Villiger defines the expression 'codification of international law' as the 'transformation of an existing rule of international law, *lex lata*, in the form of writing, of *jus non-scriptum* into *jus scriptum*'; as the author states, 'the normative material for codification is customary law.' Villiger, *Customary International Law and Treaties*, above n 22 at 102.

[24] Article 15 of the ILC Statute, above n 22.

[25] At the time of writing, the LOSC and the ECT had 166 and 54 Contracting Parties, respectively. See Appendix 2.

Paper on Telecommunications Services (Telecom Reference Paper), incorporated into some WTO Members' Schedules of Specific Commitments under the GATS as an 'additional commitment'.[26]

There could be other ways of improving WTO rules regulating pipeline gas transit. For example, the rights of access to and use of the public infrastructure (including for transit) are currently discussed in the context of the WTO negotiations on the Annex on Road Freight Transport Services. However, these other options will have to be left for further research.

2. Relationship between WTO law and regional or bilateral agreements governing energy transit

The second question raised in this study does not suggest that the WTO is the only forum that can and should regulate all aspects of cross-border gas trade. Certainly, there are other international regimes,[27] such as the ECT and regional/bilateral trade agreements that deal with energy-specific issues in a more direct way. This study posits that the multilateral and regional/bilateral modes of regulating energy trade and transit are complementary and serve different functions. WTO law sets out general principles and rules of trade acceptable to all Members. The advantage of the WTO over any bilateral or regional trade agreement lies in its compulsory dispute settlement system with the impressive record of compliance with WTO panel and the Appellate Body decisions.[28] In contrast, regional and bilateral agreements often address matters

[26] See World Trade Organization, Council for Trade in Services Special Session, 'Energy Services', Communication from the United States, S/CSS/W/24 (18 December 2000) at 4; Poretti and Rios-Herran, 'A Reference Paper on Energy Services', above n 5; Evans, 'Strengthening WTO Member Commitments in Energy Services', above n 5 at 177–185; and Nartova, above n 5 at 234–257. See GATS specific commitments explained in Chapter IV, Sections II and IV.A.

[27] The term 'international regime' has been commonly defined as: 'a set of "implicit or explicit principles, norms, rules and decision-making procedures around which actors' expectations converge in a given area of international relations"'. Vinogradov, 'Challenges of Nord Stream', above n 2 at 8 (citing Krasner).

[28] The average rate of compliance with WTO panel and the Appellate Body decisions, which is about to exceed 300, is above 80 per cent. See Pauwelyn, Joost, 'The Calculation and Design of Trade Retaliation in Context: What Is the Goal of Suspending WTO Obligations?', in C. P. Bown and J. Pauwelyn (eds.), *The Law, Economics and Politics of Retaliation in WTO Dispute Settlement* (New York: Cambridge University Press, 2010), 34 at 47–48; Davey, William, J. 'Dispute Settlement in the WTO and RTAs: A Comment', in L. Bartels and F. Ortino (eds.), *Regional Trade Agreements and the WTO Legal System* (New York: Oxford University Press, 2006) 343 at 348; and *Summary of Complaints and Resolution*, http://www.worldtradelaw.net/databases/summary.php, accessed 22 August 2014.

that stretch beyond mere economic cooperation, such as political cooperation.[29] However, the enforcement system of regional and bilateral agreements is usually significantly weaker and is used less frequently than that of the WTO.[30]

The enforceability and juridical nature of WTO obligations are the main reasons this study focuses primarily on the improvement of the international regulation of pipeline gas transit under WTO law in particular.[31] However, it will be demonstrated throughout this study that other legal frameworks, especially more specialised bilateral and regional agreements, may also play an important role in developing an international gas market and establishing cross-border pipeline infrastructure. They can be regarded as agreements that implement the basic principles and rules governing transit established under WTO law, but at a more technical and region-specific level. Given that each transit project is unique in terms of the stakeholders involved (e.g., various combinations of public-private partnerships) and applicable national and international rules,[32] it is unlikely that a multilateral treaty could provide a complete legal regime for regulating all aspects of gas transit.

B. *The Scope and Limitations of This Study, Its Structure*

This study contains a systemic analysis of WTO transit rules, which, in addition to WTO law, examines relevant principles of general international law, municipal law, the law of the sea, some aspects of energy and competition law, as well as different regional and bilateral transit regimes. Due to such a broad scope, it is not possible to delve into all aspects of international transit regulation that could potentially inform and contribute to the development of the WTO transit rules. For example, a comprehensive and cutting-edge comparative analysis of national laws regulating easements and servitudes relevant to construction of energy

[29] Cottier, Thomas and Foltea, Marina, 'Constitutional Functions of the WTO and Regional Trade Agreements', in L. Bartels and F. Ortino (eds.), *Regional Trade Agreements and the WTO Legal System* (New York: Oxford University Press, 2006) 43 at 44; and WTO Secretariat, 'World Trade Report 2011: The WTO and Preferential Trade Agreements: From Co-existence to Coherence' (Geneva, 2011) at 94–99.

[30] Davey, above n 28 at 349.

[31] The advantages of developing energy trade rules in the WTO are also discussed in Marceau, 'The WTO in the Emerging Energy Governance Debate', above n 5 at 38.

[32] Dow, Siddiky and Ahmmad, 'Cross-border Oil and Gas Pipelines and Cross-border Waterways', above n 9 at 108–115, 123.

facilities, as well as national competition laws, could be useful, but this has to be left for further research.

Furthermore, this study focuses primarily on the question of how the WTO's transit regulation can facilitate the establishment of *new* transit flows and pipeline connections between WTO Members. Another important question is how WTO law could guarantee the unimpeded transit of *the already existing* gas flows. For example, in the ECT, this goal is achieved through rules prohibiting the interruption of transit until a disagreement between the parties concerned, inter alia, over transit charges, is resolved through expedited conciliation proceedings. Importantly, the conciliator has a mandate to order interim measures with a view to ensuring the continuous gas transit.[33] WTO law does not have dispute settlement rules suitable for resolving energy disputes in emergency situations, such as a gas crisis: its dispute settlement proceedings are lengthier and do not provide for interim measures.[34] The effective regulation of the already established gas transit flows in the WTO would thus require substantial changes in WTO rules, including primarily the WTO dispute settlement mechanism. As this study, in its major part, examines how gas transit is regulated under current WTO rules, this question must be left for separate research.

This study does not discuss the on-going negotiations on the ECT Transit Protocol, although they aim to elaborate on certain aspects of third-party access and capacity establishment rights at a more technical level.[35] These negotiations have lasted for more than a decade. However, they appear to have stalled, in part, due to the announcement by Russia (the key party in these negotiations), in August 2009, of its termination of the provisional application of the ECT, as well as Russia's subsequent proposal of a new (conceptually different) legal framework regulating the

[33] See ECT Articles 7(5)-7(7). Energy Charter Secretariat, 'The Energy Charter Treaty and Related Documents: A Legal Framework for International Energy Co-operation' (Brussels: Energy Charter Secretariat, 2004). See a detailed overview of the ECT conciliation proceedings in Liesen, Rainer, 'Transit under the 1994 Energy Charter Treaty', 17 *Journal of Energy and Natural Resources Law* (1999) 56 at 65–68; and Ehring, Lothar and Selivanova, Yulia, 'Energy Transit', in Y. Selivanova (ed.), *Regulation of Energy in International Trade Law: WTO, NAFTA and Energy Charter* (The Netherlands: Kluwer Law International, 2011) 49 at 91–95.

[34] Azaria notes similar deficiencies in the WTO dispute settlement system in Azaria, Danae, *Treaties on Transit of Energy via Pipelines and Countermeasures* (Oxford: Oxford University Press, 2015) at 168–169.

[35] See the text of the Transit Protocol at: Energy Charter Secretariat, 'Transit Protocol', http://www.encharter.org/index.php?id=37, accessed 23 December 2013.

East–West energy cooperation, embodied in the Draft Convention on Energy Security.[36] Moreover, whereas the utility of the Transit Protocol for energy cooperation between the EU, on one hand, and non-EU neighbouring states (such as Ukraine and Balkan countries), on the other, was pertinent at the beginning of the new millennium, currently, this cooperation appears to have reached a much higher level outside the Energy Charter (namely within the Energy Community).[37] It remains to be seen how the negotiations on the Transit Protocol would adjust to the rapidly changing landscapes of the international energy politics. Currently, however, the ECT negotiations on the Transit Protocol do not appear to have produced results that could influence the international regulation of gas transit beyond the ECT.[38]

Finally, although this study focuses on WTO transit rules, it must be acknowledged that not all important transit and energy-exporting states are Members of the WTO, such as Algeria, Azerbaijan, Belarus, Iran, Iraq, Turkmenistan and Uzbekistan.[39] Consequently, the contribution of this study would fully extend to these countries only after their accession to the WTO.

This monograph is divided into six chapters and a Conclusion. Chapter II outlines the important characteristics of the concept of freedom of transit in international relations, including its interaction with the territorial sovereignty of transit states, and provides a more detailed overview of political and economic aspects of pipeline gas transit. Chapter III provides the context for this research by placing it within a wider framework of the international regulation of transit. Chapter IV analyses the key disciplines in WTO law relevant to transit, third-party access and capacity establishment. Chapter V examines a number of relevant non-WTO legal sources that can inform the nature of WTO transit obligations in the context of pipeline gas, by applying the principle of systemic integration. In Chapter VI, two functions

[36] Russia has never ratified the ECT, but it applied it provisionally. See Belyi, Andrei, Nappert, Sophie and Pogoretskyy, Vitaliy, 'Modernizing the Energy Charter Process? The Energy Charter Conference Road Map and the Russian Draft Convention on Energy Security', 29(3) *Journal of Energy & Natural Resources Law* (2011) 383 at 383–384, 387.

[37] The Energy Community framework is briefly addressed in Chapter III, Section III.D.

[38] On the detailed overview of the negotiations on the Transit Protocol and its key provisions, see Ehring and Selivanova, 'Energy Transit', above n 33 at 95–100; and Konoplyanik, Andrei, A., 'A Common Russia-EU Energy Space (The New EU-Russia Partnership Agreement, *Acquis Communautaire*, the Energy Charter and the New Russian Initiative)', 7 *Oil, Gas & Energy Law Intelligence* (2009), www.ogel.org, accessed 15 February 2014.

[39] See World Trade Organization, 'Members and Observers', above n 3.

that the WTO could fulfil in the regulation of gas transit are discussed: (i) enforcement of WTO transit provisions through the WTO's dispute settlement mechanism; and (ii) the further improvement of WTO transit rules.

III. Overview of Relevant Theoretical Issues

It was explained earlier that one of the important contributions of this study is the assessment of WTO rules governing pipeline gas transit in light of other relevant sources of international law, by applying the principle of systemic integration. Before concluding this chapter, this section elaborates on the meaning of the concept 'principle of general international law', discusses the key functions of the principle of 'systemic integration', and clarifies how different sources of non-WTO law are used in this study to interpret WTO transit obligations.

A. What Is a Principle of General International Law and How Can Its Existence Be Ascertained?

Different scholars have assigned different meanings to the concept 'principle of general international law'.[40] In this study, this term is used to refer to principles of customary international law and general principles of law recognised by civilised nations in the sense of the ICJ Statute Article 38(1)(b) and (c),[41] which generally apply to all nations, whether WTO Members or not.

The role of principles in a legal system, both national and international, is to represent certain values or goals, which direct human conduct towards specific behaviour.[42] Legal theorists argue that, in any

[40] See Tunkin, Grigory, 'Is General International Law Customary Law Only?', 4 *European Journal of International Law* (1993) 53; Gardiner, Richard, *Treaty Interpretation* (New York: Oxford University Press, 2008) at 262.

[41] Arguably, one legal source can take both forms when a principle of law, such as good faith or the prohibition of abuse of rights (*abus de droit*) becomes recognised as customary international law. See Statute of the International Court of Justice (ICJ) of 1945, http://www.icj-cij.org, accessed 10 January 2012.

[42] Mitchell, for example, describes principles as norms or standards used to judge or direct human conduct, derived from a set of premises, such as moral, economic and social premises. If those premises support the principles of a legal system, suggests Mitchell, this lends legitimacy to the system as a whole. Mitchell, Andrew, D., *Legal Principles in WTO Disputes* (New York: Cambridge University Press, 2011) at 10–12. Dworkin also stated that principles incline a decision towards a certain direction. Dworkin, Ronald, *Taking Rights Seriously* (London: Duckworth, 1977) at 35–36.

legal system, principles of law are general and non-conclusive norms, the application of which individually does not lead to any final or concrete result.[43] Unlike rules, which are conclusive norms and apply in an 'all or nothing' manner, principles have weight or a degree to which they are applied, and, therefore, their application requires balancing. Thus, whereas contradicting rules exclude or invalidate each other, competing principles can apply simultaneously to the extent that this is factually and legally possible.[44]

For example, in the context of treaty 'rules' regulating gas transit, if a violation of the obligation to provide freedom of transit under GATT Article V:2 is justified under a general exception relating to the protection of the environment under GATT Article XX(g), this means that, in a particular factual situation (such as an environmental hazard), a rule on 'general exceptions' excludes the application of a transit rule. By contrast, if a principle of sovereignty interacts with another competing principle (inter alia, the principle of economic cooperation or abuse of rights),[45] both principles are valid and continue to co-exist. A judge or a policy maker would have to balance them to determine which principle has more weight in light of particular facts. Only through the process of balancing, which ultimately leads to an articulation of a conditional relation of precedence between the competing principles (that is a rule on their joint application), do these principles acquire a conclusive meaning.[46]

From the normative perspective, the existence of a customary international law principle or a general principle of law must be established in light of the requirements of the ICJ Statute Article 38. In order to ascertain the existence of a principle of *customary international law*, one must demonstrate its two qualifying elements: state practice

[43] Dworkin, above n 42 at 35–36. See also Alexy, Robert, *A Theory of Constitutional Rights*, J. Rivers (trans.) (New York: Oxford University Press, 2010) at 45. Along the same lines, Koskenniemi states that principles are norms of a general character that bring out relevant arguments in support of one or another solution. Koskenniemi, Martti, 'General Principles: Reflexions on Constructivist Thinking in International Law', in M. Koskenniemi (ed.), *Sources of International Law* (Aldershot: Ashgate Darmouth, 2000) 360 at 370, 373.

[44] See Dworkin, above n 42 at 24–27; Alexy, above n 43 at 47–48; Koskenniemi, 'General Principles', above n 43 at 374.

[45] See these principles discussed in Chapter V, Section II.

[46] See Alexy, above n 43 at 51–53. See the examples of 'rules' that can be derived from the joint application of principles relating to transit, including territorial sovereignty, in Chapter VI, Section III.A.1(ii).

(material and detectable element), which must be common and consistent; and *opinio juris sive necessitatis* (immaterial and psychological element).[47] Villiger defines state practice as: 'what states do in their relations with one another' and the 'process of continuous interaction, of continuous demand and response'.[48] He explains '*opinio juris*' as 'the conviction of a State that it is following a certain practice as *a matter of law*'.[49]

The evidence of state practice is normally found in treaties, practice of international organisations (arguably, this may include reports of the ILC and WTO Decisions), non-binding legal instruments (such as UN General Assembly resolutions and declarations), municipal legislation and judgements of national courts.[50] The evidence of *opinio juris* may consist of public statements made on behalf of states, and the conduct of states in connection with the conclusion of treaties and the adoption of international resolutions or declarations (e.g. unanimous voting).[51] Moreover, judgements of international courts and tribunals as well as doctrinal sources are commonly regarded as subsidiary means for the determination of rules of law, including customary international law.[52]

This monograph does not delve into all theoretical aspects of state practice and *opinio juris*, as they are discussed elsewhere.[53] One practical question relevant to this monograph relates to the use of treaties as

[47] Article 38 of the ICJ Statute, above n 41; Shaw, Malcolm, N., *International Law*, 6th ed. (New York: Cambridge University Press, 2008) at 74–75, 77–78; D'Amato, Anthony, A., *The Concept of Custom in International Law* (London: Cornell University Press, 1971) at 49.

[48] Villiger, *Customary International Law and Treaties*, above n 22 at 16.

[49] Ibid. at 48, emphasis added.

[50] See ILC, 'Ways and Means for Making the Evidence of Customary International Law More Readily Available', above n 22; International Law Commission, 'Third Report on Identification of Customary International Law, by Michael Wood, Special Rapporteur', A/CN.4/682 (27 March 2015), paras. 31–61; D'Amato, above n 47 at 79, 85–86; Villiger, *Customary International Law and Treaties*, above n 22 at 15–29; Shaw, above n 47 at 82.

[51] Villiger, *Customary International Law and Treaties*, above n 22 at 50–52; Degan, Vladimir, D., *Sources of International Law* (The Netherlands: Kluwer Law International, 1997) at 175, 194–201; ILC, 'Third Report on Identification of Customary International Law', above n 50, paras. 43, 49; and Shaw, above n 47 at 88.

[52] See Article 38(1)(d) in the ICJ Statute, above n 41; ILC, 'Third Report on Identification of Customary International Law', above n 50, paras. 58–67.

[53] For instance, the relationship between state practice and *opinio juris* is not always clear-cut: how can a state consider a norm as law prior to its crystallisation through state practice? To resolve this theoretical puzzle, D'Amato proposes to analyse customary international law as an articulation of an objective claim of international legality – *opinio juris* (qualitative element), further endorsed by the evidence of state practice (quantitative element). D'Amato, above n 47 at 73–102.

evidence or sources of customary international law. In particular, it is not always clear whether a treaty is a mere source of contractual obligations, or whether it contains norms of general application, independent from treaty law. It will be demonstrated in Chapter III that this question has an important bearing on the status of the principle of freedom of transit in international law, which has been recognised almost universally through a web of multilateral, regional and bilateral treaties in different areas of public international law.

There are, at least, three situations in which treaty obligations may reflect customary international law. First, it appears to be commonly recognised that some 'law-making' or 'codification' conventions accepted by a great number of states may contain norms of customary international law.[54] The most illustrative examples of these conventions are the UN Charter, the 1969 Vienna Convention on the Law of Treaties (VCLT) and the LOSC.[55] The law-making nature of a given convention can be derived from its object and purpose to codify or create a norm of customary international law, confirmed, inter alia, in its preamble or negotiating history. For example, the preamble of the LOSC declares among the purposes of the Convention the establishment of a legal order for the seas and oceans and the achievement of the codification as well as the progressive development of the law of the sea.[56] Second, there is a strong view among scholars and international adjudicators that a rule of international law can become generalised and converted into customary international law if it is included in a great number of conventions setting out identical or similar obligations. This view has been expressed in particular with respect to the development of the minimum standard of

[54] See *North Sea Continental Shelf*, Judgment of 20 February 1969: I.C.J. Reports 1969, p. 3, para. 73; Degan, above n 51 at 175; Jia, Bing B., 'The Relations between Treaties and Custom', 9 *Chinese Journal of International Law* (2010) 81 at 92; and Shaw, above n 47 at 88–89.

[55] The fact that the UN Charter had given expression to certain principles of customary international law was acknowledged in *Military and Paramilitary Activities in and against Nicaragua (Nicaragua v. United States of America)*, Judgment of 27 June 1986, I.C.J. Reports 1986, p. 14, para. 181. Similarly, certain provisions of the LOSC (in particular those governing the regime of the continental shelf) were regarded as the reflection of state practice crystallised into customary international law in *Continental Shelf (Tunisia/Libyan Arab Jamahiriya)*, Judgment of 24 February 1982, I.C.J. Reports 1982, p. 18, para. 101. It is generally recognised that the principles of treaty interpretation codified by the VCLT have attained the status of customary international law. See this chapter, Section III.B.1.

[56] See United Nations Convention on the Law of the Sea, http://www.un.org/depts/los/convention_agreements/texts/unclos/unclos_e.pdf, accessed 30 December 2012.

the treatment of investment through a web of bilateral investment treaties (BITs), and the development of the customary international law on consuls.[57] In this connection, Shaw (citing Hersch Lauterpacht) states that 'one should regard all uniform conduct of governments as evidencing the *opinio juris*, except where the conduct in question was not accompanied by such intention'.[58] Finally, a treaty obligation can evolve into a rule of customary international law when a significant number of states recognises it as such. This recognition can be derived from other evidence, including the unanimous voting of states on a relevant international resolution, and unilateral statements of individual states.

Turning to *general principles of law recognised by 'civilised' nations*, these principles must be either common to most legal systems of the world[59] or they can be inductively generalised from a uniform state practice, reflected in treaties or customary international law.[60] General principles of law are usually classified as:

- principles common to all or a majority of the municipal law systems, such as *res judicata*, and equality of the parties in judicial proceedings;
- principles applicable to international legal relations, such as the principle of the independence and equality of states; and
- principles intrinsic to the idea of law and basic to all legal systems, such as *lex specialis*, 'no one can transfer to another a greater right than he himself possesses' (*'nemo plus iuris transferre potest quam ipse habet'*), and good faith.[61]

[57] See Schwebel, Stephen M., 'Investor–State Disputes and the Development of International Law: The Influence of Bilateral Investment Treaties on Customary International Law', 98 *ASIL Proceedings* (2004) 27 (citing *CME Czech Republic B.V. v. Czech Republic* and *Mondev International v. United States of America*); and International Law Commission, 'Report of the International Law Commission covering the work of its Twelfth Session', *Yearbook of the International Law Commission*, vol. 2 (1960) at 145.

[58] Shaw, above n 47, footnote 59.

[59] Ibid. at 68–72. This can be determined, inter alia, by a comparative study. Lauterpacht, Hersch, *Private Law Sources and Analogies of International Law* (USA: Archon Books, 1970, reprinted) at 85. In the contemporary legal order, there appears to be a lesser emphasis on the notion of '*civilised* nation', since, apart from a few controversial exceptions, any member of the UN is generally considered 'civilised'.

[60] Mitchell, above n 42 at 63–64; Koskenniemi, 'General Principles', above n 43 at 364–365 (citing O'Connel).

[61] Mitchell, above n 42 at 63–64; Schachter, Oscar, *International Law in Theory and Practice* (Boston: Martinus Nijhoff Publishers, 1991); Koskenniemi, 'General Principles', above n 43 at 363–364.

In Schachter's view, this classification is not set in stone and one principle can fall into one or several of these categories.[62]

Both principles of customary international law and principles of law recognised by civilised nations play an important role in interpreting international treaties by filling gaps in their provisions.[63] Precisely because of this common role, for the purposes of this study, a strict distinction between these sources is irrelevant, as they can equally inform the meaning of WTO obligations. This, however, can only be done in accordance with customary principles of treaty interpretation, including the principle of systemic integration, discussed in the following section.

B. *The Role of the Principle of Systemic Integration in Treaty Interpretation*

1. Systemic integration in treaty interpretation

WTO law, in Article 3(2) (General Provisions) of the Dispute Settlement Understanding (DSU), explicitly mandates the interpretation of WTO-covered agreements in accordance with customary rules of interpretation of public international law.[64] This reference is generally understood as a 'gateway' to VCLT Articles 31–33, which codified customary rules of treaty interpretation.[65] In particular, VCLT Article 31(1) establishes a basic principle that a treaty be interpreted 'in good faith in accordance with the ordinary meaning to be given to the terms of the treaty in their context and in the light of its object and purpose'.[66] In addition, among other important interpretation principles codified by the VCLT, there is a principle of 'systemic integration', incorporated in Article 31(3)(c) thereof.[67] This

[62] Schachter, above n 61 at 50.

[63] Mitchell, above n 42 at 9, 22; Koskenniemi, 'General Principles', above n 43 at 365.

[64] Dispute Settlement Understanding, *WTO Legal Texts*, above n 10.

[65] The customary status of the principles of treaty interpretation set out in VCLT Articles 31–33 has been recognised in a consistent line of WTO as well as non-WTO judicial decisions. See Appellate Body Reports in *US – Gasoline* at 17; *EC – Chicken Cuts*, para. 176; *US – Hot-Rolled Steel*, para. 60; and *China – Publications and Audiovisual Products*, para. 348. See also *Case Concerning Oil Platforms (Islamic Republic of Iran v. United States of America)*, Judgment of 6 November 2003: I.C.J. Reports 2003, p. 161, para. 41; and other selected international decisions confirming the customary status of the VCLT rules on treaty interpretation in Weeramantry, Romesh, J., *Treaty Interpretation in Investment Arbitration* (Oxford: Oxford University Press, 2012) at 235, 237.

[66] Vienna Convention on the Law of Treaties of 1969 (VCLT), 1155 *United Nations Treaty Series* (1980) 331.

[67] In *EC and certain member States – Large Civil Aircraft*, the Appellate Body explicitly stated that 'Article 31(3)(c) of the Vienna Convention is considered an expression of the "principle

provision requires that in treaty interpretation 'there shall be taken into account, together with the context: . . . (c) any relevant rules of international law applicable in the relations between the parties'.[68]

The idea behind systemic integration in treaty interpretation is that despite all the emphasis on the document's text as a primary indication of the parties' intention,[69] the latter can rarely be self-sufficient due to its relative indeterminacy. This indeterminacy may result from the fact that different types of legal standards have different degrees of justifying power (such as rules and principles), and normative language inherently contains ambiguous expressions and terminology.[70] Along the same lines, Wälde notes that 'it is in the nature of any negotiations . . . to break the deadlock by proposing an open-ended text and thus delegating the specification to the subsequent process of application'.[71] It has also been suggested that, due to a number of reasons, including restrictions of vocabulary, limits to the calculable capacity of linguistic conventions, different degrees of cultural diversity and the evolution of international law, legal concepts have an indeterminate nature and give only an appearance of certainty.[72]

of systemic integration"'. Appellate Body Report, *EC and certain member States – Large Civil Aircraft*, para. 845. On this principle, see also Marceau, Gabrielle, 'WTO Dispute Settlement and Human Rights', 13 *European Journal of International Law* (2002) 753 at 779–789; Sands, Philippe, 'Treaty, Custom and Cross-fertilization of International Law', 1 *Yale Human Rights and Development Law Journal* (1998) 85; McLachlan, Campbell, 'The Principle of Systemic Integration and Article 31(3)(c) of the Vienna Convention', 54 *International and Comparative Law Quarterly* (2005) 279; French, Duncan, 'Treaty Interpretation and the Incorporation of Extraneous Legal Rules', 55 *International and Comparative Law Quarterly* (2006) 281; McGrady, Benn, 'Fragmentation of International Law or "Systemic Integration" of Treaty Regimes: EC-Biotech Products and the Proper Interpretation of Article 31(3)(c) of the Vienna Convention of the Law of Treaties', 42(3) *Journal of World Trade* (2008) 589; Broude, Tomer, 'Principles of Normative Integration and the Allocation of International Authority: the WTO, the Vienna Convention on the Law of Treaties, and the Rio Declaration', The Hebrew University of Jerusalem, Research Paper No. 07–08 (August 2008); Gardiner, above n 40 at 256–291; and Weeramantry, above n 65 at 88–95.

[68] Vienna Convention on the Law of Treaties, above n 66.

[69] See Wälde, Thomas, W., 'Interpreting Investment Treaties: Experiences and Examples', in C. Binder et al. (eds.), *International Investment Law for the 21st Century: Essays in Honour of Christoph Schreuer* (New York: Oxford University Press, 2009) 724 at 752; Villiger, Mark, E., *Commentary on the 1969 Vienna Convention on the Law of Treaties* (Leiden: Martinus Nijhoff Publishers, 2009) at 426; and Appellate Body Report, *Japan – Alcoholic Beverages II* at 11.

[70] Koskenniemi, Martti, *From Apology to Utopia: The Structure of International Legal Argument* (New York: Cambridge University Press, 2009, reissue) at 37–39.

[71] Wälde, above n 69 at 736–737 and 777.

[72] Fastenrath, Ulrich, 'Relative Normativity in International Law', 4 *European Journal of International Law* (1993) 305.

Because of these reasons, there appears to be a general understanding that a purely semantic interpretation of treaties may not be sufficient, and, therefore, other interpretative means, including the principle of systemic integration, must be resorted to. In fact, such a formalistic interpretation would be contrary to the principle of holistic treaty interpretation, which envisages a harmonious application of interpretation instruments established by the VCLT with a view to rendering the treaty provision legally effective and avoiding interpretations that are mutually contradictory.[73] In this regard, Ehlermann (a former Member of the WTO Appellate Body) notes that, on one hand, the 'security and predictability'[74] of the WTO legal system should be ensured by preserving the carefully crafted balance of rights and obligations struck by WTO agreements. To this end, the Appellate Body has favoured literal interpretation over other methods. On the other hand, states Ehlermann, '[n]obody can, however, deny that the method of literal interpretation has limits, and that the recourse to other interpretative criteria may be necessary, as is recognized by Articles 31 and 32 of the Vienna Convention'.[75]

Furthermore, the principle of systemic integration reflects the idea of unity of international law sources. Amid the topical concern regarding the fragmentation of international law, the principle of systemic integration has gained prominence and has been regarded as a key remedy in the attempts to reconcile various rights and obligations created or emerging within different legal regimes.[76] In this regard, McLachlan points out that, while reference to the document as a starting point is a natural aspect of legal reasoning, in hard cases, the impact of the surrounding legal system must also be taken into account.[77] According to McLachlan, the natural liaison between treaties and other international norms follows inevitably from the very definition of the term 'treaty' in VCLT Article 2(1)(a): an agreement *governed by international law*.[78] Similarly, in Hersch Lauterpacht's view, the intention of the parties must be

[73] Appellate Body Report, *US – Continued Zeroing*, para. 268.

[74] See this principle in Article 3(2) of the DSU. Dispute Settlement Understanding, *WTO Legal Texts*, above n 10.

[75] Ehlermann, Claus-Dieter, 'Six Years on the Bench of the "World Trade Court": Some Personal Experiences as Member of the Appellate Body of the World Trade Organization', 36 *Journal of World Trade* (2002) 605 at 615–617.

[76] International Law Commission (ILC), 'Fragmentation of International Law: Difficulties arising from the Diversification and Expansion of International Law', A/CN.4/L.682 (13 April 2006) at 206.

[77] McLachlan, above n 67 at 287. [78] Ibid. at 280.

interpreted by reference to a broader system of international law, rather than a treaty itself, since the latter does not exist in a legal vacuum.[79] Higgins also noted that: 'International law is not rules. It is a normative *system*.'[80] Pauwelyn explains the principle of systemic integration by stating that '[a]ll norms [inter alia, treaties] are created in the background of already existing norms, in particular norms of general international law'.[81] In *Oil Platforms*, Judge Simma remarks in his Separate Opinion:

> The Court . . . accepts, and rightly so, the principle according to which the provisions of any treaty have to be interpreted and applied in the light of the treaty law applicable between the Parties *as well as of the rules of general international law 'surrounding' the treaty.*[82]

In line with these views, WTO adjudicators have consistently held that WTO law is not to be read in clinical isolation from public international law.[83] In *EC and certain member States – Large Civil Aircraft*, the Appellate Body clarified that the principle of systemic integration 'seeks to ensure that "international obligations are interpreted by reference to their normative environment" in a manner that gives "coherence and meaningfulness" to the process of legal interpretation'.[84] In this way, the Appellate Body recognised that not only does the principle of systemic integration serve the purpose of clarifying the ordinary meaning of the treaty terms, but it also ensures the coherence of international norms within the system of public international law.

In light of these authoritative views, the principle of systemic integration can be said to play a dual role in treaty interpretation by filling normative gaps in treaties and ensuring their harmonious co-existence

[79] Lauterpacht, Hersch, *The Function of Law in the International Community* (Hamden: Archon Books, 1966) at 109.

[80] Higgins, Rosalyn, *Problems and Process: International Law and How We Use It* (Oxford: Clarendon Press, 1994) at 1, emphasis added.

[81] Pauwelyn, Joost, *Conflict of Norms in Public International Law: How WTO Law Relates to Other Rules of International Law* (New York: Cambridge University Press, 2003) at 201.

[82] *Case Concerning Oil Platforms (Islamic Republic of Iran v. United States of America)*, Separate Opinion of Judge Simma, Judgment of 6 November 2003: I.C.J. Reports 2003, p. 161, p. 324 para. 9, emphasis added.

[83] See, inter alia, the very first decision of the Appellate Body in *US – Gasoline*, para. 158.

[84] Appellate Body Report, *EC and certain member States – Large Civil Aircraft*, para. 845. Along the same lines, Gardiner states that 'roles for the rule in article 31(3)(c) may include: . . . (d) resolving conflicting obligations arising under different treaties; [and] (e) taking account of international law developments'. Gardiner, above n 40 at 260. See also Marceau, 'WTO Dispute Settlement and Human Rights', above n 67 at 785–786.

with other sources of international law. That said, its application must conform to certain requirements stipulated in VCLT Article 31(3)(c).

2. General requirements of Article 31(3)(c)

(i) Rules of international law According to VCLT Article 31(3)(c), treaty interpretation shall take into account any relevant rules of international law applicable in the relations between the parties. This provision thus contains three requirements. The first one concerns the sources of international law that can be used in the process of interpretation. On its face, the provision only refers to '*rules* of international law'. However, in practice, this reference has been extended to cover all primary sources of international law listed in Article 38 of the ICJ Statute, namely international conventions, customary international law and general principles of law.[85] In this regard, the Panel in *EC – Approval and Marketing of Biotech Products* states:

> In considering the provisions of Article 31(3)(c), we note, initially, that it refers to 'rules of international law'. Textually, this reference seems sufficiently broad to encompass all generally accepted sources of public international law, that is to say, (i) international conventions (treaties), (ii) international custom (customary international law), and (iii) the recognized general principles of law ... [T]he Appellate Body in *US – Shrimp* made it clear that pursuant to Article 31(3)(c) general principles of international law are to be taken into account in the interpretation of WTO provisions.[86]

Arguably, this reference to international law sources can also encompass commonly recognised municipal law principles that fall under the ICJ Statute Article 38(1)(c). In *US – Shirts and Blouses*, the Appellate Body refers to general principles of civil law, common law and most other jurisdictions to determine the standard of the burden of proof in WTO law, albeit without directly mentioning Article 31(3)(c).[87] Different scholars have also regarded these sources as a useful tool for filling in gaps in international treaties.[88]

[85] See Article 38(1)(a)-(c) in the ICJ Statute, above n 41. On the role of the ICJ Statute in defining the primary sources of international law, see Gardiner, above n 40 at 261, 268; Villiger, *Customary International Law and Treaties*, above n 22 at 8; Villiger, *Commentary on the 1969 Vienna Convention on the Law of Treaties*, above n 69 at 433, and Cheng, Bin, *General Principles of Law as Applied by International Courts and Tribunals* (New York: Cambridge University Press, 2006) at 22.

[86] Panel Report, *EC – Approval and Marketing of Biotech Products*, para. 7.67. See also Appellate Body Report, *US – Shrimp*, para. 158 and footnote 157.

[87] Appellate Body Report, *US – Wool Shirts and Blouses* at 14.

[88] See Lauterpacht, *The Function of Law*, above n 79 at 115; Wälde, above n 69 at 774.

(ii) Rules that are relevant The second requirement Article 31(3)(c) establishes is that the 'rules of international law' must be 'relevant'. In *EC and certain member States – Large Civil Aircraft*, the Appellate Body clarified that '[a] rule is "relevant" if it concerns the subject matter of the provision at issue'.[89] For example, in that dispute, the Appellate Body considered whether various limitations on the government support that could be provided to the large civil aircraft sector under Article 4 of the 1992 Agreement between the then European Economic Communities (now the EU) and the United States constituted a 'relevant rule' for the interpretation of the term 'benefit' in Article 1.1(b) of the Agreement on Subsidies and Countervailing Measures (SCM Agreement). The Appellate Body held that they did not, as Article 4 did not clarify which types of government support could place the recipient in a more advantageous position in the market – the issue central for the assessment of 'benefit' under the SCM Agreement.[90] More recently, in *Peru – Agricultural Products*, the Appellate Body analysed whether the ILC Articles on State Responsibility (namely Articles 20 (Consent) and 45 (Loss of the right to invoke responsibility)) were 'relevant' for the analysis of whether additional duties Peru imposed on agricultural products qualified as 'variable import levies', 'minimum import prices' or 'similar border measures' within the meaning of Article II:1(b) of the GATT and Article 4.2 of the Agreement on Agriculture. The Appellate Body found that that was not the case, as the ILC Articles and the interpreted provisions addressed different issues, and, therefore, had different subject matters.[91]

Similarly, in *Oil Platforms*, in relation to Article 31(3)(c), Judge Higgins stated in her Separate Opinion that the possibility of applying extraneous rules should be assessed through their 'contextual' relation to the interpreted provision, whereby both instruments should have a common subject matter.[92] In this connection, she warned that Article 31(3)(c) 'is not a provision that on the face of it envisages incorporating the entire substance of international law on a topic not mentioned in the clause'.[93]

[89] Appellate Body Report, *EC and certain member States – Large Civil Aircraft*, para. 846.
[90] Ibid., paras. 847–851.
[91] Appellate Body Report, *Peru – Agricultural Products*, paras. 5.103–104.
[92] *Case Concerning Oil Platforms (Islamic Republic of Iran v. United States of America)*, Separate Opinion of Judge Higgins, Judgment of 6 November 2003: I.C.J. Reports 2003, p. 161, p. 225 para. 46.
[93] Ibid.

However, one important challenge in determining 'relevance' is that the concepts of 'contextuality' or 'subject matter' are not free from ambiguities. It is clear that few international legal instruments (especially treaties) will have identical subject matters.[94] Therefore, an overly narrow interpretation of these concepts may reduce the utility of the principle of systemic integration for treaty interpretation.

(iii) Rules that are applicable in the relations between the parties: relevance of treaties The third requirement is that the 'rules of international law' must be 'applicable in the relations between the parties'. The key element here is the term 'parties', which may relate either to parties to a dispute or a treaty. If the latter interpretation is adopted, then an external source can only be taken into account if it is applicable in the relations between *all* parties to the interpreted treaty. Thus, the integration of the WTO Agreement with other treaties (even if they are multilateral treaties concluded by the overwhelming majority of the international community members) would be highly problematic. This is especially so taking into account a large membership of the WTO (162 Members), which includes non-state Members, such as the EU and Hong Kong. The only relevant sources in the interpretation of WTO provisions would thus be customary international law and general principles of law (i.e. general international law).[95]

The meaning of the term 'parties' in Article 31(3)(c) has brought about heated debates among analysts. In *EC – Biotech Products*, the Panel defined this term pursuant to VCLT Article 2(1)(g) as meaning a *party to a treaty*.[96] This reading of Article 31(3)(c), however, resulted in harsh criticism of the Panel Report, including criticism by the ILC Study Group on Fragmentation.[97] In *EC and certain member States – Large Civil*

[94] The ILC has acknowledged the difficulty in defining common subject matters in its analysis of the principle of *lex specialis* in ILC, 'Fragmentation of International Law', above n 76 at 62.

[95] Among principles of general international law that WTO adjudicators have referred to through VCLT Article 31(3)(c) are the principle of permanent State sovereignty over natural resources and the principles of State responsibility. See Panel Report, *China – Raw Materials*, paras. 7.377–378; Panel Report, *China – Rare Earths*, para. 7.262; and Appellate Body Report, *US – Anti-dumping and Countervailing Duties (China)*, paras. 308–309.

[96] See the definition of the term 'party' in Article 2(1)(g) of the Vienna Convention on the Law of Treaties, above n 66; and Panel Report, *EC – Approval and Marketing of Biotech Products*, paras. 7.69–72.

[97] See ILC, 'Fragmentation of International Law', above n 76 at 227; Broude, above n 67, footnote 79; McGrady, above n 67 at 613; and Merkouris, Panos. 'Keep Calm and Call (no, not Batman but …) Articles 31–32 VCLT: A Comment on Istrefi's Recent Post on

Aircraft, the Appellate Body shifted the emphasis of the requirement in Article 31(3)(c) from the term 'party' to '*relevant* rules', pointing out that, to satisfy the requirement of systemic integration, a treaty that is 'relevant' could be taken into account for the purposes of interpreting a WTO-covered agreement, even if not all WTO Members have given their explicit consent to be bound by that treaty.[98] In this connection, the Appellate Body clarified that 'a delicate balance must be struck between, on the one hand, taking due account of an individual WTO Member's international obligations and, on the other hand, ensuring a consistent and harmonious approach to the interpretation of WTO law among all WTO Members'.[99] Arguably, this clarification by the Appellate Body paves the way for a closer relationship between WTO law and other treaties, leaving the door open for integrating the relevant rules under those treaties into the interpretation of WTO law.

Recall that prior to *EC and certain member States – Large Civil Aircraft*, the Appellate Body had already used non-WTO treaties (such as the relevant provisions of the LOSC), albeit for the limited purpose of interpreting the 'ordinary meaning' of terms in WTO-covered agreements within the meaning of VCLT Article 31(1), including generic terms the meaning of which had been considered to have evolved.[100] In addition, in quite a number of cases, WTO adjudicators appear to have used non-WTO legal sources, including treaties, as evidence of facts.[101] For

R.M.T. v. The UK', *EJIL: Talk!* (19 June 2014), http://www.ejiltalk.org/keep-calm-and-call-no-not-batman-but-articles-31-32-vclt-a-comment-on-istrefis-recent-post-on-r-m-t-v-the-uk/, accessed 20 June 2014.

[98] Appellate Body Report, *EC and certain member States – Large Civil Aircraft*, paras. 844–845.

[99] Ibid.

[100] See Appellate Body Report, *US – Shrimp*, paras. 130–132, in which the Appellate Body interpreted the ordinary meaning of the term 'exhaustible natural resources' in GATT Article XX(g). See also Appellate Body Report, *US – FSC (Article 21.5 – EC)*, paras. 141–148, in which the Appellate Body interpreted the ordinary meaning of the term 'foreign-source income', as used in footnote 59 of the SCM Agreement. In this regard, in *EC – Approval and Marketing of Biotech Products*, the Panel clarified: 'other relevant rules of international law may in some cases aid a treaty interpreter in establishing, or confirming, the ordinary meaning of treaty terms in the specific context in which they are used. *Such rules would not be considered because they are legal rules, but rather because they may provide evidence of the ordinary meaning of terms in the same way that dictionaries do.* They would be considered for their informative character'. Panel Report, *EC – Approval and Marketing of Biotech Products*, para. 7.92, footnote omitted, emphasis added.

[101] See Cook, Graham, *A Digest of WTO Jurisprudence on Public International Law Concepts and Principle* (Cambridge: Cambridge University Press, 2015) at 82–85.

example, in *US – Shrimp*, the Appellate Body took into account various non-WTO legal instruments, including treaties, supporting its finding that the achievement of the United States' policy objective of the protection and conservation of highly migratory species of sea turtles under GATT Article XX(g) 'demands concerted and cooperative efforts on the part of the many countries'.[102] Similarly, in *EC – Seal Products*, the Panel considered provisions of a non-WTO treaty in assessing whether the difference in purpose between commercial hunts and hunts by Inuit and other indigenous communities justifies the distinction the EU draws between these two hunts under its ban on seal products.[103] Finally, as explained in Section III.A, treaties can constitute evidence of customary international law. In this way, non-WTO treaties do not enter the process of systemic integration directly for the purpose of interpreting WTO law. Rather, they assist in identifying and clarifying other sources of public international law that can be integrated in treaty interpretation through Article 31(3)(c).

In light of the foregoing, the question can be asked whether treaties such as the ECT and the LOSC could be used in the interpretation of WTO transit obligations through VCLT Article 31(3)(c). On one hand, these treaties could provide useful guidance to a WTO panel in a dispute involving gas transit, land-locked states or restrictions on the access to transit infrastructure, and, for this reason, they are arguably 'relevant'.[104] On the other hand, some WTO Members are not Contracting Parties to these treaties.

Given that the question of whether and how non-WTO treaties can be used in WTO dispute settlement proceedings through Article 31(3)(c) has not been resolved in WTO jurisprudence,[105] in this study, the ECT and the LOSC are used mainly as: (i) 'evidence of facts' (demonstrating

[102] Appellate Body Report, *US – Shrimp*, para. 168.

[103] This issue was relevant to the Panel's analysis of whether the EU's measure was consistent with the non-discrimination obligation under Article 2(1) of the Agreement on Technical Barriers to Trade (TBT Agreement). See Panel Report, *EC – Seal Products*, paras. 7.292–295.

[104] See the detailed analysis of these treaties in Chapter III, Sections III.B.2(i) and III.C(2) as well as Chapter VI, Sections III.2 and III.3.

[105] Notably, in *Peru – Agricultural Products*, the Appellate Body appears to suggest that provisions of a regional trade agreement (especially that with an insignificant number of parties) will unlikely constitute a 'rule that is applicable in the relations between [WTO Members]', as it cannot be used in establishing the *common intention* of WTO Members underlying the interpreted provision under WTO law. In the Appellate Body's view, the contrary approach would suggest that 'WTO provisions can be interpreted differently, depending on the Members to which they apply and on their rights and obligations

common approaches among the members of the international community to operationalising freedom of transit in the context of network-bound trade); (ii) evidence of the development of principles of general international law in the area of transit; and (iii) sources elucidating the 'evolving' meaning of generic terms in WTO law, such as 'freedom' of transit under GATT Article V:2. In addition, in Chapter VI, they are discussed as illustrative examples of possible ways to improve WTO transit rules. These treaties are not used as 'rules of international law applicable in the relations between the parties' within the meaning of VCLT Article 31(3)(c).

(iv) What is the role of soft law in treaty interpretation? Strictly speaking, non-binding legal or political instruments (soft law) do not qualify as 'relevant rules of international law applicable in the relations between the parties', because of their non-binding nature.[106] However, it has often been argued that these instruments, especially when accepted by the consensus or overwhelming majority of participating states, reflect common intentions of the international community (including WTO Members) regarding the preferred way to deal with a particular problem of common concern. In this regard, Fastenrath states that certain soft law instruments, including international resolutions or declarations, may be 'effected through an extensive or restrictive application of already established legal rules'.[107] Along the same lines, in *Continental Shelf (Tunisia/Libyan Arab Jamahiriya)*, Judge Jiménez de Aréchaga stated that certain 'trends', even if not accepted as applicable law, can provide 'an indication of the direction' in which rules of international law should be interpreted.[108]

under an FTA to which they are parties'. Appellate Body Report, *Peru – Agricultural Products*, para. 5.106.

[106] Villiger, *Commentary on the 1969 Vienna Convention on the Law of Treaties*, above n 69 at 433.

[107] Fastenrath, above n 72 at 314. See also the discussion of the informative function of soft law in WTO jurisprudence in Footer, Mary, 'The Return to "Soft Law" in Reconciling the Antinomies in WTO Law', 11 *Melbourne Journal of International Law* (2010), 241 at 261–263; and a general overview of the role soft law plays in international law, including treaty interpretation, in Desta, Melaku, G., 'Soft Law in International Law: An Overview', in A. K. Bjorklund and A. Reinisch (eds.), *International Investment Law and Soft Law* (Cheltenham: Edward Elgar Publishing, 2012) 39.

[108] See *Continental Shelf (Tunisia/Libyan Arab Jamahiriya)*, Separate Opinion of Judge Jiménez de Aréchaga, Judgment of 24 February 1982: I.C.J. Reports 1982, p. 18, p. 100, paras. 33–34.

In *Nuclear Tests*, the International Court of Justice (ICJ or the World Court) held that in public international law, a unilateral act that pursues the intention to confer on a recipient a specific right (inter alia, a political promise) has the character of a legal undertaking and, therefore, is binding on the state making it. The underlying principle of general international law behind the obligatory nature of a unilateral act is the principle of good faith.[109] Like a unilateral act, soft law instruments also often contain promises to engage in a specific conduct, which must be fulfilled in good faith. Therefore, even if these instruments are not in themselves rules of international law, they cannot be disregarded in interpreting existing rules.

As explained in the previous section, WTO jurisprudence itself recognises that various legal instruments, even if not binding on all WTO Members, such as bilateral treaties, can be used for interpreting the 'ordinary meaning' of the terms in the sense of VCLT Article 31(1).[110] Furthermore, WTO adjudicators have often made references to soft law instruments to support their factual findings.[111] Finally, as in the case of treaties, non-binding international resolutions or declarations can constitute evidence of customary international law.[112]

In light of the foregoing, like non-WTO treaties, relevant soft law instruments are used in this study mainly as 'evidence of facts' (demonstrating common approaches among the members of the international community to operationalising freedom of transit in the context of network-bound trade), evidence of customary international law, and sources clarifying the 'evolving' meaning of the term 'freedom' of transit under GATT Article V:2. They are not used as 'rules of international law' in the sense of VCLT Article 31(3)(c).

(v) **Shall be taken into account** If an external legal source satisfies the requirements under VCLT Article 31(3)(c), it must be *taken into account*. The term 'shall be taken into account' determines the normative weight of a source integrated into the treaty interpretation process.[113] Notably,

[109] *Nuclear Tests (New Zealand v. France)*, Judgment 20 December 1974: I.C.J. Reports 1974, p. 457, paras. 46, 49.

[110] Panel Report, *EC – Approval and Marketing of Biotech Products*, para. 7.92.

[111] For example, the Appellate Body consulted the Rio Declaration on Environment and Development and Agenda 21 in Appellate Body Report, *US – Shrimp*, para. 168. In *EC – Seal Products*, the Panel consulted the UN Declaration on the Rights of Indigenous Peoples. See Panel Report, *EC – Seal Products*, paras. 7.292, 7.295.

[112] See Section III.A.

[113] Appellate Body Report, *EC and certain member States – Large Civil Aircraft*, para. 841.

while Article 31(3)(c) establishes an obligation to have recourse to a 'relevant rule of international law', it does not obligate the interpreter forcefully to integrate it into the interpreted treaty. The question thus arises as to what extent a 'relevant rule' must be 'taken into account', and what exactly does this term mean for the purposes of systemic integration?

Sands notes that '[i]nternational law does not provide a general definition of the term "take account"'.[114] In his view, the '[o]rdinary usage suggests that the formulation is stronger than "take into consideration" but weaker than "apply"'.[115] Sands appears to seek clarification of this term in a general presumption of coherence and harmonious interpretation of norms within the international law system.[116] The meaning of this term was to some extent clarified in *US – Anti-dumping and Countervailing Duties (China)*, where the Appellate Body held: 'Rules of international law within the meaning of Article 31(3)(c) are one of several means to ascertain the common intention of the parties to a particular agreement reflected in Article 31 of the *Vienna Convention*.'[117] Consequently, in light of the Appellate Body's ruling, a relevant non-WTO rule must be taken into account to the extent that it clarifies the common intention of WTO Members, as any other interpretative tool in Article 31.[118] This reading of the phrase 'shall be taken into account' conforms to the principle of holistic treaty interpretation, which, as already explained, envisages a harmonious application of the VCLT interpretation instruments.[119]

However, as was also noted earlier, another important function of the principle of systemic integration is to ensure coherence in the international legal system, within which WTO law is one element.[120] Thus, when treaty interpretation leads to two results, one of which is in conflict and the other of which is consistent with a non-WTO legal instrument applicable in relations between all or an overwhelming majority of WTO Members, this function can be fulfilled by selecting the second

[114] Sands, above n 67 at 103. [115] Ibid. [116] Ibid.

[117] Appellate Body Report, *US – Anti-dumping and Countervailing Duties (China)*, para. 312.

[118] See a similar view in Gardiner, above n 40 at 266; Marceau, 'WTO Dispute Settlement and Human Rights', above n 67 at 784; and Villiger, *Commentary on the 1969 Vienna Convention on the Law of Treaties*, above n 69 at 432.

[119] Appellate Body Report, *US – Continued Zeroing*, para. 268.

[120] Appellate Body Report, *EC and certain member States – Large Civil Aircraft*, para. 845; Gardiner, above n 40 at 260; Marceau, 'WTO Dispute Settlement and Human Rights', above n 67 at 785–786; Sands, above n 67 at 103.

option.[121] This approach would be in line with a generally accepted presumption against a conflict of norms in public international law,[122] and it will likely strike the delicate balance between individual WTO Members' international obligations and a consistent and harmonious approach to the interpretation of WTO law, as mentioned by the Appellate Body in *EC and certain member States – Large Civil Aircraft*.[123]

3. 'Systemic integration' beyond treaty interpretation?

So far, the principle of systemic integration has only been discussed from the perspective of treaty interpretation. However, the question can be asked whether this principle may have broader relevance outside the VCLT, for example, relating to the law-making processes within different legal regimes. In other words, does this principle fulfil an informative or guiding function in the codification and development of a coherent system of public international law, where different legal sources must interact in a harmonious manner?[124]

While for the purposes of this study it is not necessary to reach a definitive answer to this question, certain roots and expressions of the principle of systemic integration or its corollaries can be found in legal sources outside the VCLT. For example, since the creation of the League of Nations, if not earlier, the international community has made a great deal of effort to progressively develop and codify international law with a view to promoting its consolidation into an integrated normative system, aiming at achieving common objectives.[125] The preamble to the UN Charter establishes the goal of an integral relationship between international law sources by declaring the fundamental determination of nations 'to establish conditions under which justice and respect for the obligations arising from treaties and *other* sources of international law can be maintained'.[126]

[121] See a similar view in Panel Report, *EC – Approval and Marketing of Biotech Products*, para. 7.69.

[122] See Pauwelyn, above n 81 at 240–244. The conflict of norms can be defined as a situation where two or more norms (treaty or custom) contain obligations that contradict each other and, therefore, cannot be complied with simultaneously. See Appellate Body Report, *Guatemala – Cement I*, para. 65.

[123] Appellate Body Report, *EC and certain member States – Large Civil Aircraft*, paras. 844–845.

[124] As stated earlier, the objectives to codify and encourage the progressive development of international law are set out in Article 13(1) of the UN Charter, above n 22.

[125] See overview in Brölmann, Catherine, 'Law-Making Treaties: Form and Function in International Law', 74 *Nordic Journal of International Law* (2005) 383 at 383–387.

[126] UN Charter, above n 22, emphasis added. Along the same lines, the UN GA Resolution 2625 of 24 October 1970 recognises that 'every State has the duty to fulfil in good faith *its*

Broude argues that, like in VCLT Article 31(3)(c), the principle of systemic integration can be found in the concept of 'sustainable development', which incorporates different concerns of the international community, namely economic, environmental and social concerns.[127] In his view, from this perspective, the principle is not necessarily limited to systemic, legal, normative and judicial reasoning dimensions, but it also relates to the institutional realm.[128]

Finally, even if, from a purely formalistic viewpoint, the principle of systemic integration fits better within the sphere of treaty interpretation rather than a harmonious and coherent codification and progressive development of international law, while being engaged in the last two processes, all states have a duty to adhere to the common principles of international law recognised by the international community to which they belong.[129] In this regard, Degan states that general principles play an important role and are 'a stimulating factor' in the codification and progressive development of international law.[130] The failure to pay due regard to principles of general international law in international law-making may create risks of fragmentation of the system of international law into conflicting legal regimes (inter alia, trade versus the law of the sea) or levels of international governance (multilateral versus bilateral or regional).

In light of this, one could argue that the principle of systemic integration (or its corollaries) fulfils two broad functions in the international community. First, it mandates coherent and harmonious treaty interpretation, taking into account other relevant sources of public international law, including principles of general international law. Second, it encourages codification as well as further development of international law based on the commonly recognised objectives and principles of the international community. The second function must be kept in mind in improving and elaborating WTO transit rules regulating pipeline gas transit.

obligations under the generally recognised principles and rules of international law.' See United Nations, GA Resolution 2625 of 24 October 1970, 'Declaration on Principles of International Law concerning Friendly Relations and Co-operation among States in accordance with the Charter of the United Nations', in United Nations, *General Assembly Resolutions*, http://www.un.org/documents/resga.htm, accessed 15 November 2015, emphasis added.

[127] Broude, above n 67 at 34. [128] Ibid. at 37.

[129] See the principle of '*pacta sunt servanda*' codified, inter alia, in VCLT Article 26. Vienna Convention on the Law of Treaties, above n 66. This principle is discussed in Chapter V, Section II.A.

[130] Degan, above n 51 at 102 and 107.

IV. Summary and Concluding Remarks

This chapter discussed a hypothetical scenario illustrating the key obstacles to trade in gas, including transit, which hamper the development of an international gas market and undermine the idea of 'free trade' promoted by the WTO. These obstacles are: (i) the lack of adequate access to pipeline infrastructure; and (ii) restrictive business practices of state, private, or quasi-private pipeline-operating companies, hindering third-party access to their pipeline facilities. These obstacles are elaborated on in more detail in Chapter II. However, it was mentioned in this chapter that, from a practical viewpoint, they can be addressed through the regulation of what is termed in this study 'third-party access' and 'capacity establishment' rights.

This chapter further explained the key questions this study would aim to answer. These questions can be summarised as follows: (i) how does WTO law regulate particular aspects of pipeline gas transit: namely third-party access and capacity establishment rights; and (ii) how could WTO transit rules be improved to regulate these aspects better? The first question will be answered by interpreting WTO provisions relevant to gas transit in light of the principles of treaty interpretation, as set out in the VCLT, and in particular the principle of systemic integration. The second question will be answered by exploring two options: (i) codifying commonly recognised principles of general international law identified for the purposes of answering the first question; and (ii) expanding the existing GATS obligations regulating trade in energy services relevant to gas transit.

Finally, this chapter devoted special attention to the discussion of two theoretical legal concepts that will be used throughout this whole study and warrant clarification, namely 'principles of general international law' and the principle of 'systemic integration'.

II

Freedom of Transit and Pipeline Gas: Overview of Relevant Legal, Political and Economic Aspects

I. Introduction

As discussed in the previous chapter, this study analyses the principle of freedom of transit in international law, including WTO law, with particular focus on the transit of natural gas via pipelines. Cross-border transit is an important element of international trade. Since the creation of the GATT 1947 and the WTO,[1] international trade promotion policies have mainly focused on import- and export-related restrictions, rarely addressing transit.[2] This fact, however, by no means undermines the importance of transit rules for international trade. As Uprety points out, in the current world of globalisation and economic integration, in order to compete in the world market, and thereby promote economic development and reduce poverty, all countries should be assured the same level of access to this market, including via transit.[3] As GATT/WTO rules have been consolidated and developed into a rule-based legal system, supported by a binding dispute settlement mechanism,[4] administered by a WTO political body – the Dispute Settlement Body (DSB) – the regulation of transit under WTO law has increasingly been attracting the attention of its Members.

[1] The relationship between the GATT 1947 and the WTO is explained in Chapter III, Section III.B.2(ii).

[2] For instance, within the frameworks of the WTO and the GATT 1947, its predecessor, there has only been one panel review of a transit restriction – *Colombia – Ports of Entry* – out of 333 GATT-related complaints just between 1995 and 2011, although some alleged violations of freedom of transit, including one dispute over the road transit of oil and oil products between Croatia and Slovenia, were settled at a diplomatic level. See WTO Dispute Settlement Tables and Statistics, http://www.worldtradelaw.net, accessed 30 September 2011; Panel Report, *Colombia – Ports of Entry*; and World Trade Organization, 'Article V of GATT 1994 – Scope and Application', TN/TF/W/2 (12 January 2005).

[3] Uprety, Kishor, *The Transit Regime for Landlocked States: International Law and Development Perspectives* (Washington, DC: The World Bank, 2006) at 3.

[4] Lamy, Pascal, 'The Place of the WTO and Its Law in the International Legal Order', 17 *European Journal of International Law* (2006) 969 at 972.

At a diplomatic level, transit barriers were extensively discussed within the WTO negotiations on trade facilitation, launched in July 2004 as part of the Doha round. These negotiations resulted in a new Agreement on Trade Facilitation, signed in Bali.[5] This Agreement contains important technical amendments to WTO transit rules, although it does not directly regulate freedom of transit.[6]

This chapter defines the meaning of the principle of freedom of transit and outlines the essential characteristics of transit in general, as well as those of the pipeline gas transit specifically. It explains the role the transit of gas plays in modern international trade, by promoting energy security and sustainable development.

It is argued in this chapter that the most practical way to ensure the long-term energy security and sustainable development of WTO Members is through the creation of a competitive international gas market. This would require putting in place adequate cross-border infrastructure connecting gas producers and consumers. From a legal perspective, this goal can be achieved only through the adequate regulation of third-party access and capacity establishment rights under WTO transit rules – the only mechanisms to ensure the access to pipeline networks. This chapter elaborates on these mechanisms and discusses their importance for gas transit and the development of an international gas market.

II. Freedom of Transit and Territorial Sovereignty

A. *Freedom of Transit and Its Essential Characteristics*

The principle of freedom of transit is articulated differently in particular legal regimes. In WTO law, it is expressed in GATT Article V:2 as the obligation to provide 'freedom of transit through the territory of each contracting party, via the routes most convenient for international transit,

[5] The negotiations on the Agreement on Trade Facilitation were concluded by virtue of a WTO Ministerial Decision of 7 December 2013. The Agreement has not yet entered into force. As the Ministerial Decision of 7 December 2013 clarifies, it will be inserted into Annex 1A of the WTO Agreement through a Protocol of Amendment and pursuant to amendment procedures set out in Article X:3 of the WTO Agreement. See World Trade Organization, Ministerial Conference, Agreement on Trade Facilitation, Ministerial Decision of 7 December 2013, WT/MIN(13)/36, WT/L/911 (11 December 2013)/ Preparatory Committee on Trade Facilitation, Agreement on Trade Facilitation, WT/L/931 (15 July 2014). On the negotiations on trade facilitation, see World Trade Organization, 'Trade Facilitation', http://www.wto.org/english/tratop_e/tradfa_e/tradfa_e.htm, accessed 30 November 2013.

[6] See Chapter IV, Section V.C.

for traffic in transit to or from the territory of other contracting parties'.[7] The term 'traffic in transit' is defined in GATT Article V:1 as:

> [g]oods (including baggage), and also vessels and other means of transport ... when the[ir] passage across ... [the] territory [of a transit State], with or without trans-shipment, warehousing, breaking bulk, or change in the mode of transport, is only a portion of a complete journey beginning and terminating beyond the frontier of the contracting party across whose territory the traffic passes.[8]

The following figure illustrates the four major scenarios of traffic in transit captured by this definition.[9] It shows a hypothetical transit route (T) between WTO Members A and C (where applicable) through a transit state B. In scenario 2, transit takes place either between A and its enclave or between C and a land-locked country A. In scenarios 3 and 4, although a transit state B can theoretically be bypassed via alternative routes, transit through the territory of B is the most convenient transit route.

Transit has certain important characteristics that may influence its international regulation. First of all, it is a highly technical matter with particular features that vary depending on the transportation mode: road, railway, maritime, air or pipeline transit.[10] Due to this characteristic, the international regulation of transit is highly fragmented.[11]

Second, transit involves crossing the territory of a state that does not receive any direct gains from trade (that is, import or export).[12] That said, a transit state, due to its geographical location and depending on how

[7] GATT 1994, *WTO Legal Texts*, http://www.wto.org/english/docs_e/legal_e/legal_e.htm, accessed 15 February 2014.

[8] Ibid.

[9] It is noteworthy that this definition does not cover the passage of goods outside the WTO Member's *territory* but within its jurisdictional control, such as via 'exclusive economic zone' (EEZ) and the 'continental shelf'. See these terms defined in Articles 55 and 76 of the LOSC. United Nations Convention on the Law of the Sea, http://www.un.org/depts/los/convention_agreements/texts/unclos/unclos_e.pdf accessed 30 December 2012. On the international regulation of the operation of pipelines in these particular areas, see Vinogradov, Sergei, 'Challenges of Nord Stream: Streamlining International Legal Frameworks and Regimes for Submarine Pipelines', 9 *Oil, Gas & Energy Law Intelligence* (2011), www.ogel.org, accessed 15 February 2014 at 34–47.

[10] See League of Nations, 'Verbatim Reports and Texts relating to the Convention on Freedom of Transit', Barcelona Conference (Geneva, 1921) at 7.

[11] See Chapter III.

[12] The indirect gains may include additional jobs and improved infrastructure, which can be used for internal commerce. Liesen, Rainer, 'Transit under the 1994 Energy Charter Treaty', 17 *Journal of Energy and Natural Resources Law* (1999) 56 at 60–62.

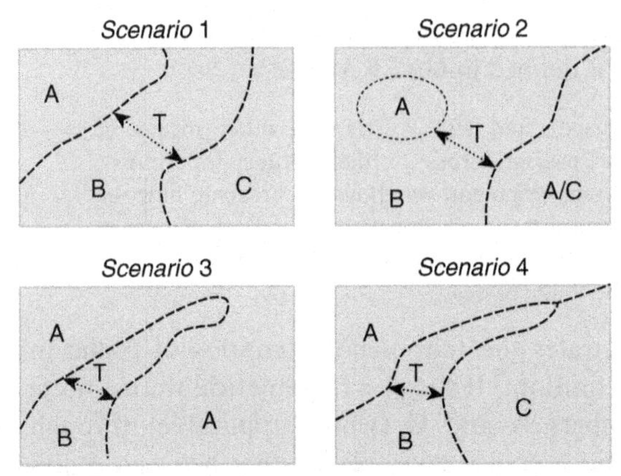

Figure 2: Transit Scenarios Regulated by WTO Law

strategic this location is (see for example scenario 2 in Figure 2), can cut off any trade flow crossing its territory or subject this flow to unreasonable demands and conditions, unless international rules may restrict this ability.

Another important characteristic of transit is its general economic and humanitarian implications, potentially extending beyond particular trade counterparts. In the circumstances in which a transit route is a vital (or even the only) channel of international or global commerce,[13] the transit of goods through that route can become an important issue for the entire international community or many of its members. This could be one of the reasons why, at different times, transit matters have been linked to international peace and friendly relations among nations. For example, the link between transit and peace was mentioned explicitly in the preparatory work of the Barcelona Convention and the Statute on Freedom of Transit of 1921 (Barcelona Convention), discussed in Chapter III of this monograph.[14] Similarly, according to Hudec, after World War II (WWII), there was a clear understanding that breaking the

[13] See, among others, the cases of maritime passage through strategic maritime canals, such as the Suez and Panama Canals.

[14] The preparatory work goes as follows: '[G]oods in transit crossing national territory but originating in and destined for places outside that territory cannot be impeded or restricted at the will of the State exercising sovereignty over such territory, without resultant injury to other States, not only inadmissible in itself, but the effects of which, in the form of reprisals, are liable to extend far beyond the States originally concerned, and, by the material inconvenience inflicted and the spirit of rivalry engendered, may contribute in no small measure to disturb the peace of the world.' League of Nations, 'Freedom of Transit', above n 10 at 283.

channels of trade, which, by implication, includes transit, made a significant contribution to the outbreak of the war.[15] Not surprising, transit-related concerns were addressed in many treaties concluded after World War I (WWI) and WWII.

The last important characteristic of transit is its peculiar interaction with territorial sovereignty of a transit state. On this account, de Visscher called the relationship between freedom of transit and sovereignty: 'the antagonism of two great ideas that, at all times, entered into conflict'.[16] While the relationship between sovereignty and various international rights (such as market access and non-discrimination) is a common characteristic of all international economic treaties (including the WTO Agreement), the underlying principle of these treaties is the principle of 'reciprocity'.[17] By contrast, the right to transit has often been claimed as a general, non-reciprocal right.[18]

B. *Sovereignty and Freedom of Transit*

The principle of territorial sovereignty has undergone a significant evolution from the old Westphalian model of international law, with the independent nation at the core,[19] to modern times, marked by globalisation and economic liberalisation. When interpreting the principle of sovereignty, the League of Nations' Permanent Court of International Justice (PCIJ), the predecessor of the ICJ, stated: 'all that can be required of a State is that it should not overstep the limits which

[15] Hudec, Robert, E., *The GATT Legal System and World Trade Diplomacy*, (New York: Praeger Publishers, Inc., 1975) at 4–5 (citing Clair Wilcox).

[16] De Visscher, Charles, *Le Droit International des Communications* (Paris: Université de Gand, 1921) at 6.

[17] See Roessler, Frieder, *The Legal Structure, Functions and Limits of the World Trade Order: A Collection of Essays* (London: Cameron May, 2000) at 95; Van den Bossche, Peter and Zdouc, Werner, *The Law and Policy of the World Trade Organization*, 3rd ed. (New York: Cambridge University Press, 2013) at 429.

[18] In the context of the negotiations on the law of the sea, whereas in initial treaties, such as the Convention on the High Seas, the right of transit for land-locked countries was subjected to the condition of reciprocity, this condition was dropped in subsequent treaties governing the same matter, such as the LOSC. See Dupuy, René-Jean and Vignes, Daniel (eds.), *A Handbook on the New Law of the Sea*. vol. 1 (Boston: Hague Academy of International Law, Martinus Nijhoff Publishers, 1991) at 514, 519; and Chapter III, Section III.B.2.

[19] See the 1648 Peace Treaty between the Holy Roman Emperor and the King of France and their respective Allies (Treaty of Westphalia), laying down the concept of sovereignty, discussed in Nussbaum, Arthur, *A Concise History of the Law of Nations* (New York: The Macmillan Company, 1954, revised edition) at 115.

international law places upon its jurisdiction; within these limits, its title to exercise jurisdiction rests in its sovereignty'.[20] Since that time, sovereignty has been often construed as an inherent right of every state to regulate (exercise jurisdiction), independent of a treaty.[21]

In the academic literature, Jackson defines sovereignty as an allocation of powers between different levels of governance: national and international.[22] This allocation has to be executed through legal means, such as treaties or general international law, which delimit the boundaries of what sovereigns can or cannot do vis-à-vis each other. As Brownlie puts it: 'If an international law exists, then the dynamics of state sovereignty can be expressed in terms of law, and, as states are equal and have legal personality, sovereignty is in a major aspect a relation to other states (and to organizations of states) defined by law.'[23] Koskenniemi, while discussing different doctrines of sovereignty, essentially describes this concept as a state's sphere of liberty determined by the normative environment with which it interacts.[24]

In light of these definitions of sovereignty, in the realm of international legal relations, the right to regulate must be balanced against other international rights and obligations. As the WTO Appellate Body has stated in a number of cases, states exercise their right to regulate by, inter alia, entering into international agreements, such as the WTO Agreement.[25] Inevitably, such international agreements reduce the scope of the independent regulatory power of each individual state by demarcating its outer limits. The duty to regulate in a manner consistent with international obligations undertaken under treaties and general international law stems from the general principle of *pacta sunt servanda*, incorporated, inter alia, in the VCLT.[26] It is further confirmed by the Declaration on Principles of International Law

[20] *'Lotus'*, Collection of Judgments, Series A. – No. 10, 7 September 1927 at 19.

[21] Appellate Body Report, *China – Publications and Audiovisual Products*, para. 222; Appellate Body Report, *China – Raw Materials*, para. 272.

[22] Jackson, John H., *Sovereignty, the WTO and Changing Fundamentals of International Law* (New York: Cambridge University Press, 2006) at 72.

[23] Brownlie, Ian, *Principles of Public International Law*, 7th ed. (New York: Oxford University Press, 2008) at 289.

[24] Koskenniemi, Martti, *From Apology to Utopia: The Structure of International Legal Argument* (New York: Cambridge University Press, 2009, reissue) at 224–233.

[25] See Appellate Body Report, *China – Raw Materials*, para. 272; Appellate Body Report, *Japan – Alcoholic Beverages II* at 15.

[26] See Article 26 in Vienna Convention on the Law of Treaties of 1969, 1155 *United Nations Treaty Series* (1980) 331. The principle of *pacta sunt servanda* is also addressed in Chapter V, Section II.A.

concerning Friendly Relations and Cooperation among States, which was recognised unanimously by the UN and adopted in the General Assembly (GA) Resolution 2625 (XXV) of 24 October 1970. Pursuant to this Declaration, each state has a duty to fulfil its obligations assumed under the generally recognised principles and rules of international law in good faith.[27] The same duty is mentioned in the preamble of the UN Charter.[28]

The principles of freedom of transit and sovereignty, while interacting in the case of transit, protect different legitimate values. Freedom of transit is associated with the ideas of interdependence and globalisation, freedom of commerce and communications, as well as, in some cases, access to the common heritage of mankind – the high seas. The principle of sovereignty guards independence and the territorial integrity of nations, including their exclusive right to regulate.

In reality, the ideas of freedom of transit and territorial sovereignty have clashed largely because they were taken to their extremes. Throughout this study, it will be examined, in the context of gas transit, whether and how these ideas could be reconciled by striking a balance between, on one hand, the right to transit gas and, on the other hand, the legitimate regulation of the practical application of this right. Similarly, de Visscher notes that this is the task of international law – to reconcile the inherently conflicting ideas of free transit and sovereignty.[29] In the area of international economic relations, this reconciliatory function can be seen in the underlying policy objectives of the GATT to provide a viable legal framework for economic cooperation that would preserve peace among nations, as well as to promote world economic welfare and sustainable development.[30] The next section discusses the particular features of pipeline gas that need to be taken into account in striking this balance.

[27] See United Nations, GA Resolution 2625 of 24 October 1970, 'Declaration on Principles of International Law concerning Friendly Relations and Co-operation among States in accordance with the Charter of the United Nations', in United Nations, *General Assembly Resolutions*, http://www.un.org/documents/resga.htm, accessed 15 November 2015. This Declaration is also discussed in Chapter V, Section II.B.

[28] Charter of the United Nations of 1945, http://www.un.org/en/documents/charter/, accessed 10 January 2014.

[29] De Visscher, above n 16 at 7.

[30] Jackson, John, H., 'History of the General Agreement on Tariffs and Trade', in R. Wolfrum et al. (eds.), *WTO – Trade in Goods* (Leiden: Martinus Nijhoff Publishers, 2011) 1 at 3–7.

III. Freedom of Transit and Pipeline Gas

A. *The Role of Gas Transit in International Trade*

1. Gas transit as a means to promote energy security and sustainable development of WTO Members

The issue of gas transit has emerged as a new and highly topical area of international trade after a number of disruptions of gas supply in Eastern Europe. Cameron describes one of the latest disruptions, the 2009 crisis between Russia (gas exporter) and Ukraine (transit state), involving the EU as a gas importer, as 'the most serious gas supply crisis to hit the EU in its history'.[31] It was a contractual dispute between Russia and Ukraine on how much 'fuel gas' Ukraine was entitled to offtake under the contractual arrangement between these countries to maintain the pressure in Ukraine's pipeline system and thereby to ensure its safe and stable operation. The disruption resulted in serious humanitarian consequences and economic losses, especially in the Eastern European part of the EU, which was fully dependent on Russian gas.[32]

This was not the only energy supply crisis in Eastern Europe. A similar crisis occurred between the same parties in 2006.[33] Among energy transit disruptions in other regions, Stevens, for example, refers to the disruption of the crude oil transit in May 1970, when a Syrian bulldozer working on a telephone cable damaged the Tapline pipeline connecting Saudi Arabia with Lebanon through Syria. Reportedly, Syria refused to allow the pipeline to be repaired without a new transit agreement being signed envisaging higher transit fees.[34]

[31] Cameron, Peter, D., 'The EU and Energy Security: A Critical Review of the Legal Issues', in A. Antoniadis, R. Shultze and E. Spaventa (eds.), *The European Union and Global Emergencies: A Law and Policy Analysis* (Oxford: Hart Publishing, 2011) 125 at 131.

[32] See Pirani, Simon, Stern, Jonathan and Yafimava, Katja, 'The Russo–Ukrainian Gas Dispute of January 2009: A Comprehensive Assessment', Fall (XVIII) *Energy Politics* (2009), 4.

[33] See Yergin, Daniel, *The Quest: Energy, Security, and the Remaking of the Modern World* (New York: Penguin Books, 2012) at 340–341; and Konoplyanik, Andrei, A., 'Russian–Ukrainian Gas Dispute: Prices, Pricing and ECT', 4(4) *Oil, Gas & Energy Law Intelligence* (2006), www.ogel.org, accessed 15 February 2014. See also other gas supply disruptions involving former Soviet Union States discussed in Yafimava, Katja, *The Transit Dimension of EU Energy Security: Russian Gas Transit across Ukraine, Belarus, and Moldova* (New York: Oxford Institute for Energy Studies, Oxford University Press, 2011) at 81–92.

[34] Stevens, Paul, 'Cross-border Oil and Gas Pipelines: Problems and Prospects' (UNDP & World Bank Energy Sector Management Assistance Programme, 2003) at 78.

As these examples of energy transit disruptions demonstrate, the issue of pipeline gas transit has a direct link with energy security. Energy security can be defined broadly as: '"Access," "Availability" and "Acceptability" of energy sources'.[35] However, in modern trade, pipeline gas transit also has an important connection with the issue of sustainable development. The concept of 'sustainable development' is generally understood as development that addresses economic, environmental and social concerns.[36] The WTO explicitly recognises the principle of sustainable development in the WTO Agreement, the WTO Ministerial Decision on Trade and Environment and the 2001 Doha Ministerial Declaration.[37]

The reason for the strong link between the effective gas transit, on one hand, and energy security and sustainable development, on the other hand, lies in the fact that natural gas is the cleanest fuel among hydrocarbons, used in a wide range of industrial processes.[38] In his bestseller,

[35] See WTO Press Release, the transcript of the video address of WTO Director-General (DG) Pascal Lamy to the 2010 World Energy Congress, www.wto.org/english/news_e/sppl_e/sppl169_e.htm, accessed 30 March 2012. For a more detailed discussion of the concept of energy security, see Haghighi, Sanam, S., *Energy Security: The External Legal Relations of the European Union with Major Oil- and Gas-Supplying Countries* (Portland: Hart Publishing, 2007) at 9–35; Pogoretskyy, Vitaliy and Melnyk, Sergii, 'Energy Security, Climate Change and Trade: Does the WTO Provide for a Viable Framework for Sustainable Energy Security?', in P. Delimatsis (ed.), *Research Handbook on Climate Change and Trade Law* (Cheltenham: Edward Elgar Publishing Ltd., 2016) 233; and Yafimava, above n 33 at 12.

[36] See Principles 4, 5 and 12 in 'Rio Declaration on Environment and Development', Rio de Janeiro (3–14 June 1992), http://www.unep.org/documents.multilingual/default.asp?documentid=78&articleid=1163, accessed 15 November 2015; *Belgium v Netherlands (Iron Rhine Arbitration)*, Permanent Court of Arbitration, 24 May 2005, www.pca-cpa.org, accessed 30 March 2012, para. 59 (in which the Arbitral Tribunal recognised the principle of 'sustainable development' as a principle of general international law); and *Gabčíkovo-Nagymaros Project (Hungary v. Slovakia)*, Separate Opinion of Vice-President Weeramantry, Judgment of 25 September 1997: I.C.J. Reports 1997, pp. 7, 88 at 89, 92–94. See also Broude, Tomer, 'Principles of Normative Integration and the Allocation of International Authority: the WTO, the Vienna Convention on the Law of Treaties, and the Rio Declaration', The Hebrew University of Jerusalem, Research Paper No. 07–08 (August 2008) at 34; Marceau, Gabrielle and Morosini, Fabio C., 'The Status of Sustainable Development in the Law of the World Trade Organization' (8 November 2011), http://ssrn.com/abstract=2547282, accessed 22 December 2015 at 64–65; and Voigt, Christina, *Sustainable Development as a Principle of International Law: Resolving Conflicts between Climate Measures and WTO Law* (Leiden: Martinus Nijhoff Publishers, 2009) at 38–42.

[37] See *WTO Legal Texts*, above n 7.

[38] See International Energy Agency, 'Tracking Industrial Energy Efficiency and CO2 Emissions' (Paris: IEA Publications, 2007).

The Quest, Yergin describes the role gas plays – and will continue to play in the coming decades – in the international manufacturing competitiveness as follows:

> Natural gas is a fuel of the future. World consumption has tripled over the last thirty years, and demand could grow another 50 percent over the next two decades. Its share of the total energy market is also growing. Three decades ago world consumption on an average-equivalent basis was only 45 percent that of oil; today it is about 70 percent. The reasons are clear: It is a relatively low-carbon resource. It is also a flexible fuel that could play a larger role in electric power, both for its own features and as an effective – and indeed necessary – complement to greater reliance on renewable generation.[39]

Pascal Lamy (the former WTO Director-General) emphasised the relationship between energy transit and energy security in his address to the 2010 World Energy Congress as follows: 'Energy security also implies that energy supply must be allowed to expand and to travel more readily from countries where there is surplus, to countries where there is demand (just as with any other natural resource).'[40] The UN GA Resolution 63/210 of 19 December 2008 'Reliable and Stable Transit of Energy and Its Role in Ensuring Sustainable Development and International Cooperation' links the goals of reliable energy transportation and transit and sustainable development by declaring that 'stable, efficient and reliable energy transportation, as a key factor of sustainable development, is in the interest of the entire international community'.[41]

2. Network dependence of gas trade as a key challenge for the development of a competitive international gas market

The development of an international gas market in which the equilibrium between supply and demand is struck in the same manner as in any other competitive market appears to be the most practical long-term solution to the problems of energy security and sustainable development

[39] Yergin, *The Quest*, above n 33 at 343. On the same subject, see also Yergin, Daniel, 'The Global Impact of US Shale' (8 January 2014), http://www.project-syndicate.org/commentary/daniel-yergin-traces-the-effects-of-america-s-shale-energy-revolution-on-the-balance-of-global-economic-and-political-power, accessed 21 January 2014; and Clark, Pilita and Oliver, Christian, 'EU Energy Costs More than Twice Those of US', *Financial Times* (21 January 2014) 1.

[40] See WTO Press Release, above n 35.

[41] United Nations, General Assembly Resolution 63/210 of 19 December 2008, 'Reliable and stable transit of energy and its role in ensuring sustainable development and international cooperation', above n 27.

experienced by net energy-importing countries. The development of such a market would also be of interest to energy-exporting states, whose budget revenues depend on energy exports. This is, however, easier said than done.

Gas resources are not evenly distributed across the world. In fact, the net gas-exporting and gas-importing countries are located long distances apart. For instance, whereas the world's largest proven gas reserves are concentrated in Russia, Central Asia and the Middle East,[42] the most appealing gas export markets for producers in those regions are in Western Europe, with its market-based gas prices, and in China, with its booming energy-intensive economy.[43]

So far, the most economically viable way to link the supply and demand sides of gas trade has been through the construction of gas pipelines, which often involves transit.[44] In 2010, global trade in pipeline gas alone was estimated at 677.6 billion cubic metres (Bcm) out of 975.2 Bcm of the world total gas trade (approximately equivalent to 611.1 million tonnes of oil).[45] Moreover, in the case of land-locked countries, having no direct access to the sea, such as most Central Asian states (including Azerbaijan, Kazakhstan and Turkmenistan), pipeline transportation is the only technical option for gas exportation.[46]

[42] This is not counting the so-called shale gas (i.e., natural gas trapped within shale formations), the abundant resources of which have recently been discovered in various countries, such as the United States. See Yergin, *The Quest*, above n 33 at 331.

[43] See British Petroleum, 'Statistical Review of World Energy' (June 2011), www.bp.com/statisticalreview, accessed 10 March 2011 at 20; Fridley, David, 'Natural Gas in China', in J. Stern (ed.), *Natural Gas in Asia: The Challenges of Growth in China, India, Japan, and Korea* (New York: Oxford Institute for Energy Studies, Oxford University Press, 2008) 7; and British Petroleum, 'Energy Outlook 2035' (January 2014), http://www.bp.com/content/dam/bp/pdf/Energy-economics/Energy-Outlook/Energy_Outlook_2035_booklet.pdf, accessed 2 April 2014 at 25.

[44] See Stevens, above n 34 at 3. Of course, this does not apply to the long-distance or intercontinental gas trade, such as between the Middle East and Japan or the United States, requiring the maritime shipping of LNG.

[45] Based on BP (2011), above n 43 at 29, and own calculations.

[46] These Central Asian states are situated around the Caspian Sea, which, despite the misleading name, is an international lake. These countries thus do not have direct access to the sea and are locked within their geographical region. See Vinogradov, Sergei, 'The Legal Status of the Caspian Sea and Its Hydrocarbon Resources', in G. Blake et al. (eds.), *Boundaries and Energy: Problems and Prospects* (London: Kluwer Law International, 1998) 137; and United Nations Office of the High Representative for the Least Developed Countries, 'Land-Locked Developing Countries and Small Island Developing States', http://unohrlls.org/, accessed 5 September 2014.

The network dependence of gas makes it a special subject of international trade. Unlike with other commodities where trade and transit merely imply a shipment of goods from one country to another via different transportation modes, the development of an international gas market is impossible without putting in place adequate infrastructure, namely cross-border pipelines.

Furthermore, this infrastructure must be sufficiently diversified to ensure competition between suppliers from different states as well as between different supply routes (including transit). This would reduce the otherwise extreme market power of any particular supplier or that of a transit state enabling them to dictate exorbitant gas prices or transit fees.[47]

As Yergin points out, 'the key to energy security has been diversification'.[48] At a political level, diversification of energy supplies and routes has been recognised as one of the global energy security principles, set out during the G8 Summit on Global Energy Security, held in St. Petersburg on 16 July 2006, and in the recently signed International Energy Charter.[49] At the regional level, it is part and parcel of the EU's energy security and solidarity action plan.[50]

3. Current state of the development of an international gas market and its short-term challenges

Currently, mainly due to the lack of sufficient and diversified pipeline routes, gas trade has not yet developed into a truly international market, governed by the laws of supply and demand, where buyers and sellers

[47] On the notion of 'pipeline-to-pipeline competition' and how it could improve the competitiveness of a gas market, see Beukenkamp, Annelieke, 'Pipeline-to-Pipeline Competition: An EU Assessment', 27 *Journal of Energy & Natural Resources Law* (2009) 5. See also the EU's investigations against the anti-competitive conduct of Russia's Gazprom (i.e., the major supplier of gas to the EU), taking the form of market partitioning, barriers to supply diversification by denying third-party access and unfair pricing, discussed in Sartori, Nicolò, 'The European Commission vs. Gazprom: An Issue of Fair Competition or a Foreign Policy Quarrel?', Istituto Affari Internazionali Working Papers (3 January 2013).

[48] Yergin, Daniel, 'Ensuring Energy Security', 85 *Foreign Affairs* (2006) 69 at 70.

[49] G8 Summit on Global Energy Security, St. Petersburg (16 July 2006), http://www.g8.utoronto.ca/summit/2006stpetersburg/energy.html, accessed 30 September 2014. See Title II (Implementation) in the International Energy Charter, http://www.energycharter.org/process/international-energy-charter-2015/, accessed 15 September 2015.

[50] European Commission, 'Second Strategic Energy Review: An EU Energy Security and Solidarity Action Plan', COM 781 final (13 November 2008).

have choices as to whom to deal with.[51] In most regions, gas is normally traded on the basis of long-term gas sales contracts between private or state companies, with gas prices linked to the value of replacement fuels (among others, heating oil, anthracite and coke). These contracts lock buyers and sellers into a bilateral gas sales agreement for an extended period of time (usually fifteen to twenty years), during which parties undertake to purchase/sell a pre-specified minimum quantity of gas, subject to penalties in the case of a shortfall. The latter obligation is stipulated in the so-called take-or-pay clauses.[52]

From a commercial perspective, the implementation of pipeline construction projects depends on 'project financing', where 'a given project is financed independently and the project sponsor(s) use the cash flow from the project output to provide for the collateral and the return on the investment made'.[53] For example, the investment needed to build the Nord Stream pipeline (a 1,224-kilometre offshore gas pipeline connecting Russia with the EU via the Baltic Sea bed) reportedly amounted to approximately €12 billion. These costs do not take into account additional investment in the development and production of natural gas, which is usually estimated at tens of billions of euros for a significant gas field. According to Riley, the development of the Yamal field above Russia's Arctic Circle would cost upward of $100 billion.[54]

Given the significant upfront investments required to create new opportunities for trade in gas and the structure of financial arrangements surrounding this process, both private and public stakeholders generally

[51] Konoplyanik, Andrei, A., 'Energy Security and the Development of International Energy Markets', in B. Barton et al. (eds.), *Energy Security: Managing Risk in a Dynamic Legal and Regulatory Environment* (New York: Oxford University Press, 2004) 47 at 52.

[52] See ibid. at 57; WTO Secretariat, 'World Trade Report 2010: Trade in Natural Resources' (Geneva, 2010) at 61–62; Energy Charter Secretariat, 'Putting a Price on Energy: International Pricing Mechanisms for Oil and Gas' (Brussels: Energy Charter Secretariat, 2007) at 234; and Roberts, Peter, *Gas Sales and Gas Transportation Agreements: Principles and Practice*, 2nd ed. (London: Sweet & Maxwell, 2008). However, see recent investigations of the EU Commission into the consistency of gas pricing by some energy-exporting companies (such as Russia's Gazprom), including long-term take-or-pay agreements, with the EU competition law in Sartori, above n 47.

[53] Talus, Kim, *Vertical Natural Gas Transportation Capacity, Upstream Commodity Contracts and EU Competition Law* (The Netherlands: Kluwer Law International, 2011) at 11.

[54] Riley, Alan, 'Can Nordstream and Southstream Survive in a Changing European Gas Market?', 7 *Oil, Gas & Energy Law Intelligence* (2009), www.ogel.org, accessed 15 February 2014 at 8; Nord Stream, http://www.nord-stream.com/, accessed 30 September 2011. On the technical and legal aspects of the development of this pipeline project, see Vinogradov, 'Challenges of Nord Stream', above n 9 at 16–29.

perceive long-term gas sales agreements as an important guarantee of energy security and the security of gas demand.[55] In other words, these agreements create conditions for a favourable investment climate in the area of energy trade and allow both energy-consuming and energy-exporting states to develop their long-term energy strategies. It is foreseeable that these agreements will continue to play this role until the international gas market reaches a mature stage of development. This will likely happen when a sufficient network of cross-border gas infrastructure (including pipelines and LNG terminals) is established and the major investments made into this infrastructure are recouped.

B. *What Role Can the WTO Play in Fostering the Development of an International Gas Market?*

The effective international regulation of pipeline gas transit must take into account its technical complexity and the unique nature of gas. Yet WTO law contains broad transit rules that at first glance may seem to be insufficient to regulate such complex technical matters.[56] This could be one of the reasons why, despite disruptions of energy transit flows having an adverse impact on WTO Members' economies, these issues have never been raised in the WTO. The only implicit reactions of WTO Members to the disruptions of gas supplies to the EU involving Russia and Ukraine, discussed in the previous section, were the conditions imposed on these states during their accession negotiations to the WTO. These conditions were namely that these states undertake an express commitment to apply

[55] See Rakhmanin, Vladimir, 'Transportation and Transit of Energy and Multilateral Trade Rules: WTO and Energy Charter', in J. Pauwelyn (ed.), *Global Challenges at the Intersection of Trade, Energy and the Environment* (Geneva: The Graduate Institute, Centre for Trade and Economic Integration, 2010) 123 at 124; G8 Summit on Global Energy Security, above n 49; and Talus, above n 53 at 12–14, 22. In the European Union, long-term gas contracts are also recognised as important instruments ensuring gas supplies to EU Member states, provided that these contracts are consistent with EU law, including competition rules. See recital 42 of the Directive 2009/73/EC of the European Parliament and of the Council of 13 July 2009 concerning Common Rules for the Internal Market in Natural Gas and Repealing Directive No. 2003/55/EC, OJ L 211, 14.8.2009, 94. On the concept of 'security of demand', see Belyi, Andrei, 'International Energy Security Viewed by Russia', 5 *Oil, Gas & Energy Law Intelligence* (2007), www.ogel .org, accessed 15 February 2014.

[56] Konoplyanik, Andrei, A., 'Russia-EU Summit: WTO, the Energy Charter Treaty and the Issue of Energy Transit', 2 *International Energy Law and Taxation Review* (2005) 30 at 32–33; and Azaria, Danae, 'Energy Transit under the Energy Charter Treaty and the General Agreement on Tariffs and Trade', 27 *Journal of Energy & Natural Resources Law* (2009). WTO transit rules are discussed at length in chapter IV.

all their laws, regulations and other measures governing transit of goods (including energy) in conformity with GATT Article V (freedom of transit) and other relevant WTO rules.[57] Certainly, some energy transit disruptions could not be remedied in the WTO because the parties involved were not WTO Members.[58] However, given the rapidly expanding membership of this international organisation, sooner or later, this barrier will be removed.[59]

The first attempt by WTO Members to specifically address trade restrictions on pipeline gas transit were made during the Doha negotiations on trade facilitation when some WTO Members suggested that the term 'goods', within the definition of 'traffic in transit' in GATT Article V:1 and the new Agreement on Trade Facilitation, would explicitly cover the movement of goods via pipelines.[60] However, this proposal faced opposition from other WTO Members. For example, Egypt and Turkey, while accepting the proposition that no products are *ex ante* excluded from the scope of the GATT, expressed doubts as to whether this Agreement was initially designed and is able to effectively regulate the energy sector, in particular pipeline transit.[61] One Member even proposed that certain rights/obligations critical to effective energy transit,

[57] World Trade Organization, Working Party on the Accession of the Russian Federation, 'Report of the Working Party on the Accession of the Russian Federation to the World Trade Organization', WT/ACC/RUS/70 (17 November 2011), para. 1161. Russia became a full Member of the WTO on 22 August 2012. World Trade Organization, 'Members and Observers', www.wto.org/english/thewto_e/whatis_e/tif_e/org6_e.htm, accessed 30 November 2015. See also World Trade Organization, Working Party on the Accession of Ukraine, 'Report of the Working Party on the Accession of Ukraine to the World Trade Organization', WT/ACC/UKR/152 (25 January 2008), para. 367. Ukraine joined the WTO on 16 May 2008. World Trade Organization, 'Members and Observers', ibid.

[58] As an example, see the aforementioned gas crisis of 2006 between Russia and Ukraine (then non-WTO Members) in Yergin, *The Quest*, above n 33 at 340–341; and Konoplyanik, 'Russian–Ukrainian Gas Dispute', above n 33.

[59] Currently, gas-exporting countries remaining outside the WTO are: Algeria, Azerbaijan, Equatorial Guinea, Iran, Iraq, Libya, Turkmenistan and Uzbekistan. See World Trade Organization, 'Members and Observers', above n 57; British Petroleum, 'Statistical Review of World Energy' (June 2013), www.bp.com/statisticalreview, accessed 21 August 2013 at 28.

[60] See World Trade Organization, 'Communication from Armenia, Canada, the European Communities, the Kyrgyz Republic, Mongolia, New Zealand, Paraguay, and the Republic of Moldova', TN/TF/W/79 (15 February 2006) at 5; and World Trade Organization, Negotiating Group on Trade Facilitation, 'Draft Consolidated Negotiating Text', TN/TF/W/165/Rev.8 (21 April 2011), Article 11(1) at 21.

[61] World Trade Organization, 'Communication from Egypt and Turkey – Discussion Paper on the Inclusion of the Goods Moved via Fixed Infrastructure into the Definition of Traffic in Transit', TN/TF/W/179 (4 June 2012) at 2, 5.

such as the right to build infrastructure of any kind in the transit state's territory and the obligation of a transit state to provide access to transit infrastructure, unless such infrastructure is open to general use by third parties, be explicitly excluded from WTO law.[62] Ultimately, under the pressure to finalise the Doha package at the Ministerial Conference in Bali (including the Agreement on Trade Facilitation), WTO Members decided to retain the status quo and excluded any specific reference to the issue of pipeline transit from the Agreement.[63]

As demonstrated throughout this study, the absence of perfect clarity in a treaty text, such as GATT Article V, does not in and of itself undermine its normativity. There could be different approaches to drafting international agreements: detailed rules do not always guarantee the best result. By way of example, despite the existence of relatively elaborate transit rules and an innovative dispute settlement mechanism under the ECT, devised specifically to resolve contractual disputes arising from gas transit, the ECT Members have never applied these instruments. Moreover, following the disruption of gas supply between Russia and Ukraine in 2009, Russia withdrew its provisional application of the ECT pending the ratification by Russia of this Treaty and announced its intention not to ratify it.[64] In this light, a WTO's principle-based approach to transit regulation, heavily relying on a successful experience of its adjudicatory system,[65] could be a good way to regulate pipeline gas transit at a multilateral level, which is a technically complex area of trade.

This is not to suggest that an international organisation, such as the WTO, should not adapt its law to new realities of trade, such as the emergence of the international gas market. This could be done through

[62] This proposal was reflected in World Trade Organization, Negotiating Group on Trade Facilitation, 'Draft Consolidated Negotiating Text', TN/TF/W/165/Rev.11 (7 October 2011) at 21.

[63] See International Centre for Trade and Sustainable Development, 'Historic Bali Deal to Spring WTO, Global Economy Ahead' (Bridges, 7 December 2013); International Centre for Trade and Sustainable Development, 'Обзор Переговоров: Краткий Обзор к Переговорам на 9-й Министерской Конференции ВТО', Мосты/Bridges (специальный выпуск – декабрь 2013). See the final version of the Agreement on Trade Facilitation, above n 5.

[64] See Cameron, Peter, D., 'The Energy Charter Treaty and East–West Transit', in G. Coop (ed.), *Energy Dispute Resolution: Investment Protection, Transit and the Energy Charter Treaty* (New York: Juris, 2011) 297 at 297–298; and Belyi, Andrei, Nappert, Sophie and Pogoretskyy, Vitaliy, 'Modernizing the Energy Charter Process? The Energy Charter Conference Road Map and the Russian Draft Convention on Energy Security', 29(3) *Journal of Energy & Natural Resources Law* (2011) 383 at 385–387.

[65] See Chapter I, Section II.A.2.

a reform, or by reconceptualising existing rules in the light of new facts. As mentioned earlier, this study examines both avenues. However, regardless of which approach WTO Members take, it must address the particular features of pipeline gas transit, which are network dependence and the lack of adequate cross-border pipeline infrastructure. As explained in the remainder of this chapter, this can be done by addressing the mechanisms described in Chapter I: (i) third-party access to existing pipelines; and (ii) capacity establishment. These mechanisms are the only practical ways to establish a new flow of gas.[66] They are not explicitly regulated by current WTO rules relevant to transit.

In light of this, a carefully crafted balance should be struck between creating an effective framework for regulating gas transit under WTO law and an overly ambitious task to establish a complete regime for gas transit taking on functions that could be discharged more effectively in regional or bilateral agreements.

C. *The Role of Third-Party Access and Capacity Establishment in Pipeline Gas Transit and the Development of an International Gas Market*

1. Third-party access and capacity establishment in a nutshell

In Chapter I, it was illustrated how the problem of the lack of adequate access to pipeline infrastructure or the absence of this infrastructure can be resolved by using third-party access and capacity establishment mechanisms (in this study called third-party access and capacity establishment rights or obligations). This section elaborates on these rights, their interaction and the role they play in gas transit and the development of an international gas market.

The **third-party access right** can be defined as the right of a third party (usually a gas seller or shipper)[67] to access a pipeline network. This network is either utilised by a pipeline owner for selling its own gas or operated by an independent pipeline operator rendering pipeline transportation services to other gas shippers. Access to the pipeline network is provided to third parties shipping (and/or selling) gas on the basis of

[66] On the important role of the third-party access mechanism in the formation of the EU internal gas market, see Talus, above n 53 at 85.

[67] While in reality gas can be sold and shipped by one and the same company, in this study, the term 'gas seller' refers to a seller of gas as a commodity, whereas the term 'gas shipper' refers to a provider of gas transportation services, including pipeline operators.

a private long-term gas-transportation or transit contract[68] or through an 'open access' system in a 'multi-shipper pipeline' (as it exists in some liberalised gas markets, such as the EU), against a reasonable fee and on practical technical terms.[69] In some cases, especially when public monopolies or complex pipeline construction projects are involved, a private contract can be accompanied by inter-governmental agreements, stipulating the basic principles of cooperation between states involved in a gas transit project, including (when relevant) those in the area of the protection of investments.[70]

The third-party access right is the first best option for establishing the transit flow of natural gas. It was demonstrated earlier that building a new pipeline requires significant investment. Cross-border pipelines are generally characterised by high fixed costs and low variable costs. As the pipeline capacity increases, its fixed costs fall. Due to these features, pipelines (especially large, long-distance pipelines) are often regarded as 'natural monopolies', the duplication of which is not always economical.[71]

[68] See as an example the Contract between Gazprom (Russian gas export monopolist) and Naftogaz Ukrayini (Ukrainian national gas company) on the Volumes and Conditions of Transit of Natural Gas through the Territory of Ukraine for the Period of 2009–2019, signed in Moscow on 19 January 2009, available at Ukrayinska Pravda, 'The Contract on Transit of Russian Gas + Additional Agreement on the Advance Payment to Gazprom' ('Контракт о транзите российского газа + Допсоглашение об авансе "Газпрома"'), http://www.pravda.com.ua/rus/articles/2009/01/22/4462733/, accessed 10 October 2012.

[69] See Wälde, Thomas, W. and Gunst, Andreas, J., 'International Energy Trade and Access to Energy Networks', 36 *Journal of World Trade* (2002) 191 at 197; Roberts, above n 52 at 179. Note that some legal systems distinguish a third-party access right from the technical connection to available pipeline networks. In these systems, third-party access may, therefore, merely imply the right of a service provider to utilise the existing pipeline system for the provision of gas transportation services. Talus, for example, writes that, in EU law, '[t]he TPA provisions impose an obligation on the Member States only in respect of "access to the system" and not in respect of "connection" to the system'. Talus, above n 53 at 84–85. It is questionable, however, whether this strict distinction between 'third-party access' and 'connection rights' can be transposed to international law. Whereas EU law creates opportunities and legal obligations for establishing energy networks within which the EU-wide third-party access functions effectively, at the international level, the only way to achieve third-party access for transit can be through connection to a pipeline network in the first place.

[70] See Agreements between Azerbaijan, Georgia and Turkey on the South Caucasus Pipeline (SCP) System, in British Petroleum, *Legal Agreements*, http://www.bp.com/en_az/caspian/aboutus/legalagreements.html, accessed 10 December 2012. See also Chapter III, Section III.E.

[71] See Dow, Stephen, Siddiky, Ishrak A. and Ahmmad, Yadgar K., 'Cross-border Oil and Gas Pipelines and Cross-border Waterways: A Comparison between the Two Legal Regimes', 6 *Journal of World Energy Law and Business* (2013) 107 at 111; and WTO Secretariat,

In national legal systems, access to such pipelines is ensured by energy regulations and competition law, including the doctrine of 'essential facilities'. 'The basic idea behind this doctrine is that essential facilities – such as . . . high voltage electricity transmission lines and gas transmission pipelines for energy companies – must be made available for use by competitors where those competitors cannot, or can only by incurring very high costs, build their own "version" of such facilities.'[72] Because of these features of energy facilities, some national laws impose compulsory third-party access rights on their operators, subject to certain conditions and requirements.[73]

However, international law, including WTO law, regulates competition to a very limited extent and, if such regulation exists, most rules are either vague or non-binding.[74] So far, the attempts to include comprehensive competition rules in the WTO framework have not been successful.[75] It will be demonstrated in this study that, with the exception

'World Trade Report 2010: Trade in Natural Resources', above n 52 at 194. In economic theory, the term 'natural monopoly' is defined as 'a monopoly that arises because a single firm can supply a good or service to an entire market at a smaller cost than could two or more firms'. See Mankiw, Gregory, N. and Taylor, Mark, P., *Economics* (USA: Thomson, 2006) at 818.

[72] Talus, Kim, 'Just What Is the Scope of the Essential Facilities Doctrine in the Energy Sector?: Third Party Access-Friendly Interpretation in the EU v. Contractual Freedom in the US', 48 *Common Market Law Review* (2011) 1571 at 1575 (citing Slot and Johnston). See also this concept discussed in Chapter VI, Section III.B.3.

[73] In this respect, some aspects of the EU's energy law are briefly addressed in Chapter III, Section III.D.

[74] See, inter alia, GATS Articles VIII (Monopolies and Exclusive Service Suppliers) and IX (Business Practices). GATS, *WTO Legal Texts*, above n 7. The only exception is the WTO Telecom Reference Paper, which, however, in its essence is a plurilateral rather than a multilateral agreement. See World Trade Organization, Reference Paper on Telecommunications Services (incorporated into some WTO Members' GATS Schedules of Specific Commitments) (Telecom Reference Paper), www.wto.org/english/tratop_e/serv_e/telecom_e/tel23_e.htm, accessed 30 March 2012.

[75] During the GATT period, GATT Contracting Parties established a group of experts examining the impact of restrictive business practices of private actors on international trade, as well as whether those practices should be tackled through new rules or the existing dispute settlement mechanism. However, in the Decision of 18 November 1960, the Contracting Parties merely agreed that this matter should be dealt with through bilateral consultations between governments and that no Party could undertake control over the restrictive business practices. GATT, Consultative Group of Eighteen, 'Restrictive Business Practices', Note by the Secretariat, CG.18/W/44 (10 October 1980), paras. 5–9, Annex 2. In the WTO, trade and competition was one of the issues WTO Members agreed to explore at the 1996 Singapore Ministerial Conference. However, due to a wide array of disagreements even among the supporters of a WTO competition law framework, this issue was dropped from the WTO agenda of the Doha round of negotiations. See International

of EU law, international law does not impose explicit compulsory third-party access obligations on pipeline operators.[76]

Against this background, a **capacity establishment right** comes as a second-best option for establishing gas transit in cases in which third-party access is either technically not feasible or cannot be practically enforced against a private pipeline operator. Capacity establishment is defined in this study as the creation of new pipeline capacity through its construction or expansion. It may involve different phases: determining the respective pipeline system corridor; procurement of all the necessary permits and technical documentation; land acquisition in accordance with the transit state's legislation; making technical surveys; carrying out the actual construction works; and the appointment of a subsequent transmission system operator responsible for the management of the pipeline.[77]

The interaction between third-party access and capacity establishment rights in the context of pipeline gas transit can be illustrated by a hypothetical scenario similar to that discussed in Chapter I. Figure 3 describes a situation in which private companies in WTO Members A (gas exporter) and C (gas importer) negotiate a gas sales agreement whereby gas would have to be delivered to the market of C via the pipeline network traversing the territory of a transit Member and gas exporter B. While C may also import some of its gas from B, it may be interested, for instance, due to its energy security concerns, in diversifying supply sources. For A, which may be a land-locked country, establishing the pipeline transit through B could be the only way to export its gas. In this situation, the best option for A and C is if B (or its domestic pipeline operator) would grant the right of third-party access to its pipeline for a certain amount of gas supplied by an exporter of A. However, if there is no idle capacity in B's pipeline network, or

Centre for Trade and Sustainable Development, 'The 1996 Singapore Ministerial Declaration', Vol. 1 No. 6 (February 2003).

[76] See also Wälde and Gunst, 'International Energy Trade and Access to Energy Networks', above n 69 at 209; and Cossy, Mireille, 'Energy Trade and WTO Rules: Reflexions on Sovereignty over Natural Resources, Export Restrictions and Freedom of Transit', in Herrmann, C. and Terhechte J. P. (eds.), 3 *European Yearbook of International Economic Law* (2012) 281 at 298–299. See Chapter IV, Section III.A.1, and Chapter V, Section III.B.

[77] See these phases addressed in Agreements between Azerbaijan, Georgia and Turkey on the South Caucasus Pipeline (SCP) System, in British Petroleum, *Legal Agreements*, above n 70; and Energy Charter Secretariat, 'Model Intergovernmental and Host Government Agreements for Cross-border Pipelines', 2nd ed. (Brussels: Energy Charter Secretariat, 2007).

Scenario 1 (Third-party Access)

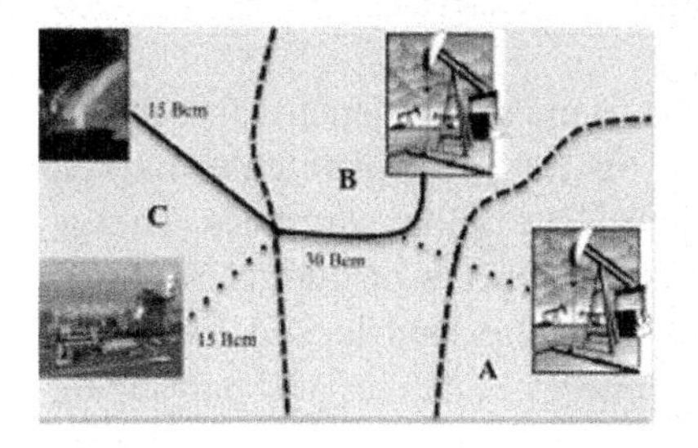

Scenario 2 (Capacity Establishment)

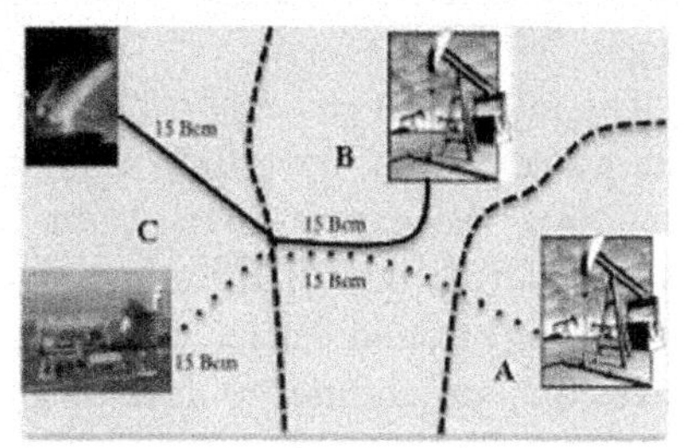

Solid line: existing pipeline
Round dot line: planned pipeline

Figure 3: Third-Party Access and Capacity Establishment

alternatively if there is no pipeline connection between these three countries at all, the establishment of a new pipeline capacity in the territory of B can be the second-best option for A, B and C.[78]

This section explained the meaning of third-party access and capacity establishment rights in the context of the transit of pipeline gas and how these rights interact. In the following sections, it will be examined whether third-party access and capacity establishment rights have been compatible with the interests of all of the stakeholders in a typical pipeline gas transit project.

2. Third-party access and capacity establishment in light of the interests of different stakeholders in a pipeline gas transit project

A typical pipeline gas transit project implies a tripartite relationship between gas-importing, gas-exporting and transit states, whose positions can be driven by different interests.[79] The interests of *gas-importing states* have already been partly addressed; they are linked to energy security, which in turn was described as access to and availability of affordable energy. Moreover, amid the globally recognised objective to mitigate

[78] The precise scope of legal rights and obligations that arise from invoking the third-party access and capacity establishment rights is elaborated in Chapters IV and V. However, it must be mentioned at the outset that a capacity establishment right invoked by A or C does not necessarily imply the entitlement to build a pipeline in the territory of B. As will be further explained, there could be different ways to establish gas transit between A and C via the territory of B.

[79] However, see transit involving only two states in Figure 2, scenarios 2 and 3.

climate change, the goal of sustainable development can prompt energy-intensive nations to shift their economies to cleaner energy sources, such as gas.[80]

The interest of *gas-exporting states* is to maximise benefits from their energy resources, by selling their gas on the most profitable market, including the export market. In some developing countries, energy export can be one of the most significant sources of budget revenue. Yet, as mentioned earlier, not all states, such as land-locked countries, have equal capabilities to develop their energy resources and participate in global energy trade.

The interests of *transit states* have not been fully addressed as yet. They can be linked to various non-economic and economic concerns. The former may relate to environmental or social implications of a given pipeline gas transit project. Unlike oil pipelines, gas networks do not pose the same risk of hazardous spills.[81] Nevertheless, damage to or the explosion of a pipeline can cause fires; in some politically volatile areas, the existence of a pipeline can attract terrorist attacks. In addition, the construction of pipelines can require tunnels to be dug, may affect the traditional lives of indigenous communities by impeding the ordinary migration of herds, local grazing or farming habits, and may impair the authentic environment of national parks. From a more technical viewpoint, the transit state's regulations can relate to gas specification requirements as a prerequisite for third-party access or the enforcement of domestic competition or property laws. Provided that all these concerns are reasonable, they can duly be accommodated by the cooperation of the pipeline project participants, including domestic authorities and local communities of the transit state, with a view to designing a commonly acceptable transit route and ensuring compliance with legitimate energy transportation standards.[82]

[80] Natural gas is the least carbon-intensive energy resource among fossil fuels. International Energy Agency, 'Energy Efficiency', above n 38; Roberts, above n 52 at 14.

[81] Vinogradov notes that '[i]n terms of technology and possible environmental impacts, pipelines have proved to be a relatively secure, safe and reliable form of transportation'. Vinogradov, 'Challenges of Nord Stream', above n 9 at 1. Dow et al. explain that '[i]n the event of any leak, the pressure of the stored gas tears the pipeline apart and the gas is released into the air causing minimal damage, assuming that the gas does not ignite upon discharge.' Dow, Siddiky and Ahmmad, 'Cross-border Oil and Gas Pipelines and Cross-border Waterways', above n 71 at 109.

[82] These issues, for example, are discussed in Zillman, Donald, N. et al. (eds.), *Human Rights in Natural Resource Development: Public Participation in the Sustainable Development of Mining and Energy Resources* (New York: Oxford University Press, 2002); and Roberts,

However, from an economic viewpoint, a transit state may not always be interested in gas transit and may not be willing to grant third-party access or capacity establishment rights. While, in principle, a transit state may receive certain indirect benefits from creating or improving its transit infrastructure[83] and will obtain a reasonable compensation for any inconveniences that transit entails, in reality, transit states are often interested in more direct benefits, both political and economic.[84]

Moreover, if a transit (or potential transit) state is also a gas exporter (such as Russia or Azerbaijan), the security of its own gas supply projects (security of demand) can also be one of its strategic economic interests, which conflicts with acting as a transit state. Consequently, opening a transit route through its territory would expose its domestic gas producers to gas-to-gas competition and may undermine the security of demand.[85] Recall that at the current initial stage of the development of an international gas market, attracting sufficient investment in energy production and transportation is directly linked to investors' ability to recoup their investment during the project span. This could also be an important reason why the pipeline capacity in the existing network of a transit state is fully utilised or not designed for a mandatory third-party access.[86]

Nevertheless, if the goal of creating an international gas market is pursued and is taken seriously, there should be some limitations imposed by international law on the discretion of a transit state as to whether to allow or reject gas transit through its territory. However different the interests of gas-importing, gas-exporting and transit states could be in a pipeline gas project, as stated earlier, international law has the important role of reconciling political and economic differences among states.[87]

As will be explained throughout this study, WTO law together with other applicable principles of international law provides sufficient autonomy to a transit state to address its regulatory concerns. Furthermore, tools, which are compatible with WTO obligations, are available in private international gas contracts to ensure the security of gas demand. These tools include long-term gas supply contracts that have a duration that matches the duration of particular investment projects. These

above n 52. The examples of possible solutions to various regulatory concerns can be found in Energy Charter Secretariat, 'Model Intergovernmental Agreements', above n 77.

[83] See this chapter, Section II.A. [84] See this chapter, Section III.C.3.

[85] Liesen, above n 12 at 60. [86] Rakhmanin, above n 55 at 124.

[87] De Visscher, above n 16 at 7. See this chapter, Section II.B.

contracts can provide sufficient security of demand for a gas seller located in a transit state, thus alleviating the impact of international competition on the transit state's ability to utilise its own natural resources, as well as the ability of its domestic energy producers to recoup investments.[88]

In the following section, a few examples are provided of pipeline gas transit projects that depend on third-party access or capacity establishment rights and that have faced challenges or resulted in a political deadlock between a transit state and transit-dependent states. These examples illustrate the importance of third-party access and capacity establishment rights for ensuring the effective international regulation of pipeline gas transit and will place the ensuing discussion of these rights in a real practical context.

3. Third-party access and capacity establishment in real practical contexts

The problem of pipeline connections between energy-exporting and energy-importing states has continuously been raised in the context of gas trade between Western European countries and Ukraine (net energy importers), on one hand, and Russia and Central Asian States (net energy exporters), on the other hand.[89] Gas trade between these countries is extremely dependent on various pipelines traversing Russia and Ukraine, as well as other Eastern European countries.[90] In this context, there has been a long-standing attempt of the EU and Ukraine to overcome their energy security challenges by importing a share of their gas supplies from Central Asia (in particular Turkmenistan) via the Russian pipeline network. However, most negotiations on this issue stalled due to Gazprom's[91] unwillingness to grant third-party access to

[88] See this chapter, Section III.A.3.

[89] Liesen, above n 12 at 60; Wälde and Gunst, above n 69 at 212.

[90] See Appendix 3 to this monograph. See the discussion of different Eastern-European pipeline routes to the European market in Yafimava, above n 33; and Vinogradov, 'Challenges of Nord Stream', above n 9 at 3. On energy politics in this region, see also Yergin, *The Quest*, above n 33 at 336–343.

[91] Gazprom is a Russian pipeline operator who owns the Unified Gas Supply System of Russia (the world's largest gas transmission system for long-distance gas transportation); it is also the world's largest gas trading company and gas export monopoly in Russia, where the state-controlled stake is 50.002 per cent of the shares. During Russia's negotiations on its accession to the WTO, Russia committed to notify Gazprom as a state-trading enterprise (STE). See Gazprom, http://www.gazprom.com/about/, accessed 10 January 2013; Mitrova, Tatiana, 'Natural Gas in Transition: Systemic Reform Issues', in S. Pirani (ed.), *Russian and CIS Gas Markets and Their Impact on Europe* (New York: Oxford Institute for Energy Studies, Oxford University Press, 2009) 13 at 23; and World Trade

its pipelines.[92] Had it granted third-party access, the Central Asian gas would enter into competition with Russian gas, thereby weakening Russia's geopolitical position vis-à-vis its neighbours.[93] As one Russian commentator noted, the possibility that the geographical disadvantage of Central Asian states could be corrected by applying the transit provisions of the ECT, in particular by claiming third-party access to Russia's pipeline network, has been one of the reasons that Russia has not yet ratified this Treaty.[94] Instead, Gazprom's policy has been to import relatively cheap Turkmen gas into Russia and sell its own gas (or that mixed with the imported Turkmen gas) to the Western European market at a high European price.

For example, under a 2005 deal between Turkmenistan and Gazprom, Turkmenistan agreed to sell its gas to Russia at $44 per 1,000 cubic metres (mcm). After adding Uzbek and Kazakh transit charges (between $10.5–21/mcm, depending on different sources), the total price for Turkmen gas amounted to $54.5–65/mcm. In the same year, the EU market price for gas was, on average, $220/mcm. After deducting the transportation charges from the latter price (approximately $57/mcm to the Czech–German border in 2004), the difference between the Russian and the EU gas prices at that time was about $100/mcm.[95] Nonetheless,

Organization, Working Party on the Accession of the Russian Federation, above n 57, paras. 76, 88.

[92] Liesen, above n 12 at 60; Wälde and Gunst, above n 69 at 212.

[93] In the region of the Commonwealth of Independent States (CIS), Russia has used its natural resources, mainly relatively cheap gas, to attract neighbouring states (such as Armenia and Ukraine) into the Eurasian Economic Union that it recently formed with Armenia, Belarus, Kazakhstan and the Kyrgyz Republic and where it informally plays a dominant role. See Pogoretskyy, Vitaliy and Beketov, Sergey, 'Bridging the Abyss? Lessons from Global and Regional Integration of Ukraine', 46(2) *Journal of World Trade* (2012) 457 at 459–460; 'Armenia Intends to Join the Customs Union' ('Армения намерена вступить в Таможенный союз'), 3 September 2013, http://ria.ru/economy/20130903/960485143.html, accessed 10 September 2013; and McElroy, Damien, 'Ukraine Receives Half Price Gas and $15 Billion to Stick with Russia', *The Telegraph* (17 December 2013), http://www.telegraph.co.uk/news/worldnews/europe/ukraine/10523225/Ukraine-receives-half-price-gas-and-15-billion-to-stick-with-Russia.html, accessed 17 December 2013.

[94] Konoplyanik, Andrei A., 'A Common Russia-EU Energy Space (The New EU–Russia Partnership Agreement, *Acquis Communautaire*, the Energy Charter and the New Russian Initiative)', 7 *Oil, Gas & Energy Law Intelligence* (2009), www.ogel.org, accessed 15 February 2014 at 18–19 (citing the former chairman of the Energy Committee of the Russian State Duma, Valery Yazev, and the former Gazprom CEO, Rem Vyakhirev). See the detailed analysis of the ECT transit rules in Chapter III, section III.C.2 and Chapter VI, Section III.A.2.

[95] Own calculations based on Stern, Jonathan, *The Future of Russian Gas and Gazprom* (New York: Oxford Institute for Energy Studies, Oxford University Press, 2005) at 78;

given the complete dependence of Turkmenistan on the Russian and Ukrainian gas transportation systems, the only option it had to sell its gas was to Russian or Central Asian consumers.

The inability of the EU to access Central Asian gas through Russia has urged it to develop other transit routes for its gas importation from this region, including via the so-called Southern Gas Corridor traversing Turkey.[96] Turkmenistan also had to look for other export opportunities, such as gas exports to China, India and Pakistan.[97] As stated earlier, the only development relevant to the improvement of gas transit conditions in Russia was the commitment Russia made during its accession negotiations to the WTO that it would apply all its laws, regulations and other measures governing the transit of goods (including energy) in conformity with GATT Article V (freedom of transit) and other relevant WTO rules.[98] This commitment, however, does not explicitly mention third-party access or capacity establishment rights.

Within the framework of the Southern Gas Corridor, the development of gas pipeline projects has also faced a number of geopolitical and economic challenges, which significantly delayed the negotiations on the construction of certain cross-border pipelines. For example,

Energy Charter Secretariat, 'Gas Transit Tariffs in Selected Energy Charter Treaty Countries' (Brussels: Energy Charter Secretariat, 2006) at 66; and British Petroleum, 'Statistical Review of World Energy' (June 2009), www.bp.com/statisticalreview, accessed 30 March 2010 at 31.

[96] The Southern Gas Corridor (or the Fourth Corridor) is a political initiative of the European Commission for the supply of natural gas to the EU from the Caspian and Middle Eastern basins. Within this initiative, Turkey (a main transit corridor) is involved in a number of pipeline projects, such as Nabucco, the Trans-Anatolian Pipeline, the Trans-Adriatic Pipeline, the Turkey–Greece Interconnector, and the Turkey-Greece-Italy Interconnector (see Appendix 3). The overall objective of these pipeline projects for the EU and Turkey is strengthening their energy security and loosening their dependence on Russian gas. See Pogoretskyy, Vitaliy, 'The Transit Role of Turkey in the European Union's Southern Gas Corridor Initiative: an Assessment From the Perspective of the World Trade Organization and the Energy Charter Treaty', 2(2) *Journal of International Trade and Arbitration Law* (Istanbul, 2013), 121 (also available at www.ogel.org, accessed 15 February 2014) at 126–128; and Winrow, Gareth M., 'Problems and Prospects for the "Fourth Corridor": The Positions and Role of Turkey in Gas Transit to Europe' (Oxford Institute for Energy Studies, June 2009).

[97] See Appendix 3 to this monograph; and Stern, Jonathan and Bradshaw, Michael, 'Russian and Central Asian Gas Supply for Asia', in J. Stern (ed.), *Natural Gas in Asia: Challenges of Growth in China, India, Japan, and Korea*, 2nd ed. (New York: Oxford Institute for Energy Studies, Oxford University Press, 2008) at 220.

[98] World Trade Organization, Working Party on the Accession of the Russian Federation, above n 57, para. 1161.

Turkey, realising its importance as a gas transit route between the EU and energy resources in Central Asia, Iraq and the Mashreq countries, has often linked the negotiations on gas transit with the EU, or transit pipeline construction projects in its territory, to its own political and economic interests, such as aspirations to accede to the EU, its own demand for cheap gas and its interest in gas re-export rather than transit.[99] Despite these challenges, the transit of pipeline gas via Turkey is still regarded by the EU as the highest priority of its external energy policy.[100]

In other regions, the establishment of pipeline networks has also faced challenges. One can mention, by way of examples, Qatar's failed gas export project to Bahrain and Kuwait via Saudi Arabia and the long-discussed Iran Peace Pipeline project to India through Pakistan, dead-locked due to political disagreements between the participating states.[101] Reportedly, one of the reasons underlying the recent case of Bolivia (an energy-endowed land-locked country) against Chile at the ICJ to reclaim sovereign access to the sea through Chile, which Bolivia lost during the War of the Pacific more than a century ago, was its intention to gain access to the international gas market by building an LNG export facility on the Pacific Ocean coast.[102] Formally, Bolivia claims in the ICJ proceedings that Chile failed to comply with its obligation, under general international law, bilateral agreements, diplomatic practice and a series of declarations attributable to its highest-level representatives, to negotiate in good faith and effectively with Bolivia in order to reach an agreement granting Bolivia fully sovereign access to the Pacific Ocean.[103]

[99] Pogoretskyy, 'The Transit Role of Turkey', above n 96 at 125–129; Özen, Erdinç, 'Turkey's Natural Gas Market Expectations and Developments' (Deloitte, April 2012) at 68; Winrow, above n 96 at 19–23.

[100] Pogoretskyy, 'The Transit Role of Turkey', above n 96 at 125–129.

[101] See Flower, Andy, 'Natural Gas from the Middle East', in J. Stern (ed.), *Natural Gas in Asia: The Challenges of Growth in China, India, Japan, and Korea* (New York: Oxford Institute for Energy Studies, Oxford University Press, 2008) 330.

[102] Jacobs, Justin, 'Bolivia's Fight for the Sea', *Petroleum Economist*, 3 May 2013, http://www.petroleum-economist.com/Article/3201020/Bolivias-fight-for-the-sea.html, accessed 30 August 2013.

[103] Application of Bolivia Instituting Proceedings before the International Court of Justice, http://www.icj-cij.org, last visited 30 August 2013, para. 31. At the time of writing, this case just started with parties' memorials being due on 17 April 2014 (Bolivia) and 18 February 2015 (Chile). International Court of Justice, Order of 18 June 2013, http://www.icj-cij.org/docket/files/153/17392.pdf, visited 30 August 2013.

IV. Summary and Concluding Remarks

This chapter defined the meaning of the principle of 'freedom of transit' as it is expressed in WTO law and the term 'traffic in transit'. It discussed the essential characteristics of transit, including its economic and general humanitarian implications, as well as its peculiar interaction with the territorial sovereignty of transit states.

This chapter further discussed the importance of gas transit in modern trade, including its relationship with energy security and the sustainable development of WTO Members. In this respect, the chapter explained how the effective regulation of third-party access and capacity establishment rights under WTO transit rules could address these concerns by providing favourable conditions for the development of an international gas market. Given, however, the abundant restrictions on these rights imposed by transit states, the improvement of the international regulation of pipeline gas transit under WTO law will naturally entail challenges. These challenges will have to be overcome through either legal or diplomatic means: namely a judicial clarification of the existing transit rules or Members' negotiations on new rules.

III

General Overview of the International Regulation of Transit

I. Introduction

This chapter provides an overview of the regulation of transit in different areas of public international law. Although, during the past two centuries, the principle of freedom of transit has been employed in different international treaties, including the WTO Agreement, it has been known since times immemorial. Therefore, to appreciate fully the role this principle plays in modern international relations, a historical analysis of the evolution of this principle in a broad environment of public international law must be conducted.

Furthermore, it is important to note that, depending on its legal status, the principle of freedom of transit could influence the interpretation of rules governing transit in particular treaty regimes, such as WTO law. As was explained in Chapter I, general international law sources can be used in treaty interpretation as rules applicable to all WTO Members. This raises the question of whether the principle of freedom of transit has entered the corpus of general international law, and, therefore, applies to the relations between states independently from treaties.

It will be demonstrated in this chapter that there are contradicting views among scholars and adjudicators regarding the true status of the principle of freedom of transit in public international law in the sense of the right to pass through foreign territory. Certainly, it would be an exaggeration to state that, based on general international law, this right can be exercised without obtaining the permission of a transit state. Indeed, such a right has been consistently disputed by many states as well as by individual judicial authorities and scholars.

Nevertheless, an argument that, in a time of peace, a transit nation has absolute discretion as to whether to allow or deny transit via its territory to the commerce of another nation is becoming increasingly outdated, especially if transit does not pose any threat to the national security, or does not undermine the legitimate regulatory objectives, of the transit nation. The principle of freedom of transit, together with other principles linked to this freedom (namely the prohibition on levying customs duties on traffic in transit and imposing charges and regulations that are unreasonable and discriminatory), has been recognised in different international legal instruments, including multilateral, regional and bilateral treaties.

It is argued in this chapter that the principle of freedom of transit is recognised as a matter of general international law as an 'imperfect right' of any member of the international community, and particularly landlocked states.[1] In complex cases, the implementation of this principle may require the conclusion of an agreement between a transit state and a transit-dependent state setting out the concrete terms and modalities of transit. This can be done at a bilateral or regional level. In the context of pipeline gas, such implementation agreements would necessarily have to regulate third-party access or capacity establishment rights.[2]

This chapter is divided into three substantive sections. The following section examines the major practical contexts in which the principle of freedom of transit has been applied. Section III analyses the development of this principle in different areas of international law, including the transit of network-bound energy. Section IV discusses whether the principle of freedom of transit is a principle of general international law. Section V concludes this chapter.

[1] The term 'imperfect right' is used in this study in the sense of a right that cannot be exercised directly without removing administrative obstacles to its implementation, by obtaining the necessary permits or complying with certification requirements or licensing conditions. See this term mentioned in Westlake, John, *International Law*, Part I (Cambridge: Cambridge University Press, 1904) at 157; and Lauterpacht, Elihu, 'Freedom of Transit in International Law', 44 *Transactions of the Grotius Society* (1958–1959) 313 at 346–347. Melgar appears to understand this term in a narrower sense as denoting a right that 'cannot be asserted by force … a law without sanction'. Melgar, Beatriz H., *The Transit of Goods in Public International Law* (Leiden: Brill Nijhoff, 2015) at 159. This monograph does not suggest that the principle of freedom of transit is unenforceable in international law. On the contrary, it argues that this principle creates specific obligations for transit states, subject to certain limitations. In this sense, the meaning of the principle of freedom of transit advocated in the monograph does not appear to be conceptually different from that Melgar promotes, namely that it creates 'the right of transit as a conditional right'. Melgar, ibid. at 316–332.

[2] On the meaning of the term 'implementation agreement', see Kojima, Chie and Vereshchetin, Vladlen S., 'Implementation Agreements', in *Max Planck Encyclopaedia of Public International Law* (2010), www.mpepil.com, accessed 30 March 2012.

II. Transit and Its Historical Contexts – General Overview

Freedom of transit or passage is not a new concept in international law and, throughout the history of nations, it has been regulated in different ways. In ancient times, restriction of the passage of troops by a transit nation was regarded as a just cause for war.[3] Some bilateral or even regional transit arrangements, as well as commercial routes traversing different kingdoms, existed in the Middle Ages.[4] During the Renaissance, Gentili associated the idea of freedom of transit with international communication and commerce.[5] De Vattel provides the example of a transit dispute brought before the Holy Roman Emperor in the fifteenth century with respect to transit restrictions imposed on merchandise by the count of Lupfen. The council called by the emperor condemned those restrictions as inequitable, giving the order to restore transit rights and pay costs and damages.[6]

The difference between the concepts of freedom of transit today and hundreds of years ago lies in their particular contexts and the modes of their international regulation. It is noteworthy that, while commercial transit has always been among the concerns, before the twentieth century, it was part and parcel of the then greater political interests. For example, at all times, the right to transit has been linked to international peace and security. This is precisely why legal theorists, such as Gentili and Grotius, discussed it in connection with the law of war ('*de iure belli*').[7] In their time, military action was still the main remedy for

[3] Phillipson, Coleman, *The International Law and Custom of Ancient Greece and Rome*, vol. 2 (London: The Macmillan Company, 1911) at 88.

[4] In the twelfth century, the members of the Lombard League (an alliance formed at that period among the communes of Lombardy (Italy) to resist the Holy Roman Empire), such as Venice, kept roads open and safe for their allies, such as Milan (a land-locked kingdom), in exchange for their allies' undertaking to keep the waterways of the Po open. Lauterpacht, 'Freedom of Transit in International Law', above n 1 at 326. See also Silk Route in *The New Encyclopaedia Britannica*, vol. 10, 15th ed. (Chicago: Encyclopaedia Britannica, Inc., 1993) at 810, and Varangian Route through Kievan Russia to the Greeks in *The Soviet Encyclopaedia*, 3rd ed. (Moscow, 1985) at 1080.

[5] Gentili, Alberico, *De Iure Belli Libri Tres*, vol. 2, Scott, J. B. (ed.), Rolfe, J. C. (trans) (London: Clarendon Press, 1933, reprinted) at 88–89.

[6] De Vattel, Emer, *The Law of Nations, Or Principles of the Law of Nature, Applied to the Conduct and Affairs of Nations and Sovereigns, with Three Early Essays on the Origin and Nature of Natural Law and on Luxury*, vol. 2, Kapossy, B. and Whatmore, R. (eds.) (Indianapolis: Liberty Fund, 2008), chapter X, para. 132.

[7] Gentili, above n 5; and Grotius, Hugo, *The Rights of War and Peace*, vol. 2, Tuck, R. and Barbeyrac, J. (eds.) (Indianapolis: Liberty Fund, 2005), chapter 2. See also Chapter II, Section II.A.

violations of international obligations. Three cases of the World Court dealing with transit – *S.S. 'Wimbledon'*, *Corfu Channel* and *Right of Passage over Indian Territory* – concerned transit restrictions on military munitions and the navy.[8]

Another important concern linked to transit has been access to the high seas. The problem of free access to the high seas emerged in the seventeenth century when Portugal and England claimed sovereignty over this common heritage of mankind. However, Grotius (a citizen of the naval rival of those nations – the Dutch Empire) successfully countered these claims in his landmark study '*Mare Liberum*'.[9] In modern international law, there is no question that all states enjoy equally the freedom of the high seas, which has been declared a fundamental principle of the world order established after WWII. This principle subsumes freedom of navigation in the high seas.[10]

Grotius also postulated a general theory of state property advocating a natural right of transit through foreign territories.[11] He based his theory on natural law that looks at the nature of men as social and rational beings, building on the ideas of uniform justice, reasonableness, theological beliefs and the ancient practice of nations.[12] Albeit not shared by many authors of the nineteenth century,[13] the Grotian theory of state property is generally regarded as a cornerstone upon which the principle of freedom of transit began its development. In the twentieth century, the relation between transit and access to the sea drew the attention of the international community to the particular situation of land-locked countries, which, due to their disadvantaged geographical position, could

[8] *S.S. 'Wimbledon'*, Collection of Judgments, Series A. – No. 1, 17 August 1923; *The Corfu Channel Case*, Judgment of 9 April 1949: I.C.J. Reports 1949, p. 4; and *Case Concerning Right of Passage over Indian Territory*, Judgment of 12 April 1960: I.C.J. Reports 1960, p. 6.

[9] Grotian work laid down the foundations for the commonly accepted principle of freedom of the high seas. Nussbaum, Arthur, *A Concise History of the Law of Nations* (New York: The Macmillan Company, 1954, revised edition) at 111.

[10] See a review of the major legal sources supporting the customary nature of the principle of freedom of the high seas, including the Atlantic Charter adopted in 1941 by Great Britain and the United States and later agreed to by all the Allies, in *The M/V 'SAIGA' (No. 2) Case (Saint Vincent and the Grenadines v. Guinea)*, ITLOS, Separate Opinion of Judge Laing, 1 July 1999, http://www.itlos.org/index.php?id=64, accessed 15 May 2014, paras. 20–31.

[11] Grotius, above n 7, chapter 2.

[12] See the Grotian theory of natural law explained in Degan, Vladimir D., *Sources of International Law* (The Netherlands: Kluwer Law International, 1997) at 26; and Nussbaum, above n 9 at 102–114.

[13] See the criticism of Grotius discussed in the *Faber* case, in Venezuelan Arbitrations of 1903, prepared by Jackson H. Ralston (assisted by W. T. H. Doyle) (Washington, DC: Government Printing Office, 1904) 600 at 629.

benefit fully neither from international communication and commerce, nor from participation in the development of offshore natural resources.[14]

With the emergence of new states in the twentieth century, the discussion of freedom of transit was revived in the context of territorial disputes and delimitation agreements. From this perspective, the concept of freedom of transit was intertwined with the private law doctrine of servitude (originated in Roman law) or easement (a similar common law concept).[15] Some international lawyers relied on servitudes in their attempt to prove that certain real rights in foreign territories (such as the right of passage via rivers, international straits or land) could be derived from territorial separation.[16]

Two other areas of international law that have a historical connection to transit (as well as access to the sea) are river navigation rights and the right of passage through international straits and canals.[17] With the

[14] Uprety counted thirty-eight such countries in the world. Uprety, Kishor, *The Transit Regime for Landlocked States: International Law and Development Perspectives* (Washington, DC: The World Bank, 2006) at 3. Notably, this author does not take into account Azerbaijan, Kazakhstan, Turkmenistan and Uzbekistan, which are also landlocked countries. See Chapter II, Section III.A.2.

[15] Servitude is a 'real' right or a right *in rem* (a right relating to a thing, but which is less than ownership). This right can be created by an agreement, usage (*longi temporis praescriptio*) and necessity (when upon conveyance or severance of a piece of land it becomes landlocked – lacking access to the public way). After its creation, a servitude becomes a perpetual right attached to land rather than to the owner or user of this land. Buckland, W. W., *A Text-Book of Roman Law from Augustus to Justinian*, P. Stein (ed.) (New York: Cambridge University Press, 1963); Prichard, A. M., *League's Roman Private Law Founded on the Institutes of Gaius and Justinian*, 3rd ed. (London: Macmillan & Co Ltd, 1964) at 215–217.

[16] See Lauterpacht, Hersch, *Private Law Sources and Analogies of International Law* (USA: Archon Books, 1970, reprinted) at 121–124; Reid, Helen D., *International Servitudes in Law and Practice* (Chicago: University of Chicago Press, 1932); Váli, F. A., *Servitudes of International Law: A Study of Rights in Foreign Territory*, 2nd ed. (London: Stevens & Sons Limited, 1958); O'Connell, D. P., 'A Re-consideration of the Doctrine of International Servitude', 30 *The Canadian Bar Review* (1952) 807; Farran d'Olivier, C., 'International Enclaves and the Question of State Servitudes', 4 *International and Comparative Law Quarterly* (1955) 294 at 304; Sinjela, Mpazi A., 'Freedom of Transit and the Right of Access for Land-Locked States: The Evolution of Principle and Law', 12 *Georgia Journal of International and Comparative Law* (1982) 31 at 51–52; and Shaw, Malcolm N., *International Law*, 6th ed. (New York: Cambridge University Press, 2008) at 540–541.

[17] International straits and canals are the channels of maritime passage between two seas or oceans. Whereas straits are natural waterways, canals are artificial facilities, such as the Suez or Panama Canals. See Brownlie, Ian, *Principles of Public International Law*, 7th ed. (New York: Oxford University Press, 2008) at 264–271.

proliferation of international commerce and increased interdependence of nations, transit also became associated with economic development.[18]

Finally, more recent concerns that have emerged with technological progress in the past two centuries are various special areas of transit, such as aerial transit, telecommunication technologies and energy transit. These areas are regulated by general principles of international law, multilateral, regional and bilateral treaties establishing their own balances between freedom of transit and territorial sovereignty. Due to the technical complexity of each such area of transit regulation, in this study only a few selected areas are discussed, which have historically contributed to the development of the principle of freedom of transit in international law. Among the new areas, the focus will be on energy transit and in particular pipeline-bound gas transit.[19] The issue of international trade in telecommunication services is addressed in Chapter VI, in which possible improvements to WTO transit regulation are analysed.

III. Freedom of Transit in Different Areas of International Law

A. *Developments in Fluvial Law, the 'Innocent Passage' through Straits and Canals*

1. Freedom of transit in fluvial law

Freedom of transit is widely recognised in international river law (also known as 'fluvial law'), which emerged in the nineteenth century,[20] and in the area of inter-canal or straits communication. In the area of fluvial law, the core ideas behind the right to transit (also represented as freedom of river navigation) have been the equitable sharing and use of

[18] See the importance of freedom of navigation for the economic global order discussed in *SAIGA*, Separate Opinion of Judge Laing, above n 10, paras. 23–31.

[19] In general, see an overview of different transit conventions in Einhorn, Talia, 'Transit of Goods over Foreign Territory', in *Max Planck Encyclopaedia of Public International Law* (2010), www.mpepil.com, accessed 30 November 2011; Vasciannie, Stephen C., *Land-Locked and Geographically Disadvantaged States in the International Law of the Sea* (Oxford: Clarendon Press, 1990); and Melgar, above n 1. For a good overview of the international regulation of aerial transit, which is not discussed here, see World Trade Organization, Council for Trade in Services, 'Developments in the Air Transport Sector since the Conclusion of the Uruguay Round', Background Note by the Secretariat, S/C/W/163/Add.3 (13 August 2001); and Melgar, ibid. at 78–105.

[20] De Visscher, Charles, *Le Droit International des Communications* (Paris: Université de Gand, 1921) at 7.

watercourses by all states, the freedom of the seas and the interdependence of states in the international community.[21]

The trans-European liberalisation of river navigation started at the end of the eighteenth century during the First French Republic, when the Scheldt and Meuse rivers were declared open on the ground that 'a nation cannot without injustice pretend to the right of exclusively occupying the channel of a river, and hinder the neighbouring peoples who border on its higher shores from enjoying the same advantage'.[22] In North America, US representatives relied on the concepts of interdependence and necessity in substantiating their claims against Great Britain and Spain to the right of navigation over the St. Lawrence River and the Mississippi, respectively. These claims were then duly accommodated by the conclusion of bilateral agreements with Spain in 1795 and Great Britain in 1871.[23]

The first multilateral treaty that established important common principles of river navigation was the Final Act of the Congress of Vienna of 1815. These principles were: (i) freedom of navigation, subject to the policing by a transit state; (ii) the obligation of transit states to apply regulations in a manner non-discriminatory and as favourable as possible to the commerce of all nations; and (iii) the prohibition on imposing duties or other charges that relate solely to navigation.[24] Similar principles were further reiterated in subsequent multilateral treaties, such as the 1921 Convention on the Regime of Navigable Waterways of International Concern (Convention on Navigable Waterways).[25]

Back in the beginning of the twentieth century, in *River Oder*, the PCIJ clarified some general principles of international law applicable to riparian states. In that dispute, the World Court was called on to determine the jurisdiction of the International Commission of the Oder over certain sections of this river situated in Polish territory, based on the Treaty of

[21] Ibid. at 6–9; Uprety, above n 14 at 28–29, 37.

[22] Lauterpacht, 'Freedom of Transit', above n 1 at 327. See also Uprety, above n 14 at 39 (citing the Executive Council of the Convention that decreed the liberalisation of the Scheldt and the Meuse on 20 November 1792).

[23] Lauterpacht, 'Freedom of Transit', above n 1 at 320–321.

[24] See Articles CIX–CXI of the Final Act of the Congress of Vienna of 1815, http://www.fao.org/docrep/005/w9549e/w9549e02.htm, accessed 5 September 2013; Zeilinger, Anton F., 'Danube River' in *Max Planck Encyclopaedia of Public International Law* (2014), www.mpepil.com, accessed 30 May 2014, para. 3.

[25] See Article 4 of the Statute of the 1921 Convention on the Regime of Navigable Waterways of International Concern, http://www.fao.org/docrep/005/w9549e/w9549e02.htm, accessed 5 September 2013.

Versailles and the Convention on Navigable Waterways, which was ultimately found to be inapplicable. However, being unable to resolve the dispute by relying solely on the text of the Treaty of Versailles, the Court decided to fall back on 'the principles governing international fluvial law *in general*', including the principle of freedom of navigation on international waterways, thereby recognising this principle as a principle of general international law.[26]

With respect to the nature of freedom of navigation, the Court admitted that 'the desire to provide the upstream States with the possibility of free access to the sea played a considerable part in the formation of the principle of freedom of navigation on so-called international rivers'.[27] Yet, in its view, a solution to this problem was not to be sought 'in the idea of a *right of passage* in favour of upstream States, but in that of a *community of interest* of *riparian States*', which the Court interpreted as '*the perfect equality of all riparian States* in the user of the whole course of the river and *the exclusion of any preferential privilege* of any one riparian State in relation to the others'.[28]

Unfortunately, the PCIJ did not elaborate on differences, if any, between the concept of the 'community of interest of riparian States' and the right of passage of an upstream state. Given, however, that the former concept incorporates the obligation of a transit state to afford equal treatment to *all* riparian states (arguably including the transit state itself), any such difference would not be significant. This view would also be supported by the fact that the PCIJ appears to have derived the concept of the 'community of interest of riparian States' from the Final Act of the Congress of Vienna of 1815, which, in the Court's view, had laid down the principles of international river law.[29] Interestingly, the Act explicitly recognised the freedom of navigation in the sense of the right of passage as a separate, fundamental principle of river law.[30]

In any event, the basis for freedom of navigation and access to the sea over international rivers appears to have expanded beyond the mere idea

[26] *Territorial Jurisdiction of the International Commission of the River Oder*, Collection of Judgments, Series A. – No. 23, 10 September 1929 at 22, 26, emphasis added. The term 'international waterways' has been defined as '[rivers] which are navigable from the open sea and at the same time either separate or pass through several states between their sources and their mouths'. Melgar, above n 1 at 12 (citing Oppenheim).

[27] *River Oder*, above n 26 at 26–28. [28] Ibid. at 27, emphasis added. [29] Ibid.

[30] Article CIX of the Act provides: 'The navigation of the rivers, along their whole course . . . from the point where each of them becomes navigable, to its mouth, *shall be entirely free*, and *shall not, in respect to commerce, be prohibited to any one*.' The Final Act of the Congress of Vienna of 1815, above n 24, emphasis added.

of the 'community of interest of *riparian* States' in modern international law, since modern treaty regimes grant freedom of navigation to *all* – not only *riparian* – states.[31] This evolution of fluvial law appears to have strengthened the relationship between the concept of 'freedom of navigation' and the right of passage of all states via an international river. This is especially so, given a significant development of concepts similar to 'freedom of navigation' (such as freedom of transit) in other areas of international law.

In addition to these basic principles of fluvial law, international river navigation has also been liberalised through bilateral and regional treaties for almost all major rivers on different continents.[32] Those treaties set out more detailed and technical rules on river navigation, including rules governing the division of competences between different territorial authorities regulating river navigation, administration of river traffic, harmonisation of national regulations and maintenance of navigation facilities.[33]

2. Transit via international straits and canals

In the area of inter-canal (straits) communication, the right of passage through straits used for international navigation is internationally regulated by the regime of 'innocent passage', defined as passage that is not prejudicial to the peace, good order or security of the coastal state.[34] This regime is now part and parcel of customary international law.[35] The right

[31] See McCaffrey, Stephen C., 'International Watercourses' in *Max Planck Encyclopaedia of Public International Law* (2014), www.mpepil.com, accessed 30 May 2014, para. 5; and Article 4 of the Statute of the Convention on Navigable Waterways, above n 25.

[32] See Uprety, above n 14 at 37–44; McCaffrey, above n 31.

[33] See the overview of the river regimes for the Rhine and the Danube as well as the EU law applicable to those rivers in Tiroch, Katrin, 'Rhine River' in *Max Planck Encyclopaedia of Public International Law* (2014), www.mpepil.com, accessed 30 May 2014; and Zeilinger, above n 24, respectively.

[34] See this right established in Article 17 and defined in Article 19 of the LOSC. United Nations Convention on the Law of the Sea, http://www.un.org/depts/los/convention_agreements/texts/unclos/unclos_e.pdf, accessed 30 December 2012.

[35] See *Corfu Channel*, above n 8 at 28. Although in the main judgment the ICJ referred specifically to the passage of warships, Judge Alvarez, in his Separate Opinion, clarifies as follows: 'it may be accepted that, to-day, the passage through the territorial sea of a State, or through straits situated therein, and also through straits of an international character, is not a simple tolerance but is a *right* possessed by *merchant ships* belonging to other States.' See *Corfu Channel*, Separate Opinion of Judge Alejandro Alvarez, Judgment of 9 April 1949: I.C.J. Reports 1949, pp. 4, 39 at 46, emphasis added. The ICJ has subsequently confirmed a general right of innocent passage via international straits belonging to all states in *Maritime Delimitation and Territorial Questions between Qatar and*

of 'innocent passage' translates into an obligation for a coastal (transit) state not to impose on foreign ships, inter alia, charges by reason only of their passage through the territorial sea, and regulations that are discriminatory or have the practical effect of denying or impairing the right.[36] Furthermore, the right of passage through international straits is also subject to a separate and more elaborate treaty-based regime of 'transit passage' established by the LOSC, which provides for narrower regulatory rights of a coastal state.[37]

A canal normally becomes subject to the previously described regime of innocent passage by means of its permanent 'dedication' to the international community (i.e. internationalisation), which is normally done through treaties.[38] In this way, most significant canals have been opened to all states under individual treaty regimes without distinctions based on the flags of vessels.[39]

Bahrain, Merits, Judgment of 16 March 2001, I.C.J. Reports 2001, p. 40, para. 223. See also Jia, Bing B., *The Regime of Straits in International Law* (Oxford: Clarendon Press, 1998) at 213–214; Brownlie, above n 17 at 268; Melgar, above n 1 at 182; and Article 10(3) of the Resolution of the Institute of International Law of 1894, 'Règles sur la Définition et le Régime de la Mer Territoriale', Session de Paris, 1894, http://www.idi-iil.org, accessed 30 September 2011.

[36] See Articles 24 and 26 of the LOSC, clarifying the right of innocent passage under the Convention. United Nations Convention on the Law of the Sea, above n 34.

[37] Those rights include the right to regulate the safety of navigation and the right to establish sea lanes or traffic separation schemes, subject to agreement with other states that border the strait and after the submission to and adoption of the scheme by the competent international organisation i.e. the International Maritime Organization. Ibid., Part III, section 2. On the regime of transit passage and its differences from 'innocent passage', see Mahmoudi, Said, 'Transit Passage', in *Max Planck Encyclopaedia of Public International Law* (2010), www.mpepil.com, accessed 15 May 2011.

[38] In *S.S. 'Wimbledon'*, the PCIJ states: 'when an artificial waterway connecting two open seas has been permanently dedicated to the use of the whole world, such waterway is assimilated to natural straits.' *S.S. 'Wimbledon'*, above n 8 at 28. See also McNair, Arnold D. 'So-Called State Servitudes', 6 *British Year Book of International Law* (1925) 111 at 122. The doctrine of 'dedication' is also discussed in this chapter, Section IV.B.1.

[39] See, for instance, the Constantinople Conventions of 1888 in relation to the Suez Canal and the Hay-Bunau-Varilla Treaty of 1903 in relation to the Panama Canal, discussed in Brownlie, above n 17 at 264–266. The 'international' character of the Suez Canal was highlighted in a number of resolutions of different UN bodies, including the UN Security Council, responding to Egypt's nationalisation of the Suez Canal Company operating the canal in 1956, and the subsequent invasion of the canal zone by Great Britain, France and Israel. Those resolutions set out the key principles governing navigation through the canal, including freedom of navigation without discrimination and Egypt's territorial sovereignty over the canal. See United Nations, Security Council Resolution 118(S/3675) of 13 October 1956, in United Nations, *Security Council Resolutions*, http://www.un.org /en/sc/documents/resolutions/1956.shtml, accessed 15 February 2014; and United Nations, GA Resolution 997(ES-I) of 2 November 1956, in United Nations, *General*

In light of the foregoing, it can be argued that passage through international rivers, straits and canals is regulated by several layers of international norms, namely (i) general principles of international law, some of which have been codified in multilateral 'framework' agreements, such as the Final Act of the Congress of Vienna and the Convention on Navigable Waterways; and (ii) the more specific bilateral/regional treaty regimes. While the former establish general standards of treatment of foreign vessels, the latter provide detailed rules on how to implement those standards.

B. Freedom of Communications and Transit, and Access to the Sea for Land-Locked Countries

1. Freedom of communications and transit: from a bilateral to a multilateral approach

With the expansion of international commerce, transit of goods has been liberalised across different regions through treaties of friendship, commerce and navigation or peace, since the beginning of the nineteenth century. To mention a few examples, in Europe, freedom of trade communication, with duties exemptions, was established by the treaty between Sardinia and the Canton of Geneva back in 1816.[40] In Africa, in 1885, as a result of the Berlin Conference, a number of colonial powers signed the General Act on the administration of Africa's Congo region and circumjacent areas, establishing the complete freedom of trade of all nations, albeit subject to some territorial exemptions.[41] Great Britain and Portugal made arrangements to respect mutual freedom of transit without the payment of transit duties within their areas of influence in Africa and India.[42] Favourable transit regimes were established, inter alia, between Great Britain and South Africa in 1890 in relation to railway and waterway transit across Swaziland and between Ethiopia, on one hand, and Great Britain and Italy, on the other.[43] In Latin America,

Assembly Resolutions, http://www.un.org/documents/resga.htm, accessed 15 November 2015.

[40] United Nations Secretariat, 'Question of Free Access to the Sea of Land-Locked Countries', A/CONF.13/29 (14 January 1958) at 312.

[41] See Chapter I of the General Act of the Berlin Conference in *Protocoles et Acts Général de la Conférence de Berlin* (1884–1885), Annexe au Protocole No. 10, available at the Peace Palace Library, http://www.peacepalacelibrary.nl/, accessed 10 March 2014.

[42] Lauterpacht, 'Freedom of Transit', above n 1 at 328; and *Right of Passage over Indian Territory*, above n 8 at 37–38.

[43] UN Secretariat, 'Question of Free Access to the Sea of Land-Locked Countries', above n 40 at 312.

international transit of goods was also significantly liberalised since the nineteenth century by a range of bilateral treaties. Treaties were concluded, inter alia, between Bolivia (a land-locked country), on one part, and Argentina, Brazil, Chile, Paraguay (a land-locked country but also a possible transit country for Bolivia) and Peru, on the other. The transit regime established between those states broadly covered transit via the sea, river ports, river navigation, road, railways and even the transport of petroleum.[44]

Although during wars in the nineteenth and twentieth centuries belligerents destroyed trade and transit channels, there appears to be a strong opinion that in peace times the violation of transit was an immoral if not an internationally wrongful act.[45] In some treaties, freedom of transit was granted in perpetuity,[46] often as a basic right to be later elaborated in subsequent implementation agreements or regulations. However, the absence or lapse of those agreements would not suspend or restrict the application of the principle of freedom of transit.[47]

In the twentieth century, the development of the principle of freedom of communications and transit was characterised by its recognition in multilateral treaties. The 1919 Versailles Treaty, bringing about the end of WWI, was the first legal instrument declaring this principle at a multilateral level.[48] Part I of the Covenant of the League of Nations, Article 23(e), laid down the following general obligation:

> Subject to and in accordance with the provisions of international conventions existing or hereafter to be agreed upon, the Members of the League:
>
> ...

[44] Ibid. at 315–318.

[45] See League of Nations, 'Verbatim Reports and Texts Relating to the Convention on Freedom of Transit', Barcelona Conference (Geneva, 1921) at 283.

[46] See Article 1 of the Treaty of Commerce between Bolivia and Brazil of 1910 and Article 6 of the Treaty of Peace between Bolivia and Chile of 1904 in UN Secretariat, 'Question of Free Access to the Sea of Land-Locked Countries', above n 40 at 316–317.

[47] See Article 5 of the Treaty of Friendship and Non-aggression between Bolivia and Peru of 1936 in ibid. at 318.

[48] Here the author does not count the 1648 Treaty of Westphalia, where the very idea of then a new concept of 'sovereignty' was subject to the obligation to ensure '*a secure passage* by sea and land'. Article LXX of the Treaty of Westphalia of 1648, http://avalon.law.yale.edu/17th_century/westphal.asp, accessed 30 September 2011, emphasis added. Grotius, in his natural law theory of state property, also argued that the initial division of territories among sovereigns had assumed the reservation of some common rights, including the right of harmless passage. Grotius, above n 7, chapter 2, sections VI:2 and VI:4.

e. will make provision to secure and maintain freedom of communications and of transit and equitable treatment for the commerce of all Members of the League . . .[49]

In addition, Part XII of the Treaty, although referring to Germany specifically, contained a statement of principles that the Allied and Associated Powers reportedly intended to adopt as a general solution to the problem of freedom of transit.[50] On this basis, it was suggested that these principles were accepted 'by all European States and, if possible, by all the nations in the world'.[51] Among the principles established by Part XII, Article 321 sets out: (i) the principle of freedom of transit for persons, goods, vessels, carriages, wagons and mails on the routes most convenient for international transit; (ii) non-discrimination; (iii) the obligation to impose charges on transport in transit that are reasonable, having regard to the conditions of the traffic; (iv) the prohibition of regulatory distinctions based directly or indirectly on the ownership or on the nationality of the ship or other means of transport; and (vi) the exemption of traffic in transit from customs duties.[52]

Further, with a view to implementing Article 23(e) of the Covenant, the League convened a General Communication and Transit Conference for the first time in 1920 in Barcelona (Barcelona Conference). For this event, it was agreed to establish the Commission of Enquiry on Freedom of Communications and Transit (the Enquiry Commission) to prepare draft documents, including draft international conventions.[53] While understanding that transit may involve complex and technical matters, in its report to the Conference participants, the Enquiry Commission stated that the Conference would be 'inspired by just those principles of freedom in the loftiest sense, and of equal respect for the rights and interests of every nation, which the Commission . . . has never failed to maintain and assert'.[54] Hence, it can be inferred from this statement that

[49] The 1919 Treaty of Versailles, http://avalon.law.yale.edu/subject_menus/versailles_menu.asp, accessed 30 May 2010. See the interpretation of this provision in Chapter V, Section III.A.

[50] See UN Secretariat, 'Question of Free Access to the Sea of Land-Locked Countries', above n 40 at 321. It is noteworthy that the Treaty also contained provisions on freedom of transit with reference to other states, such as Article 86, referring to the Czecho-Slovak state, and Article 93 – Poland. Treaty of Versailles, above n 49.

[51] UN Secretariat, 'Question of Free Access to the Sea of Land-Locked Countries', above n 40 at 321.

[52] Treaty of Versailles, above n 49.

[53] UN Secretariat, 'Question of Free Access to the Sea of Land-Locked Countries', above n 40 at 321–22.

[54] Ibid. at 322.

the intention of the Conference was to codify only basic principles and state practice that at that time were already in existence.[55]

As a result of the Conference, two conventions were concluded, namely the Barcelona Conventions on Freedom of Transit and the Convention on Navigable Waterways.[56] Following the Barcelona Conference, at the second Conference on Communications and Transit, which met in Geneva in 1923, Members of the League signed three other important conventions: Conventions on the International Regime of Railways, on Maritime Ports and on the Transmission in Transit of Electric Power (Convention on Electric Power).[57] All of these conventions reiterate, in their own way, the principle of freedom of transit as well as the other fundamental principles linked to transit mentioned earlier in this section. For example, without delving into those special areas, it suffices to quote Article 2 of the Statute of the Barcelona Convention on Freedom of Transit, which provides:

> [M]easures taken by Contracting States for regulating and forwarding traffic across territory under their sovereignty or authority shall facilitate free transit on routes in use convenient for international transit. No distinction shall be made which is based on the nationality of persons, the flag of vessels, the place of origin, departure, entry, exit or destination, or on any circumstances relating to the ownership of goods or of vessels, coaching or goods stock or other means of transport.[58]

It has been observed in this section that the process of the development of international transit rules started from various bilateral and regional economic agreements and was later systematised by the League of

[55] League of Nations, 'Freedom of Transit', above n 45 at 286.

[56] UN Secretariat, 'Question of Free Access to the Sea of Land-Locked Countries', above n 40 at 322–326; Convention and Statute on Freedom of Transit of 1921, 8 *American Journal of International Law, Supplement* (1924) 118; and Convention on Navigable Waterways, above n 25.

[57] See Convention and Statute on the International Regime of Railways of 1923, *League of Nations Treaty Series*, No. 1129 (1926) 57; Convention and Statute of the International Regime of Maritime Ports of 1923, *League of Nations Treaty Series*, No. 1379 (1926–1927) 287; and Convention relative to the Transmission in Transit of Electric Power of 1923, 22 *American Journal of International Law Supplement* (1928) 83.

[58] The preamble of this Convention recognises more generally that 'it is well to proclaim the right of free transit and to make regulations thereon as being one of the best means of developing cooperation between states without prejudice to their rights of sovereignty or authority over routes available for transit'. See Barcelona Convention on Freedom of Transit, above n 56; and Appendix 1 to this monograph. For a more detailed analysis of the negotiating history and substantive obligations under the 1921 Barcelona Convention on Freedom of Transit, see Melgar, above n 1 at 50–78.

Nations into a number of multilateral legal instruments, pronouncing certain important principles relevant to freedom of transit, including the principle of freedom of transit itself. While, even before the League of Nations, the international regulation of transit in different treaties was already based on similar principles or ideas, the League of Nations made an important initial contribution to their gradual codification and progressive development, which has continued throughout the second half of the twentieth century.

2. Freedom of transit in modern multilateral treaties

(i) The law of the sea During the post-WWII period, the issue of freedom of transit got its second wind in multilateral talks during the codification and development of the law of the sea in the time between UN Conferences on the Law of the Sea (UNCLOS I and III). These Conferences lasted between 1956 and 1982 and resulted in a range of conventions, including the 1958 Geneva Convention on the High Seas, the 1965 New York Convention on Transit Trade of Land-Locked States, and the LOSC.[59] At these conferences, a general view of land-locked countries was that freedom of transit was inextricably linked to their right of access to the sea and the principle of sovereign equality of nations. On this basis, some land-locked countries claimed that freedom of transit was a general principle of international law. Not surprising, a group of transit states contested the existence of a general international law obligation to provide freedom of transit even to land-locked countries, arguing that this freedom could only be granted by treaties on the basis of reciprocity.[60]

Despite this clash of positions, the UNCLOS negotiations resulted in the ultimate endorsement of the principle of freedom of transit with certain caveats. The LOSC, which ended the long process of codification and development of the law of the sea,[61] in its Part X, devotes special attention to the problem of transit barriers that land-locked countries

[59] See these conventions discussed in Uprety, above n 14 at 65–75; Melgar, above n 1 at 154–161.

[60] On the details of negotiations, see Uprety, above n 14 at 59–96; Nandan, Satya N. and Rosenne, Shabtai (eds.), *United Nations Convention on the Law of the Sea 1982: A Commentary*, vol. 3 (The Hague: Centre for Oceans Law and Policy, Martinus Nijhoff Publishers, 1995) at 371–457; and Dupuy, René-Jean and Vignes, Daniel (eds.), *A Handbook on the New Law of the Sea*. vol. 1 (Boston: Hague Academy of International Law, Martinus Nijhoff Publishers, 1991) at 512–523.

[61] See the preamble of the LOSC. United Nations Convention on the Law of the Sea, above n 34.

face. The LOSC is a multilateral treaty regime consisting of 166 states.[62] The regulation of transit under the LOSC is discussed in greater detail in Chapter VI.[63] At this stage, however, it is important to outline some relevant aspects of this regulation, since the LOSC is an important milestone in the overall development of the principle of freedom of transit in public international law. Moreover, this Convention recognises and addresses the problem of the lack of adequate transit facilities, as well as access to such facilities, for the effective implementation of transit obligations.

In Article 125(1), the Convention recognises the right of access to and from the sea for land-locked states and, to this end, establishes freedom of transit of land-locked states through the territory of transit states.[64] Further, Article 125(3) recognises that transit states may have a legitimate sovereign interest to regulate transit.[65] Article 125(2) strikes a balance between the interests of land-locked and transit states, setting out the obligation for those states to agree on '[t]he terms and modalities for exercising freedom of transit ... through bilateral, sub-regional or regional agreements'.[66] Nandan and Rosenne explained the meaning of the obligations under Article 125 as follows:

> Article 125 ... *balances conflicting concerns regarding the transit regime for land-locked States.* One concern is represented by the claim of the land-locked States for freedom of access to the sea across the territory of other States. The second concern entails the assertion of the transit States that complete [absolute] freedom of transit would compromise their territorial sovereignty.[67]

Importantly, the balance between the interests of land-locked and transit states that Article 125(2) establishes does not mean that freedom of transit depends on the conclusion of an agreement on the terms and modalities of transit. While the conditions of transit may be the subject of this agreement, the right to transit per se is recognised by the LOSC and cannot be denied at the discretion of a transit state. Indeed, the denial of transit, without just reasons, would appear to be inconsistent with the principle of the prohibition of abuse of rights (*abus de droit*), incorporated in the LOSC through Article 300.[68]

[62] Appendix 2. [63] See Chapter VI, Section III.A.3.
[64] United Nations Convention on the Law of the Sea, above n 34. [65] Ibid. [66] Ibid.
[67] Nandan and Rosenne (eds.), above n 60 at 409, emphasis added.
[68] See United Nations Convention on the Law of the Sea, above n 34; Dupuy and Vignes, above n 60 at 517–518; Melgar, above n 1 at 197; and Chapter VI, Section III.A.3(ii). The principle of *abus de droit* is discussed in Chapter V, Section II.B.3.

The approach adopted in the LOSC is not different from that taken in the Treaty of Versailles, Article 23(e), or other public international law regimes regulating transit. The general trend observed in this chapter is that the principle of freedom of transit, while being recognised as such, is always balanced against the legitimate sovereign concerns of the transit state. Taking into account particular technical complexities of the regulation of transit in different areas of international law, this balance is usually struck at two or more levels of regulation: (i) the multilateral (general) level; and (ii) the level of bilateral, sub-regional or regional agreements, at which the detailed conditions and modalities of transit are specified.[69]

Like other treaties discussed earlier, the LOSC establishes additional legal guarantees that aim to facilitate transit and complement the principle of freedom of transit. They include: (i) reasonableness and proportionality of charges; (ii) the prohibition on imposing customs duties on traffic in transit; (iii) the obligation to afford non-discriminatory treatment to traffic in transit with respect to taxes or charges levied in connection with the use of transit facilities/means of transport or use of ports; and (iv) the obligation to take all appropriate measures to avoid delays or other difficulties of a technical nature in traffic in transit.[70]

Importantly, as elaborated in Chapter VI, the LOSC aims to resolve the problem of the lack of transit facilities and adequate access to the existing transit facilities in some transit countries, thereby recognising the inherent importance of these issues for the effective implementation of freedom of transit. In particular, in Article 129, it requires that the transit states and land-locked states engage in cooperation on the construction or improvement of the means of transport in circumstances where the existing means are inadequate to give effect to the freedom of transit or where they do not exist.[71]

(ii) UN Charter and international economic agreements: ITO Charter, GATT 1947 and GATT 1994 Apart from the LOSC, freedom of transit has been addressed in other multilateral treaties. The UN

[69] Among examples relevant to pipeline gas transit, see agreements between Azerbaijan, Georgia and Turkey on the South Caucasus Pipeline (SCP) System, discussed in this chapter, Section III.E. See also various regional economic agreements regulating energy transit discussed in this chapter, Section III.D.

[70] See LOSC, Articles 127, 130(1) and 131. United Nations Convention on the Law of the Sea, above n 34. These provisions are discussed in greater detail in Chapter VI, Section III.A.3.

[71] See Article 129 in United Nations Convention on the Law of the Sea, above n 34.

Charter mentions it, albeit in a limited and particular context – the maintenance of international peace and security.[72] In addition, the UN Charter indirectly addresses the principle of freedom of transit through the requirement of co-operation in international economic matters 'based on respect for the principle of equal rights and self-determination of peoples'.[73] Other, more detailed aspects of economic transit were left to international organisations and agreements specialised in international economic relations, such as the International Trade Organisation (ITO) and the GATT 1947. This could be the main reason the UN Charter itself does not address this matter.

During the negotiations on the ITO Charter (also known as the Havana Charter) in the 1940s, the principle of freedom of transit was enshrined in Article 33 (Freedom of Transit). Although the intentions to create the ITO were ultimately not fruitful, the principle of freedom of transit in Article 33 was also incorporated almost verbatim into the GATT 1947 (Article V), which was created as a separate agreement to be applied provisionally until the establishment of the ill-fated ITO. On 1 January 1995, the GATT 1947 was incorporated into the newly established WTO legal framework as one of the covered agreements (GATT 1994).

Both Article 33 of the ITO Charter and Article V of the GATT were based on the principles developed by the Barcelona Convention on Freedom of Transit. As Jackson notes: 'The preparatory work [of Article V] reflects the view that this article was not inconsistent [with the Barcelona Convention]. . . . Thus GATT contracting parties who were also parties to the convention were protected from the embarrassment that might otherwise occur.'[74] This explains why Article 33(2)

[72] Article 43(1) of the Charter establishes the common obligation of all UN Members 'to make available to the Security Council, on its call and in accordance with a special agreement or agreements . . . assistance, and facilities, *including rights of passage*'. UN Charter, http://www.un.org/en/documents/charter/, accessed 10 January 2012, emphasis added.

[73] See UN Charter, Article 55 in ibid. The principle of economic cooperation is discussed separately in Chapter V, Section II.B.1 of this study.

[74] See Jackson, John H., *World Trade and the Law of GATT* (Indianapolis: The Bobbs-Merill Company, Inc., 1969) at 507. See also Economic and Social Council, 'Preparatory Committee of the International Conference on Trade and Employment', E/PC/T/C.II/54/Rev.1 (28 November 1946) at 7; and World Trade Organization, Secretariat, 'Article V of GATT 1994 – Scope and Application', Negotiating Group on Trade Facilitation, TN/TF/W/2 (12 January 2005) at 3.

of the ITO Charter and Article V:2 of the GATT closely mirror provisions of Article 2 of the Statute of the Barcelona Convention (quoted in Section III.B.1) and in particular its principle of freedom of transit and the prohibition on discrimination (distinctions) against traffic in transit. In addition, Article 33 of the ITO Charter and Article V of the GATT incorporated other major principles that had been recognised by previous treaties dealing with transit (including the Barcelona Convention), such as reasonableness and proportionality of charges and regulations applied in connection with transit, and the exemption of traffic in transit from customs duties.[75]

The GATT, including Article V, contains a number of provisions safeguarding sovereign interests of a transit state, namely the standards of reasonableness and necessity in paragraphs 3 and of 4 of Article V,[76] and general and security exceptions in Articles XX and XXI of the GATT.[77]

Interestingly, Article 33(6) of the ITO Charter established the mechanism pursuant to which the ITO would undertake studies, make recommendations and promote international agreements relating, inter alia, to the equitable use of transit facilities and other measures designed to promote the objectives of Article 33. This provision required ITO Members to cooperate in fulfilling this task of the ITO.[78] A similar provision, however, was not incorporated into the GATT 1947, arguably because, during its initial existence, the GATT lacked a formal institutional structure. Nevertheless, to some extent, the new Agreement on Trade Facilitation appears to fill this gap by establishing the Committee on Trade Facilitation. The Agreement delegates to the Committee the function of, inter alia, facilitating the discussions among Members of specific matters related to its implementation and requires Members to 'cooperate and coordinate

[75] See Articles V:2, V:3 and V:4 of the GATT 1994, *WTO Legal Texts*, http://www.wto.org/english/docs_e/legal_e/legal_e.htm, accessed 5 March 2014; Articles 33(2)-33(4) of the International Trade Organization Charter, http://www.wto.org/english/docs_e/legal_e/havana_e.pdf, accessed 10 January 2014; and Appendix 2.

[76] See Articles V:3 and V:4 of the GATT 1994, *WTO Legal Texts*, above n 75. These provisions are further elaborated in Article 11 of the Agreement on Trade Facilitation. World Trade Organization, Ministerial Conference, Agreement on Trade Facilitation, Ministerial Decision of 7 December 2013, WT/MIN(13)/36, WT/L/911 (11 December 2013)/ Preparatory Committee on Trade Facilitation, Agreement on Trade Facilitation, WT/L/931 (15 July 2014). The obligations under Article V of the GATT and Article 11 of the Agreement on Trade Facilitation are discussed at length in Chapter IV of this study.

[77] See these provisions discussed in Chapter IV, Section V.A.2.

[78] The International Trade Organization Charter, above n 75.

with one another with a view to enhancing freedom of transit'.[79] These discussions in the Committee and cooperation/coordination efforts could address the lack of adequate transit infrastructure in some states.

C. *Freedom of Transit and Network-Bound Energy*

1. The emergence of network-bound energy transit as a separate area of the international transit regulation

It should be noted at the outset that transit of network-bound energy (such as gas, electricity and pipeline crude oil) is a relatively new area of international law. At the bilateral or regional level, certain aspects of such transit were already regulated in the early twentieth century.[80] However, at the multilateral level, until recently, there were scant rules governing this subject. The only exception is the League of Nations' Convention on Electric Power, which to some extent, in Articles 4 and 5, dealt with capacity establishment rights.[81] Nevertheless, this attempt to regulate network-bound transit was rather modest. Apart from Article 6, which prohibits levying special dues or charges based on whether the power transmission was in transit, it is difficult to find any 'hard-law' obligation in the Convention. Moreover, in accordance with Article 1 of this document, transit states can avoid any obligation at all if they show that transmission of electric power would be detrimental to their national economy or security.[82]

Furthermore, it is often argued that neither the ITO Charter nor the GATT 1947 initially intended to specifically regulate energy trade, including network-bound transit.[83] In the area of trade in oil and gas, prior to the creation of the WTO, the GATT's role was in any event insignificant, given that the majority of producing countries were

[79] See Articles 11(16) and 23(1) in World Trade Organization, Agreement on Trade Facilitation, above n 76. This Agreement is discussed in Chapter IV, Section V.C.

[80] See the Montevideo Convention on Construction of Oil Pipelines of 1941 concluded at the Regional Conference of the River Plate States (that is, Argentina, Bolivia, Brazil, Paraguay and Uruguay) discussed in Váli, above n 16 at 293.

[81] Convention on Electric Power, above n 57. [82] Ibid.

[83] See Lamy, Pascal, *The Geneva Consensus: Making Trade Work for All* (New York: Cambridge University Press) at 116–117; Marceau, Gabrielle, 'The WTO in the Emerging Energy Governance Debate', 5 *Global Trade and Customs Journal* (2010) 83 at 83. While this general proposition is sound, it is important to note that the GATT drafters did think of trade in natural resources at least in the context of the definition of the term 'state-trading enterprise' for the purposes of GATT Article XVII (State-Trading Enterprises). See AD Note to Article XVII:1(a), GATT, *WTO Legal Texts*, above n 75.

outside this Treaty[84] and that any energy-related complaint could effectively be blocked through a positive consensus rule of the adoption of panel reports that then existed in the GATT dispute settlement system.[85]

In the LOSC, despite Afghanistan's 1976 proposal to include pipelines and gas lines as 'the means of traffic' in the definition of 'traffic in transit', this issue was ultimately excluded from the general regulation of transit in Part X of the Convention, and it was made subject to possible special agreements between transit and land-locked states.[86] To some extent, the cross-border movement of energy is directly regulated by the LOSC in its provisions addressing the rights of states to lay submarine cables and pipelines on the continental shelf and exclusive economic zone (EEZ), which is not transit per se.[87]

In light of the foregoing, while the principle of freedom of transit was recognised in most areas of international law, before the early 1990s, it does not appear to play a significant role in the field of transit of network-bound energy, including pipeline gas. This situation started to change after the negotiations began on the Energy Charter and the ECT in 1991. The substantive content of the ECT transit rules is discussed at greater length in Chapter VI.[88] However, given that the ECT is the only multilateral treaty specialised in energy[89] and, therefore, constitutes a significant contribution to the development of international rules governing energy transit, the following section explains its origin and outlines its key principles relevant to our discussion. The ECT can provide certain guidance to the interpreters of WTO transit rules on the aspects of the international transit regulation that are of particular

[84] Among developing countries producing oil and gas, Mexico joined the GATT 1947 in 1986; Venezuela and Bolivia joined in 1990. Others, such as Russia and Saudi Arabia, became Members after the creation of the WTO. See Desta, Melaku G., 'The GATT/WTO System and International Trade in Petroleum: An Overview', 21 *Journal of Energy and Natural Resources Law* (2003) 385 at 392; World Trade Organization, 'Members and Observers', www.wto.org/english/thewto_e/whatis_e/tif_e/org6_e.htm, accessed 30 November 2015.

[85] Currently, the recommendations and rulings of the panels and the Appellate Body are automatically adopted by the DSB, unless Members decide by consensus not to adopt a report. After the report is adopted, the parties to the dispute must unconditionally accept it. Dispute Settlement Understanding, Article 17(14), *WTO Legal Texts*, above n 75.

[86] See Nandan and Rosenne (eds.), above n 60 at 404; and Article 124(2) of the LOSC. United Nations Convention on the Law of the Sea, above n 34. The relevance of this exclusion is addressed in Chapter VI, Section III.A.3(i).

[87] See Chapter VI, Section III.A.3(i). [88] See Chapter VI, Section III.A.2.

[89] See Vinogradov, 'Challenges of Nord Stream', above n 9 at 6.

importance to trade in pipeline gas, as well as general trends in the international regulation of this area of transit.[90]

2. Freedom of network-bound energy transit and the Energy Charter

(i) Origins of the Energy Charter The Energy Charter is an international organisation consisting of forty-seven full Members that signed and ratified the ECT, and Australia, Belarus, Iceland, Norway and Russia who have not yet ratified this treaty.[91] The history of this Organisation can be traced back to the dissolution of the Soviet Union when, after the end of the Cold War, at the Summit of the European Council in Dublin, in June 1991, the Dutch Prime Minister, Ruud Lubbers, proposed a 'European Energy Charter' (a political declaration), which was adopted the same year in The Hague.[92] This document laid down the foundations for further negotiations on a more comprehensive legal framework – the ECT – which regulates most aspects of energy cooperation between its Members, including investment, transit and trade.[93]

Among the key interests underlying the Charter as well as the ECT negotiations were to:

- draw Western capital into the oil and gas industry of the former Soviet Union states;
- open up the possibility for the investment flows from the former Soviet Union states to Western markets;
- connect energy-rich Central Asian states with Western consumers by improving energy transit conditions, including via Russia and Ukraine;
- ensure energy security of the EU energy-importing Members; and
- support political and economic reforms in newly established Eastern European nations.[94]

[90] See Chapter I, Section III.B.2(iii).

[91] Energy Charter Secretariat, http://www.encharter.org/, accessed 30 November 2013.

[92] Doré, Julia, 'Negotiating the Energy Charter Treaty', in T. W. Wälde (ed.), *The Energy Charter Treaty: An East–West Gateway for Investment & Trade* (London: Kluwer Law International, 1996) 137 at 138.

[93] Ibid. at 140. Article 2 of the ECT sets out its purpose to establish 'a legal framework in order to promote long-term cooperation in the energy field, based on complementarities and mutual benefits, in accordance with the objectives and principles of the Charter'. Energy Charter Secretariat, 'The Energy Charter Treaty and Related Documents: A Legal Framework for International Energy Co-operation' (Brussels: Energy Charter Secretariat, 2004).

[94] Originally the negotiations were designed as Western–Eastern European cooperation. However, later on, other non-European countries were invited to join, including the

These interests were to be secured by a number of heavily negotiated international obligations, such as investment protection, including non-discrimination with respect to the making of investments,[95] and transit obligations.[96] Negotiations on energy trade rules (another element of the ECT) appear to have been less intense at that time, given that the negotiating period overlapped with the then on-going GATT Uruguay round talks and, in general, all negotiating states (including non-GATT Members) intended to incorporate GATT's trade rules into the ECT.[97] Among other issues discussed during the negotiations, but irrelevant to our subject, were energy efficiency and a nuclear protocol.[98]

Particular attention of negotiators was drawn to the issue of transit, since, from a business perspective, the lack of effective transit rules in the ECT would not provide a sufficient signal to the private sector to invest in the upstream energy market in the new and unfamiliar environment of the former Soviet Union. As Jenkins writes: 'It is not an exaggeration to say that the success of all western oil and gas investment in the former Soviet Union effectively hangs on the reliable provision of economically viable transit routes from point of production to hard currency

United States and other non-European OECD Members. Doré, above n 92 at 138–140. See Russian position in negotiations in Konoplyanik, Andrei A. and von Halem, Freidrich, 'The Energy Charter Treaty: A Russian Perspective', in T. W. Wälde (ed.), *The Energy Charter Treaty: An East–West Gateway for Investment & Trade* (London: Kluwer Law International, 1996) 156.

[95] Article 1(8) of the ECT defines the terms 'Make Investments' or 'Making of Investments' as 'establishing new Investments, acquiring all or part of existing Investments or moving into different fields of Investment activity'. Energy Charter Treaty, above n 93. As Doré states, the right to make an investment was the principal interest of the United States in the negotiations. The ultimate failure of the ECT to establish this right was one of the major reasons the United States did not sign the Treaty. Doré, above n 92 at 151–152.

[96] Doré, above n 92 at 140–152.

[97] In this regard, Doré writes: 'the Energy Charter Treaty negotiations were directly linked to the GATT talks, since – based on a US proposal – trade in energy materials under the Treaty was to be governed by GATT rules.' Ibid. at 145. Along the same lines, the negotiating history of the ECT indicates that the ECT drafters intended to build the Treaty upon the basic trade principles developed under the auspices of the GATT. At the same time, their intention was to expand on these principles when they were deemed insufficient to address matters specific to energy. See European Energy Charter Conference Secretariat, 'Note from the Chairman of Working Group II', 14/91 (BP 3), Brussels, 11 October 1991 at 2; European Energy Charter Conference Secretariat, 'Note from the Chairman of Working Group II', 36/93 (BA-40), Brussels, 6 May 1993 at 7; European Energy Charter Conference Secretariat: Comments on Issues Assigned to the Legal Sub-group, No. 1396, 14 May 1993 at 3. See also Rakhmanin, above n 55 at 124; and Wälde, Thomas W. and Gunst, Andreas J., 'International Energy Trade and Access to Energy Networks', 36 *Journal of World Trade* (2002) 191 at 213.

[98] Doré, above n 92 at 141.

markets.'[99] Moreover, the importance of transit for energy cooperation was already recognised at that time by the Energy Charter, which declared that '[t]ransit should take place in economic and environmentally sound conditions'.[100]

In the context of transit, while the negotiations concerned various issues – the prohibition of the interruption of the existing energy flows, conduct of private pipeline operators, dispute settlement rules, and some technical matters, including the territorial scope of the application of the ECT – the most heavily negotiated issues were the conditions of access to existing pipelines (third-party access) or establishing the new pipeline infrastructure (capacity establishment). These issues appear to have permeated the discussions on almost all transit provisions of the ECT. Without delving into the details of the ECT's transit rules, which are analysed in Chapter VI, the drafters discussed these issues in the context of an overarching obligation in ECT Article 7(1) (Transit) to take the necessary measures to facilitate energy transit;[101] a non-discrimination obligation relating to energy transport and use of energy transport facilities under Article 7(3);[102] the obligation not to place obstacles in the way of new capacity being established under Article 7(4);[103] as well as exceptions and qualifications to the latter obligation under Articles 7(5) and 7(9).[104] In addition, ECT Article 7(2) established a Members'

[99] Jenkins, David, 'An Oil and Gas Industry Perspective', in T. W. Wälde (ed.), *The Energy Charter Treaty: An East–West Gateway for Investment & Trade* (London: Kluwer Law International, 1996) 187 at 191.

[100] See Energy Charter Secretariat, 'The Energy Charter Treaty and Related Documents', above n 93 at 211, 217.

[101] European Energy Charter Conference Secretariat, 'Room Document 10', Plenary Session, 24–26 March 1993 at 2–3.

[102] European Energy Charter Conference Secretariat, 'Basic Agreement for the European Energy Charter (Draft)', 31/92 (BA 13), 19 June 1992 at 39; European Energy Charter Conference Secretariat, 'Revised Draft', Annex, 15/93 (BA-35), Brussels, 9 February 1993 at 37, 39.

[103] European Energy Charter Conference Secretariat, 'Submission from European Communities', Brussels, 1 April 1992 at 2; European Energy Charter Conference Secretariat, 'Room Document 10', above n 101 at 3; European Energy Charter Conference Secretariat, 'Memorandum to Legal Sub-group, Agenda for 21–25 June Legal Sub-group Meeting', 47/93 (LEG-9), Brussels, 9 June 1993 at 1; and Organisation for Economic Co-operation and Development (OECD), 'Memorandum', 27 May 1993 at 2.

[104] European Energy Charter Conference Secretariat, 'Room Document 2 (Attached to the Basic Agreement)', Working Group II, Brussels, 1 June 1992 at 2; European Energy Charter Conference Secretariat, 'Room Document 6', Working Group II, Brussels, 22–23 April 1993; and General Secretariat of the Council, 'European Energy Charter of Negotiations: Meeting of EC Ad Hoc Group (High Level) on 19 May 1993', Note to Delegations by the Presidency, 17 May 1993 at 3 and Annex II.

best-endeavour commitment to encourage relevant entities to cooperate, inter alia, in: (i) modernising energy transport facilities necessary for transit; (ii) developing and operating facilities serving more than one contracting party; and (iii) facilitating the interconnection of energy transport facilities.[105]

The key questions the negotiating parties raised with respect to these provisions were whether the ECT implies obligations for Members to: (i) provide compulsory third-party access to pipelines; (ii) allow 'making of investment'; or (iii) construct or improve their pipeline facilities, and, if so, what the scope was of these obligations.[106] Thus, it can be inferred from the ECT negotiating history that access to pipeline infrastructure through some sort of third-party access and/or capacity establishment was considered the fundamental pillar on which the effective freedom of gas transit would rest.

(ii) Key transit principles under the ECT The final provisions of the ECT reiterate the basic transit principles of the GATT, while introducing a number of important clarifications and amendments to these principles pertinent to network-bound energy trade. The ECT, in Article 7(1), expressly links its transit obligations to the principle of freedom of transit, thereby recognising the application of this principle in the area of network-bound energy trade.[107] The ECT recognises other basic principles relating to transit, such as: (i) the prohibition of distinctions based on the origin, destination or ownership of energy; and (ii) the requirement not to impose unreasonable delays, restrictions or charges on energy transit.[108]

[105] Energy Charter Treaty, above n 93.

[106] See Doré, above n 92 at 142–143. In addition, see, inter alia, European Energy Charter Conference Secretariat, 'Basic Agreement for the European Energy Charter (Draft)', above n 102 at 39; European Energy Charter Conference Secretariat, 'Revised Draft', above n 102 at 37, 39; European Energy Charter Conference Secretariat, 'Room Document 10', above n 101 at 3; and European Energy Charter Conference Secretariat, 'Memorandum to Legal Sub-group, Agenda for 21–25 June Legal Sub-group Meeting', above n 103 at 1.

[107] Energy Charter Treaty, above n 93. This link is further strengthened by the Transit Protocol, which is being negotiated by the Energy Charter Members and aims to further elaborate on ECT transit rules. The Protocol declares the objective of ensuring secure, efficient, uninterrupted and unimpeded transit as a means of promoting the economic growth and security of energy supply of its Members. See the text of the Protocol at: Energy Charter Secretariat, Transit Protocol, http://www.encharter.org/index.php?id=37, accessed 23 December 2013.

[108] See Article 7(1). Energy Charter Treaty, above n 93.

Importantly, the ECT expressly addresses third-party access and capacity establishment rights. Although the Treaty does not require a mandatory third-party access right, opting instead for a negotiated one, it provides that in the event that the energy transit cannot be achieved by means of existing transit facilities, transit states must not place obstacles in the way of new capacity being established, in accordance with their applicable legislation, consistent with the principle of freedom of transit.[109] In this way, the ECT strikes a balance between the interests of transit-dependent states and the regulatory autonomy of a transit state in the area of network-bound energy trade. As explained in the following chapters, given the particular features of gas transit, a mandatory third-party access right would not always be technically feasible.[110] That said, the ECT recognises clearly the importance of third-party access and capacity establishment rights for the effective regulation of freedom of transit in this area, by requiring the transit state to operationalise its transit obligations through invoking either of these rights and regardless of whether energy transit is technically feasible via existing transit facilities.

Before concluding on the ECT, it should be mentioned that most recently, in May 2015, at the Ministerial Conference hosted by the Government of The Netherlands ('The Hague II'), a number of states signed a declaration of political intention (the International Energy Charter), which aims at strengthening further energy cooperation between the Members of the Energy Charter, as well as involving additional parties in this cooperation, such as China, the Republic of Korea and the Economic Community of West African States (ECOWAS). This document reiterates the key principles set out in the European Energy Charter and the ECT, including those promoting freedom of transit and access to energy networks for international transit purposes.[111]

3. Other international instruments recognising freedom of network-bound energy transit

Freedom of transit of network-bound energy has also gained importance in a more general economic law context. It is now commonly accepted

[109] The latter obligation is subject to certain exceptions, including cases in which a transit state can demonstrate that energy transit would endanger the security or efficiency of its energy systems, including the security of supply. See Understanding IV:1(b)(i), Articles 7(4), 7(1) and 7(5) of the ECT, ibid.

[110] See Chapter V, Section III.B.1

[111] The International Energy Charter, http://www.energycharter.org/process/international-energy-charter-2015/, accessed 15 September 2015.

that WTO law, including its transit regulation, applies to network-bound energy.[112] Freedom of transit has been linked to particular energy-related concerns, such as energy security, in a range of non-binding international resolutions and declarations adopted under the auspices of the UN. For example, the UN GA Resolution 63/210 of 19 December 2008 declares that stable, reliable and efficient energy transportation is in the interest of the entire international community. This resolution refers to reliable energy transit as an important factor in ensuring sustainable development and international cooperation.[113]

In the area of economic cooperation between land-locked and transit states, the UN Millennium Declaration, among its developmental goals, established an objective to devise practical solutions to the barriers land-locked countries face when exercising their right of transit and access to the sea.[114] In 2003, the UN convened an international ministerial conference in Almaty, Kazakhstan (the Almaty Conference) to devise the means to implement this objective. The Conference was attended by eighty-two UN Members, including land-locked and developing transit states and addressed, among other things, the pipeline transit problems land-locked countries experience. The result of this Conference, the Almaty Declaration, reaffirmed 'the right of access of land-locked countries to and from the sea and freedom of transit through the territory of transit countries by *all* means of transport [including pipelines], in accordance with applicable rules of international law'.[115] In 2014, the UN convened a follow-up conference in Vienna, Austria, which reiterated the main principles of the Almaty Declaration. In particular, among the objectives of the Vienna Programme of Action that the participants agreed to implement in the Vienna Declaration adopted at the Conference are: the 'promot[ion] [of] unfettered, efficient and cost-effective access to and from the sea by all means of transport', and the 'develop[ment] [of] adequate transit transport infrastructure networks [including energy

112 The details of WTO transit rules are discussed in Chapter IV.

113 United Nations, GA Resolution 63/210 of 19 December 2008, 'Reliable and stable transit of energy and its role in ensuring sustainable development and international cooperation', above n 39.

114 See United Nations, 'Millennium Declaration, adopted by the GA Resolution at 8th plenary meeting', A/55/L.2 (8 September 2000).

115 See United Nations, 'The UN Report of the International Ministerial Conference on Land-Locked and Transit Developing Countries and Transit Transport Co-operation', and the attached Almaty Declaration, A/CONF.202/3 (28 and 29 August 2003) at 25, emphasis added.

infrastructure] and complet[ion of] missing links connecting landlocked developing countries'.[116]

D. *Freedom of Transit in Regional Economic Agreements*

In addition to multilateral treaties, the principle of freedom of transit is established through a web of regional economic agreements. These agreements can be seen as a means to implement the basic transit obligations laid down at a more general level in WTO law as well as other multilateral treaties governing transit. They set out the modes and modalities of energy transit that are appropriate for a given region, doing so in a more direct and specific manner than multilateral negotiations could achieve.

In Europe, among Members of the EU, freedom of transit falls within the scope of free movement of goods, established by Part III, Title II of the EU Treaty.[117] In *Case 159/94 Commission v France*, the European Court of Justice held that the EU regime of the free movement of goods applies to trade in gas.[118] The EU Treaty also addresses particular aspects of energy trade within the Union relating to gas transit. In Part III, Title XVI, it assigns to the Union the tasks to: (i) contribute to the establishment and development of trans-European networks in, inter alia, the area of energy infrastructure; and (ii) promote the interconnection and interoperability of national networks as well as access to such networks, by establishing guidelines, technical standards, and supporting infrastructure projects of common interest among EU Members.[119] In Part III Title XXI (Energy), the Treaty defines the EU policies in the context of the establishment and functioning of the internal market for energy, which include ensuring security of energy supply in the Union and promoting the interconnection of energy networks.[120]

[116] See United Nations, GA Resolution 69/137 of 12 December 2014, 'Programme of Action for Landlocked Developing Countries for the Decade 2014–2024', above n 39, Annexes I (Vienna Declaration), and II (Vienna Programme of Action), sections IV and V.

[117] See the most recent consolidated versions of the Treaty on European Union and the Treaty on the Functioning of the European Union and the Charter of Fundamental Rights of the European Union (EU Treaty) at http://europa.eu/lisbon_treaty/full_text/index_en.htm, accessed 5 September 2013. The existence of the principle of freedom of transit within the EU was explicitly recognised in *Case 266/81 Siot*, [1983] ECR 731, para. 16; and *Case 115/02 Rioglass & Transremar*, [2003] I-12719, paras. 17–18.

[118] *Case 159/94 Commission v France* [1997] ECR I-5819, paras. 31–42.

[119] EU Treaty, above n 117. [120] Ibid.

Other legislative acts of the EU provide more specific rules on the functioning of the EU's internal gas market, which EU Members implement at their national levels. These rules require: (i) third-party access to pipeline and LNG facilities on the basis of published and regulated tariffs; (i) access to storage facilities when technically and/or economically necessary for gas supply to customers; and (iii) the unbundling of vertically integrated undertakings by separating pipeline operation from other activities, such as supply and production.[121] These rules are enforced at both the *ex ante* level by national regulatory authorities and the *ex post* level through EU competition law.[122]

At the level of energy cooperation between the EU and third-party countries, the EU has successfully 'exported' its *acquis communautaire* relating to the regulation of energy and competition policies to a number of neighbouring states, such as the non-EU-Member Balkan states, Moldova and Ukraine, through the framework of the so-called Energy Community Treaty.[123] This Treaty aims to establish a stable regulatory and market framework for energy cooperation between the EU and its neighbours aspiring to join the EU and, to that end, requires its non-EU parties to implement certain relevant provisions and principles of the EU energy law in national legislation.[124]

[121] See the third energy package of the EU, in particular, Articles 9, 32 and 33 of Directive 2009/73/EC of the European Parliament and of the Council of 13 July 2009 concerning Common rules for the Internal Market in Natural Gas and Repealing Directive No. 2003/55/EC, OJ L 211, 14.8.2009, 94; and Regulation (EC) No. 715/2009 of the European Parliament and the Council of 13 July 2009 on Conditions for Access to the Natural Gas Transmission Networks and Repealing Regulation, OJ 2009 L 211, 36. In addition, see the analysis of the EU second energy package in Cameron, Peter D., 'Completing the Internal Market in Energy: An Introduction to the New Legislation', in P. Cameron (ed.), *Legal Aspects of EU Energy Regulation: Implementing the New Directives on Electricity and Gas across Europe* (New York: Oxford University Press, 2005) 7 at 12–19. With respect to the issues of third-party access to gas infrastructure, the second energy package does not appear to have undergone significant changes in the third energy package, apart from new rules on access to gas storage facilities.

[122] See Talus, Kim, *Vertical Natural Gas Transportation Capacity, Upstream Commodity Contracts and EU Competition Law* (The Netherlands: Kluwer Law International, 2011) at 199–219; and Slotboom, Marco, 'Recent Developments of Competition Law and the Impact of the Sector Inquiry', in M. M. Roggenkamp and U. Hammer (eds.), *European Energy Law Report VII* (Antwerp: Intersentia, 2010) 97 at 106–108.

[123] See Energy Community Secretariat, 'Energy Community Treaty', http://www.energy-community.org, accessed 12 September 2013.

[124] See the 2005 Treaty Establishing the Energy Community and in particular Articles 2 and 3 thereof, ibid.

In Eastern Europe,[125] the principle of freedom of transit is established by the 2011 Free Trade Agreement of the CIS, which, in Article 7(1), incorporates provisions of GATT Article V.[126] Notably, however, Article 7 of the Agreement does not regulate pipeline transit, which will be subject to a separate agreement.[127] Furthermore, within the Eurasian Economic Union (a customs union between Armenia, Belarus, Kazakhstan, the Kyrgyz Republic and Russia), the Treaty on the Eurasian Economic Union, in section XX (Energy), calls for energy cooperation and the phased development of the common energy market among its Members, based on the following principles and objectives: (i) market-oriented energy pricing; (ii) competition; (iii) joint development of common energy infrastructure (including construction and reconstruction of pipelines and storage facilities); (iv) non-discrimination with respect to transportation charges; (v) harmonisation of technical standards and rules, including those on the functioning of energy facilities; (vi) priority of internal energy demand; and (vi) environment protection.[128] The Treaty requires Members to ensure non-discriminatory third-party access to gas pipeline infrastructure, on the basis of commercial contracts between the entities concerned and taking into account available pipeline capacities. This obligation is conditioned on the fulfilment of certain technical arrangements, such as the development of the indicative balance of gas demand in the Union, maintenance of market prices and

[125] On the relationship between different forms of regional integration within the area of former Soviet Union states, see Pogoretskyy, Vitaliy and Beketov, Sergey, 'Bridging the Abyss? Lessons from Global and Regional Integration of Ukraine', 46(2) *Journal of World Trade* (2012) 457.

[126] Commonwealth of Independent States, Free Trade Agreement, http://www.cis.minsk.by/, accessed 30 November 2011.

[127] See Articles 7(3) and 7(4), ibid.

[128] See Article 79 and Annex 22 of the Treaty on the Eurasian Economic Union of 29 May 2014 (Договор 'О Евразийском Экономическом Союзе'), http://www.economy.gov.ru/wps/wcm/connect/economylib4/mer/about/structure/depsng/agreement-eurasian-economic-union, accessed 4 July 2014. Pursuant to Annex 33 thereof, as of the date of its entry into force on 1 January 2015, the Treaty repealed old agreements governing energy trade between the Members of the Eurasian Economic Union, such as the 2010 Agreement between Belarus, Russia and Kazakhstan 'On the Rules of Access to Services of Natural Monopolies in the Area of Gas Transportation through Gas Transportation Systems, including the Foundations of Pricing and Tariff Policy' (Соглашение от 9 декабря 2010 'О правилах доступа к услугам субъектов естественных монополий в сфере транспортировки газа по газотранспортным системам, включая основы ценообразования и тарифной политики'), http://www.economy.gov.ru/minec/activity/sections/formuep/agreement/doc20110404_08, accessed 5 September 2013.

the harmonisation of technical standards. Importantly, this obligation does not extend to entities of non-constituent Members and is limited to internal gas traffic only.[129]

In the Association of Southeast Asian Nations (ASEAN), freedom of transit is recognised as a legal principle and is enshrined in the Framework Agreement on the Facilitation of Goods in Transit.[130] In the context of gas transit, this principle is being implemented through cooperative efforts of ASEAN Members to establish interconnecting arrangements for natural gas including, among others, the Trans-ASEAN Gas Pipeline Project.[131]

In Africa, various regional economic frameworks explicitly recognise the principle of freedom of transit.[132] These frameworks have also established various regional agreements promoting cooperation among their members in the area of gas transit and the construction of pipeline infrastructure. For example, both the EAC Treaty and COMESA Treaty require their Members to cooperate in the development of pipeline transport and in the utilisation of existing pipeline facilities.[133]

In North America, the North American Free Trade Agreement (NAFTA) does not appear to establish an explicit freedom of transit obligation, although there are transit routes passing various US states and Canadian provinces.[134] In the context of intra-NAFTA trade, the

[129] See Article 83(4) and Annex 22 of the Treaty on the Eurasian Economic Union above n 128.

[130] Association of Southeast Asian Nations, http://www.asean.org, accessed 30 November 2011.

[131] See Association of Southeast Asian Nations, Memorandum of Understanding on the Trans-ASEAN Gas at http://www.asean.org/communities/asean-economic-community/item/the-asean-memorandum-of-understanding-mou-on-the-trans-asean-gas, accessed 23 March 2014.

[132] See the Protocol on Transit Trade and Transit Facilities of the Common Market for Eastern and Southern Africa (COMESA), http://www.comesa.int/, accessed 30 November 2011; the Treaty of the East African Community (EAC), Article 7(1)(c), http://www.eac.int/, accessed 30 November 2011; the Treaty of the ECOWAS, Article 45.2, http://www.ecowas.int/, accessed 30 November 2011; the Agreement of the Southern African Customs Union, Article 24, http://www.sacu.int, accessed 30 November 2011; and Article 15 of the Protocol on Trade of the Southern African Development Community (SADC), http://www.sadc.int, accessed 30 November 2011.

[133] See EAC Treaty, Articles 89(b) and 101.2(e), above n 132; and COMESA Treaty, Article 90, above n 132.

[134] The term 'transit' is mentioned in the NAFTA only in passing, namely with respect to: (i) the obligation in Annex 311 (Country of Origin Marking) to exempt a good of another Party that is in transit from a country of origin marking requirement; and (ii) an exception from the obligation to enforce intellectual property rights in Article 1718.

recognition of the principle of freedom of transit can be deduced from certain general objectives of the Agreement, such as to eliminate barriers to trade in, and facilitate the cross-border movement of, goods and services between the territories of the Parties, which are set out in Article 102.[135]

Furthermore, while chapter 6 of NAFTA establishes a number of specific rules and principles governing energy, such as national treatment and import/export restrictions, it does not mention transit.[136] In Article 601, paragraphs 2 and 3, the Parties recognise in general the importance of having viable and internationally competitive energy and petrochemical sectors (including gas) to further their individual national interests and express their desire to enhance the role trade in energy plays in the free trade area through sustained and gradual liberalisation.[137]

In any event, NAFTA imposes certain important limitations on its obligations with respect to energy trade. First of all, in the very first provision of chapter 6 (Article 601.1), the Parties confirm their full respect for their constitutions.[138] In the case of Mexico (a significant energy exporter within NAFTA), for example, this meant that the exclusive right to own and exploit mineral resources would belong to the nation.[139] This exclusive right is further confirmed in Annex 602.3 to NAFTA, in which Mexico reserved for itself a wide array of activities relating to trade in gas, including exploration, exploitation, refining or processing, foreign trade, transportation, storage and distribution, which, in practice, are performed by its state-owned petroleum company Pemex.[140] Nevertheless, despite these limitations of NAFTA, its Members have engaged in energy cooperation at the bilateral level, including various pipeline gas transit projects.[141]

See North American Free Trade Agreement (NAFTA), http://www.worldtradelaw.net/fta/agreements/nafta.pdf, accessed 5 September 2013.

[135] Ibid. [136] Ibid. [137] Ibid. [138] Ibid.

[139] Watkins, G. C., 'Constitutional Imperatives and the Treatment of Energy in the NAFTA', 7 *Oil and Gas Law and Taxation Review* (1994) 199 at 201–203. At the time of writing, Mexico appears to have changed its constitutional law towards opening up its petroleum industry to private and foreign investors. See Eljuri, Elisabeth and Johnston, Daniel, 'Mexico's Energy Sector Reform', 17 *Journal of World Energy Law and Business* (2014) 168; Agren, David, 'Mexico opens up petroleum business to private and foreign companies', Special to *USA Today* at http://www.usatoday.com/story/news/world/2013/12/20/mexico-petroleum-oil-foreign/4152521/ (20 December 2013), accessed 21 December 2013.

[140] Watkins, above n 139 at 201–203; North American Free Trade Agreement (NAFTA), above n 134.

[141] See, inter alia, a project to import gas from Asia, Australia and New Zealand to regasification LNG terminals in Mexico and then supply it to northern Mexico and

Finally, the principle of freedom of transit is also recognised in Latin America,[142] although, in the area of network-bound energy, cooperation among Latin American states has been fraught with challenges, such as political differences and various national constitutional limitations on private participation in energy business. These challenges have significantly constrained the development of legal frameworks regulating pipeline gas transit and trade.[143] The only achievements in this area appear to have been made on a general political level.[144]

This overview of regional economic agreements demonstrates that freedom of transit is generally recognised as a principle. Some of these agreements even set forth elaborate rules regulating network-bound energy transit, although this is not always the case.

E. *Freedom of Transit and Cross-border Gas Transit Projects*

In certain circumstances, the successful implementation of a gas transit project may require the involvement of states either in facilitating the conclusion of a transit contract between commercial entities addressing

California in Hufbauer, Gary, C. and Schott, Jeffrey, J., *NAFTA Revisited: Achievements and Challenges* (Washington, DC: Institute for International Economics, 2005) at 415. See also a long-standing project to build a gas transit pipeline running from Alaska to Canada and other US states in 'Alaska Pipeline Project', http://thealaskapipelineproject.com/, accessed 5 September 2013; and Ballem, John B., 'International Pipelines: Canada – United States', 18 *The Canadian Yearbook of International Law* (1980) 148.

142 For example, this principle appears to be listed as one of the objectives of the Southern Common Market (MERCOSUR) in the Treaty of Asuncion. Barrera-Hernández, Lila, 'South American Energy Networks Integration: Mission Possible?', in M. M. Roggenkamp et al. (eds.), *Energy Networks and the Law: Innovative Solutions in Changing Markets* (New York: Oxford University Press, 2012) 61 at 68.

143 Currently, Latin American countries are split into different trade blocs with different economic agendas, such as MERCOSUR, the Andean Community and the Pacific Alliance (a recent initiative). Due to political disagreements between Chile and Bolivia over Bolivia's right to access to the sea via Chile's territory, Bolivia (an energy-exporting, land-locked state) has often cut its energy supplies to Chile. The constitutional laws of Bolivia, Venezuela and Ecuador appear to require the participation of their national companies in any energy-related projects, such as pipeline transportation. See ibid. at 63 and 66; and Casas de las Peñas del Corral, Amalia, 'Regional Energy Integration: A Wide and Worthy Challenge for South America', 5 *Journal of World Energy Law & Business* (2012), 166 at 169.

144 See, inter alia, the 2007 Declaration of Margarita and various strategies, outlines and action plans adopted by the Union of South American Nations discussed in Barrera-Hernández, above n 142 at 65–67. In the context of the MERCOSUR, see also a 1999 Memorandum of Understanding on gas integration, addressing the issues of access to excess transport capacity and refraining from imposing restrictions on the free movement of gas. Ibid. at 68.

third-party access to an existing cross-border pipeline, or in the construction of a new pipeline. For example, the 2009 gas crisis between Russia and Ukraine was resolved when Gazprom and Naftogaz concluded, inter alia, a new gas transit contract, which was accompanied by an inter-governmental agreement signed by Prime Minister Vladimir Putin of Russia and Prime Minister Yulia Tymoshenko of Ukraine.

Transit contracts would normally regulate the terms and conditions of transit (namely transit volumes, transportation charges, gas quality specification, technical and payment conditions). Inter-governmental transit agreements accompanying these contracts contain political commitments. They rarely address the technical aspects of transit.[145]

The inter-governmental agreements on the development of the SCP system (a pipeline transporting gas from the Shah Deniz field in Azerbaijan's sector of the Caspian Sea through Georgia to the Georgia–Turkish border), concluded between Azerbaijan, Georgia and Turkey, is a good example of how the terms and modalities of gas transit are regulated at the level of specific pipeline construction projects. It also illustrates technical matters that must be addressed in implementing such projects. These Agreements establish a comprehensive legal framework, comprising:

- the 2001 Agreement between Georgia and Azerbaijan relating to Transit, Transportation and Sale of Natural Gas in and beyond the Territories of Georgia and Azerbaijan through the South Caucasus Pipeline System;
- the 2001 Agreement between Azerbaijan and Turkey concerning the Delivery of Azerbaijan Natural Gas to Turkey; and

[145] See Contract between Gazprom and Naftogaz Ukrayini (Ukrainian national gas company) on the Volumes and Conditions of Transit of Natural Gas through the Territory of Ukraine for the Period of 2009–2019, signed in Moscow on 19 January 2009, available at Ukrayinska Pravda, 'The Contract on Transit of Russian Gas + Additional Agreement on the Advance Payment to Gazprom' ('Контракт о транзите российского газа + Допсоглашение об авансе "Газпрома"'), http://www.pravda.com.ua/rus/articles/2009/01/22/4462733/, accessed 10 October 2012. See the discussion of contracts/agreements concluded during the 2009 gas crisis between Russia and Ukraine in Yafimava, Katja, *The Transit Dimension of EU Energy Security: Russian Gas Transit across Ukraine, Belarus, and Moldova* (New York: Oxford Institute for Energy Studies, Oxford University Press, 2011) at 190–193. On gas transportation contracts more generally, see Roberts, Peter, *Gas Sales and Gas Transportation Agreements: Principles and Practice*, 2nd ed. (London: Sweet & Maxwell, 2008).

- Host Government Agreements between states hosting the project (namely Azerbaijan and Georgia) and the private investors involved.[146]

The inter-governmental agreement between Georgia and Azerbaijan governs, inter alia, the transit and transportation of gas through the SCP system in the territory of the Contracting Parties and is the most relevant to our discussion. The Agreement, in its preamble, notes the desire of each state to ensure the principle of freedom of transit of natural gas in accordance with international law and that, in this respect, the Agreement furthers the principles of the ECT and other international agreements to which each state is a party.[147] The preamble recognises other objectives of the Agreement, such as: (i) providing for gas transit and the development of gas transportation infrastructure in the territory of each state; (ii) ensuring the protection of investments; and (iii) preserving the environment. The preamble also acknowledges that gas sales in and beyond the territories of the Contracting Parties will contribute to the improvement and development of additional exploration, production, transportation and gas sales opportunities. It thus emphasises the important implications of the project for the economic development of the states concerned.

Article II of the Agreement (Mutual Representations, Warranties and Covenants) deals with the issues of the prompt implementation and effective application of the Agreement in national legal systems of the Contracting Parties, including: (i) steps taken to ratify or adopt the Agreement; (ii) acquisition of rights to land in the territory of the Contracting Parties, which must be done under clear commercial terms and conditions; (iii) avoidance of administrative, regulatory or other similar procedural delays which might adversely affect the design, construction, ownership, operation, capacity expansion and maintenance of the SCP system; (iv) authorisation and facilitation of imports/exports/re-exports as well as utilisation of foreign currency by entities

[146] See agreements between Azerbaijan, Georgia and Turkey on the construction and operations of the South Caucasus Pipeline (SCP) System, in British Petroleum, *Legal Agreements*, http://www.bp.com/en_az/caspian/aboutus/legalagreements.html, accessed 10 December 2012. These agreements are discussed as examples of how 'freedom of gas transit' could be implemented at the level of bilateral agreements in Willems, Arnoud R. and Li, Qing, 'Using WTO Rules to Enforce Energy Transit and Influence the Transit Fee', 4 *European Energy Journal* (2014) 34 at 41–42.

[147] While both Georgia and Azerbaijan are Members of the ECT, Azerbaijan is not yet a WTO Member.

involved in the SCP project; (v) subject to the terms and conditions set out in applicable Host Government Agreements and the provisions of Article III of the Agreement (i.e. immigration, customs and criminal law matters), the right to move freely goods, materials, supplies, technology and personnel to and among SCP facilities located in the territories of the Contracting Parties (free of all taxes, customs duties and other restrictions); (vi) cooperation and support for all financing efforts and activities conducted by project investors; and (vii) cooperation in negotiations on other international agreements with other states or international institutions necessary for the implementation of the SCP project.[148]

Article II also addresses matters relevant to securing the established gas flows and transit. It states that, in mutual recognition that the project will involve substantial, capital intensive environmentally sound infrastructure development, each state will at all times fulfil and perform its respective obligations under this Agreement, including the obligations: (i) not to interrupt or impede the freedom of gas transit and gas transportation flows moving into, within or through its territory by use of all parts of the SCP system; and (ii) to take all necessary and lawful measures and actions required to eliminate threats of any such interruption of or impediment to freedom of transit, except to the extent that the interruption has been effected to address health, safety or environmental concerns.[149]

Article II further establishes the principles of gas sales activities, conducted by the project investors and other entities involved in the project (namely authorised gas sellers and shippers) in the territory of the Contracting Parties. To this end, subject to the provisions of applicable Host Government Agreements, this provision requires the Contracting Parties to: (i) support and make it possible for the aforementioned entities to sell natural gas to licensed buyers on mutually agreeable commercial terms on a non-discriminatory basis (set forth in relevant gas purchase and sale agreements); and (ii) secure the interconnection of the SCP facilities with pipelines owned by other persons. The latter requirement must be fulfilled by, inter alia, granting the right of access to non-SCP system pipeline infrastructure established in the territory of Azerbaijan on the terms and conditions no less favourable than those offered or provided by or to other persons.[150]

[148] See Articles II(1), II(4)(vii), II(4)(ix), II(4)(x), II(4)(xi), II(xii) and II(xiii) of the Agreement, in British Petroleum, *Legal Agreements*, above n 146.

[149] See Article II(4)(iii) as well as Article VII(4), ibid.

[150] See Articles II:4(iv) and (v), ibid.

Interestingly, Article II exempts the SCP project investors and other entities involved in the project from national laws regulating competition, restraints of trade and establishing other similar requirements. In this respect, Article II clarifies that the SCP project will not be regulated as a public utility, does not involve the supply of services for the public benefit, nor does it constitute a concession.[151] This provision thus limits the possibility of the Contracting Parties to impose on project investors third-party access rights of other entities or universal service obligations. The Agreement makes clear that the SCP project is a commercial one, pursuing the private interests of its stakeholders. This understanding is further reinforced by investment protection obligations, including the obligation of the Contracting Parties to ensure the uniform, non-discriminatory application of international law standards of investment protection as set out in bilateral and multilateral agreements to which each state is a party as well as applicable Host Government Agreements. Pursuant to Article VII:2, Contracting Parties acknowledge and agree that title to or ownership of all gas transported through the SCP system will remain vested in or on behalf of project investors and/or gas shippers concerned and will, therefore, not claim this title.[152]

Among other relevant provisions of the Agreement, Article III (Security and Access) establishes obligations for states to ensure the safety and security of project implementation. Article IV (Technical, Safety, and Environmental Standards) requires cooperation and coordination between the Contracting Parties in the development and establishment of uniform technical, safety and environmental standards for the construction, operation and maintenance of the SCP system (including interconnection rights and further capacity expansion). Article V (Taxes) deals with taxes, including on goods, services and technology provided with respect to the SCP project. The Agreement also contains institutional provisions setting up the implementation commission, laying down dispute settlement rules, and clarifying its relationship with Host Government Agreements, attached thereto as appendices.[153]

The inter-governmental agreement between Azerbaijan and Turkey regulates the terms and conditions of the purchases and sales of natural

[151] See Articles II(4)(vi) and II(8), ibid.

[152] See Articles II:4(xiv), VII(2), VII(5) and VIII(5), ibid.

[153] See Articles II(2), VI, VII(5) and VIII, ibid.

gas delivered through the SCP system, including re-export rights, and the allocation of responsibilities for the financing, construction and operation of the pipeline in the territories of the Contracting Parties. This Agreement also recognises in its preamble that it aims to further the principles set forth in relevant international trade and investment agreements to which the participating states are parties. It provides that gas sellers, authorised by Azerbaijan, will bear the responsibility for the financing, construction and operation of the part of the pipeline located between Azerbaijan to the Turkish–Georgian border, whereas the buyer authorised by Turkey will bear similar responsibilities with respect to the part of the pipeline within Turkish territory. Interestingly, the Agreement makes a reference to another pipeline project pursued by Turkey and Turkmenistan (the Trans-Caspian natural gas pipeline project), involving gas transit through the territories of the participants of the SCP (namely Azerbaijan and Georgia). In this respect, the Agreement establishes a best-endeavour obligation of the Contracting Parties to undertake all necessary actions and efforts to provide assistance for the implementation of that other project.[154]

Finally, the Host Government Agreements attached to the Agreement between Georgia and Azerbaijan, which contain more than 100 pages each, provide for the detailed regulation of rights and obligations of each participating state and project investors at various stages of the development of the SCP project, from the determination of a land corridor for the construction of the SCP system to the operation of this system.[155]

F. Final Remarks on the Regulation of Freedom of Transit in Different Areas of International Law

It has been demonstrated that the principle of freedom of transit, together with certain other principles linked to transit, has been recognised in various multilateral treaties, namely the Treaty of Versailles, the 1921 Barcelona Convention on Freedom of Transit, the LOSC, WTO law and the ECT,[156] regional and bilateral treaties, and is iterated in international resolutions and declarations promoting energy trade and transit.

[154] See preamble and Articles 1–4 of the Agreement in British Petroleum, *Legal Agreements*, ibid.

[155] See the Host Government Agreements between Azerbaijan and Georgia respectively and the private investors involved in ibid.

[156] See a summary of common principles relevant to transit, as incorporated in selected multilateral treaties discussed in this chapter in Appendix 2.

These instruments jointly cover almost all the nations of the world. Moreover, some of these instruments acknowledge that, in certain circumstances (such as network-bound energy trade), freedom of transit cannot be operationalised without giving effect to other, inherent ancillary rights (such as the right to access pipeline infrastructure). The enforcement of such rights can be compelled by a multilateral treaty itself. This is the case, to some extent, under the ECT. However, often, it is left for implementation agreements, which take the form of a regional or bilateral economic agreement or an inter-governmental agreement governing a particular gas transit project.

IV. Is the Principle of Freedom of Transit a Principle of General International Law?

It has been demonstrated throughout this chapter that the principle of freedom of transit, together with certain other principles linked to transit, is recognised in different areas of international law, including multilateral, regional and bilateral treaties. Yet, despite this almost universal recognition, in this section, it will be seen that the status of the principle of freedom of transit as a principle of general international law, applicable across all of its branches, independently from treaties, has been contested by some scholars and international tribunals. This section provides an overview of scholarly arguments, as well as different legal sources, supporting or undermining the existence of the principle of freedom of transit in general international law.

A. *Freedom of Transit as a General Principle in Doctrine*

1. Opinions supporting the existence of the principle of freedom of transit in general international law

In the doctrine, a general right to transit (passage) was initially discussed by naturalists: Grotius and later de Vattel and Pufendorf.[157] These

[157] Some aspects of the Grotian theory of natural law were explained in this chapter, Section II. While natural law does not have a direct legal force in the contemporary system of international law, it appears to have played a significant role in shaping its development and crystallisation, including transit regulation. As Shelton writes: 'For all the emphasis on positive law, many authors revealed a belief that such law derives from and is inferior to international morality or *natural law* precepts. The existence of a common reservoir of universal principles governing "civilized nations" and on which the positive law is based seems to have been taken for granted.' Shelton, Dinah, 'Normative Hierarchy in

scholars advocated for the right of passage based on 'necessity' (a natural law concept); in their view, an unjustified restriction on transit was contrary to the nature of a 'reasonable' man and the law of humanity.[158] Grotius' particular attention was devoted to transit of goods and merchandise, which fell within the category of 'necessity', since trading was considered to be in the interest of society in general.[159] It is noteworthy that all of the aforementioned scholars balanced freedom of transit against what was considered a legitimate interest of a transit nation. For example, on one hand, in the case of certain inconveniences caused by transit, a transit nation, in their view, had the right to request reasonable compensation, covering the maintenance of infrastructure, such as canals or roads. On the other hand, they condemned disguised or arbitrary restrictions on transit, such as offering a circuitous and impractical route for transit, or rejecting transit without substantial reasons. Back in that time, Pufendorf suggested that the particular terms and conditions of transit should be settled in bilateral conventions.[160]

These views of natural law theorists laid the foundation for the contemporary development of the principle of freedom of transit in international law.[161] Some of those views were espoused and adjusted to contemporary realities by scholars of the twentieth century. In particular, Elihu Lauterpacht has argued that the interests of the international community require the acknowledgement of a general right to freedom of transit, which, in his view, could be derived from the notion of 'necessity', which Lauterpacht defined as a *bona fide* and legitimate interest.[162] Lauterpacht has also supported this right by referring to other legal instruments, namely the principle of the prohibition of abuse of rights (*abus de droit*), the doctrine of prescription, the

International Law', 100 *American Journal of International Law* (2006) 291 at 295, emphasis added, footnote omitted. The negotiating history of the PCIJ Statute discussing the 'principles of law' within the meaning of Article 38(1)(c) also reveals that these sources were intended to embrace the fundamental principles of morality and justice. Fastenrath, Ulrich, 'Relative Normativity in International Law', 4 *European Journal of International Law* (1993) 305 at 328.

[158] Grotius, above n 7, chapter 2; S. Von Pufendorf, Samuel, *De Jure Naturae Et Gentium Libri Octo*, vol. 2, Scott J. B. (ed.), Oldfather C. H. and Oldfather W. A. (trans) (Oxford: Clarendon Press, 1934) at 354.

[159] Grotius, above n 7, chapter 2, section XIII:5.

[160] See Grotius, above n 7 chapter 2, sections XIV, XIII:4; de Vattel, above n 6, chapter X, sections 131, 134; and von Pufendorf, above n 158 at 354–357.

[161] UN Secretariat, 'Question of Free Access to the Sea of Land-Locked Countries', above n 40 at 310–311.

[162] Lauterpacht, 'Freedom of Transit', above n 1 at 332–333 and 351–352.

concept of *estoppel* and 'the way of necessity' (servitude).[163] While calling freedom of transit an 'imperfect right', the enforcement of which would depend on an international agreement stipulating specific terms and modalities of transit, for Lauterpacht, this freedom is not a 'soft law' instrument. His arguments go as follows:

> [B]y virtue of their physical juxtaposition one to another, States are not free arbitrarily to deny to each other the use of convenient or necessary routes of communication. ... States, far from being free to treat the establishment or regulation of routes of transit as a substantial derogation from their sovereignty which they are entirely free to refuse, are bound to act in this matter in the fulfilment of an obligation to the community of which they form a part.[164]

Other scholars have also supported the right to transit as a general principle of international law. Armstrong linked transit to the principle of 'sovereign equality of nations'.[165] He wrote: '[A]t the present time, when international trade and commerce are of such vast importance, the doctrine of equality must have an economic implication, namely, that on highways or in zones, international or under international control, nations must have equal passage and trade.'[166] A number of scholars have expressed their opinion that a general right to transit was established by the Barcelona Convention and other subsequent multilateral transit regimes.[167] In some special areas of transit, Westlake, after a detailed assessment of international fluvial law, concluded that 'a sufficient consent of states exists to warrant the assertion that the right of navigation [separate from treaties] ... exists as an imperfect right on navigable rivers traversing or bounding the territories of more than one state.'[168] A general right of

[163] Ibid. at 335–338. [164] See ibid. at 319–320, 322, 346.

[165] This fundamental principle of the UN is recognised in Article 2(1) of the UN Charter, above n 72.

[166] Armstrong, S. W., 'The Doctrine of the Legal Equality of Nations in International Law and the Relations of the Doctrine to the Treaty of Versailles', 14 *American Journal of International Law* (1920) 540 at 551.

[167] See de Visscher, above n 20 at 12; McNair, Arnold D., 'The Functions and Differing Legal Character of Treaties', 11 *British Year Book of International Law* (1930) 100 at 105–106; and Fawcett, J., 'Trade and Finance in International Law', 123 *Recueil des Cours* (1968) 214 at 267. McNair referred to those treaties as the 'law-making treaties' or 'objective regimes' the effect of which extends beyond the original parties. See McNair, ibid. at 105–106. Similarly, Reid regarded 'the newer type of multilateral collective agreements' (law-making treaties) as instruments capable of creating international servitudes, guaranteeing, inter alia, transit rights. Reid, above n 16 at 17–19.

[168] Westlake, above n 1 at 157 (see also Westlake, ibid. at 159). For a recent assessment with a similar conclusion, see Uprety, above n 14 at 37–44 and 149–150.

transit has also been advocated in the contexts of the access to the sea for land-locked countries, the right of 'innocent passage', and it was even linked to the 'right of survival'.[169] Furthermore, Farran argued that the right of access to enclaves exists as a 'general principle of law recognised by civilised states'; in his view, without the right of access to the enclave, the very right to this piece of land would be inapplicable and nugatory.[170] Bedjaoui's discussion of the elementary right of each state to its own development, and the duty of states to contribute to the balanced expansion of the world economy, has an implicit logical connection to freedom of transit, although the author did not explicitly refer to this freedom.[171]

Other authors have supported a general right to transit by reference to servitudes, which, in their view, could establish a transit corridor through a foreign territory for the benefit of transit-dependent states or the international community as a whole. Such arguments were based on the idea of solidarity and interdependence of nations, whereby through multilateral cooperation (embodied in treaties or other instruments of international law) the collective interests of the international community (such as international trade and intercommunication) would gradually replace individual interests.[172]

A general right to transit has also been advocated in more recent academic works.[173] For instance, Benvenisti derives it from the so-called

169 Caflisch, Lucius C., 'Land-Locked States and Their Access to and from the Sea', 49 *British Year Book of International Law* (1978) 71 at 78; Uprety, above n 14 at 149–150; Jia, above n 35 at 213–214; Rivier, Alphonse, *Principes du Droit des Gens*, vol. 1, A. Rousseau (ed.) (Paris: Librairie Nouvelle de Droit et de Jurisprudence, 1896) at 265. On the right of 'innocent passage', see this chapter, Section III.A.2.

170 Farran, above n 16 at 304, footnote omitted. See also the expert opinion of Professor Rheinstein submitted to the ICJ in *Right of Passage over Indian Territory* arguing that the doctrine of servitude is recognised in almost all legal systems of the world, and, therefore, crystallised into a general principle of law applicable in the relations between all states. Rheinstein, Max, 'Observations et Conclusions du Gouvernement de la République Portugaise sur les Exceptions Préliminaire du Gouvernement de l'inde (Annexe 20), Étude Comparative sur le Droit D'Accès aux Domaines Enclaves par le Professeur Rheinstein', in *Case Concerning Right of Passage over Indian Territory*, Judgment of 12 April 1960: I.C.J. Reports 1960, p. 6, 714.

171 Bedjaoui, Mohammed, 'The Right to Development', in M. Bedjaoui (ed.), *International Law: Achievements and Prospects* (Dordrecht: Martinus Nijhoff Publishers, 1991) 1177 at 1182 and 1186.

172 See O'Connell, above n 16 at 811–812; Rivier, above n 169 at 258–259; Reid, above n 16 at 31; Sinjela, above n 16 at 51; Uprety, above n 14 at 33; Lauterpacht, 'Freedom of Transit', above n 1 at 335.

173 See Benvenisti, Eyal, 'Sovereigns as Trustees of Humanity: On the Accountability of States to Foreign Stakeholders', 107 *American Journal of International Law* (2013) 295 at 320, 322, 327; Melgar, above n 1 at 12–13, 206, 304–332; and Ehring, Lothar and

restricted version of the 'Pareto principle'. Pursuant to this principle, each sovereign, as a trustee of humanity, has an obligation to accommodate the interests of others when it sustains no loss.[174] Benvenisti derives the Pareto principle from a number of legal sources, including the underlying commitment of states to long-term cooperation, the doctrine of innocent passage and decisions of international courts.[175] He contends that '[a]ny legal system that perceives itself as reflecting the common enterprise of a "human society" and that allocates shared resources among its members must endorse *at least* a restricted Pareto criterion as a principle for regulating the interactions between group members'.[176] Melgar provides a comprehensive review of state practice and *opinio juris* promoting the principle of freedom of transit (both for land-locked countries and other states) to a principle of customary international law. She argues that the right of transit exists in general international law as a 'conditional right', that is a right subject to limited conditions applied on a case-by-case basis, such as the protection of *ordre public*, respect of national laws and regulations and security interests.[177]

The existence of a general principle of freedom of transit has been advocated by different bodies of international organisations. It has already been mentioned that the Enquiry Commission, appointed by the League of Nations, intended to base the draft of the Barcelona Convention on Freedom of Transit on the fundamental principles and state practice that, at that time, had already been established by nations.[178] In 1958, during the attempts to codify and progressively develop the law of the sea, the UN Secretariat came to the conclusion that the practice of states had evolved a number of principles relevant to transit:

- the principle of freedom of transit as such;
- the principle of non-discrimination, irrespective of the origin and destination of the goods and passengers in transit;

Selivanova, Yulia, 'Energy Transit', in Y. Selivanova (ed.), *Regulation of Energy in International Trade Law: WTO, NAFTA and Energy Charter* (The Netherlands: Kluwer Law International, 2011) 49 at 52 (citing Elihu Lauterpacht).

[174] Benvenisti, above n 173 at 320. Benvenisti suggests that a more general Pareto outcome, envisaging the allocation of resources among sovereigns or foreign and domestic citizens (such as in the case of catastrophes or trans-boundary damages), would require a more robust institutional infrastructure that sovereigns could trust to make impartial and competent decisions. Ibid. at 320.

[175] Ibid. at 321–325. [176] Ibid. at 321, emphasis original.

[177] Melgar, above n 1 at 12–13, 206, 304–332.

[178] See League of Nations, 'Freedom of Transit', above n 45 at 286.

- the principle that persons and goods in transit should not be subjected to any vexatious formalities; and
- the principle that charges payable by or in respect of persons and goods in transit should be the same as those payable by other users.[179]

The UN Secretariat has noted that these principles apply to 'traffic by whatever means of communication are chosen – rail, air or river'.[180] The recent technical note on transit of the UN Conference on Trade and Development (UNCTAD) also provides:

> It is recognised that based on existing international law, freedom of transit and the freedom of access to the sea cannot be absolutely restricted by the transit state. Absolute restrictions are only considered lawful if they are applied on a temporary and exceptional basis – justified by war, civil unrest.[181]

2. Opinions contesting the existence of the principle of freedom of transit in general international law

Nevertheless, a significant number of scholars contested the existence of the principle of freedom of transit in general international law.[182] For example, a group of authors argued that, before the creation of transit treaties in the twentieth century, transit states regularly restricted transit over land and rivers and through straits within their jurisdictions. Mowat stated that '[t]he law of nations has never recognised any international right of way as attaching naturally to the land of an independent state.'[183] Toulmin expressed a similar opinion.[184] Hyde also had doubts regarding a general right to transit. He believed that claims of transit over foreign territory must always be regarded as subordinate to the requirements of the sovereign state exercising jurisdiction over this territory. It is noteworthy that Hyde, nonetheless, reserved an important forthcoming role for the principle of freedom of transit. He stated that in the time of peace a transit restriction,

[179] UN Secretariat, 'Question of Free Access to the Sea of Land-Locked Countries', above n 40 at 327–328.

[180] Ibid.

[181] See United Nations Conference on Trade and Development, Trust Fund for Trade Facilitation Negotiations, 'Freedom of Transit', Technical Note 8 (February 2009) at 3.

[182] For a review of older literature sources criticising a general right to transit, published before the twentieth century, see *Faber Case*, above n 13 at 627–630.

[183] See Mowat, R. B., *The Concert of Europe* (London: Macmillan and Co., 1930) at 110.

[184] Toulmin, G. E., 'The Barcelona Conference on Communications and Transit and the Danube Statute', 3 *British Year Book of International Law* (1922–1923) 167 at 169–170.

especially if transit involves the vitally important channel of commerce, 'is likely, as time goes on, to be increasingly regarded as inequitable'.[185] Makil argued, 'while the principle of freedom of transit is universally recognized, this right is subject to agreement by the states concerned'.[186]

In the context of fluvial law, Hall observed that this area of transit in most regions was regulated by international agreements before which most rivers had been either closed or subject to the riparian states' regulation. He argued that the principle of freedom of transit through international watercourses had not been instituted by usage and was merely a contractual right.[187] Schwarzenberger also questioned the existence of customary international law or general principles governing transit on rivers. He derives this conclusion from the lack of a uniform regulation of river navigation between the Congress of Vienna of 1815 and the 1921 Barcelona Convention on Waterways.[188]

Some objections were also expressed in relation to freedom of transit for land-locked countries. For example, Vasciannie argued that to ignore the fact that, during the UNCLOS I and III negotiations, transit states consistently refused to recognise the access to the sea for land-locked countries over their territory as a matter of general international law would mean to disregard the sovereign will of states.[189] Caflisch also noted that the oft-occurring denials of transit by transit states and the existing practice between countries of regulating transit by concluding agreements at the bilateral level provide evidence against a general right of transit.[190] In the context of energy trade, Liesen expressed doubts as to whether there is a principle of freedom of transit in public international law.[191]

[185] Hyde, Charles C., *International Law Chiefly as Interpreted and Applied by the United States*, vol. 1, 2nd ed. (Boston: Little Brown and Company, 1951) at 618.

[186] Makil, A., 'Transit Rights of Land-Locked Countries: An Appraisal of International Conventions', 4 *Journal of World Trade Law* (1970) 35 at 12.

[187] Hall, William E., *A Treatise on International Law*, 3rd ed. (Oxford: Clarendon Press, 1890) at 130–139.

[188] Schwarzenberger, Georg, *International Law*, vol. 1, 3rd ed. (London: Stevens & Sons Limited, 1957) at 216–223.

[189] See Vasciannie, above n 19 at 217. See also Makil, above n 186 at 50–51; Churchill, R. R. and Lowe, A. V., *The Law of the Sea*, 3rd ed. (Manchester: Manchester University Press, 1999) at 440–441; and Brownlie, above n 17 at 271.

[190] Caflisch, above n 169 at 78–79.

[191] Liesen, Rainer, 'Transit under the 1994 Energy Charter Treaty', 17 *Journal of Energy and Natural Resources Law* (1999) 56 at 64.

There could be other arguments supporting or contesting the existence of a general principle of freedom of transit.[192] This overview of the literature, however, sufficiently demonstrates the existing debate on this subject. The next section looks at legal instruments and judicial decisions supporting or rejecting freedom of transit as a general right.

B. *Legal Instruments and Judicial Decisions Supporting or Rejecting the Existence of a General Principle of Freedom of Transit*

1. Legal instruments and judicial decisions supporting freedom of transit as a general principle

A number of legal instruments attest to the promotion of the principle of freedom of transit to a principle of general international law. First, it was mentioned earlier that, in an analogous situation of necessity in private law, the principle of servitude (or easement in common law) could create the right of passage through a neighbour's land. In *Right of Passage over Indian Territory*, Professor Rheinstein argued in his comparative study submitted to the Court as an expert opinion that servitudes/easements were recognised as a general principle of municipal law in the sense of the ICJ Statute Article 38(1)(c) in almost all jurisdictions, including civil law and common law.[193] Other comparative

[192] For example, some commentators objected to the alleged establishment of the principle by the Barcelona Convention, since this treaty covers only transit by rail and waterways. See Makil, above n 186 at 40; and Einhorn, above n 19 at 3. This argument overlooks some important facts. The preparatory work of the Barcelona Convention indicates that at that time those means of communication were the most crucial channels of transit, some of which were distorted and damaged by war, and, consequently, requiring urgent international solutions. Other areas were not included in the Convention due to their particular technical complexities, and they were left to be dealt with by other specialised legal regimes. Therefore, a limited scope of the Convention would not appear to undermine the recognition by this treaty of a broad principle of freedom of transit, explicitly declared in its preamble. See League of Nations, 'General Transport Situation in 1921. Statements submitted by the States which took part in the First General Conference on Communications and Transit, held in March–April 1921, With an Introduction by Proof Tajani', vol. 1 (Geneva, 1922) at LXXII-IV; League of Nations, 'Freedom of Transit', above n 45 at 3 and 284; Barcelona Convention on Freedom of Transit, above n 56.

[193] Professor Rheinstein states that, among the sixty-one systems of national law investigated during his comparative study, '[n]ot one system was found in which the compulsory right of passage would be denied or where even reasonable doubts could be entertained as to its existence'. Rheinstein, above n 170 at 714.

studies also confirm the wide recognition of the principles of servitude/easement in national laws.[194]

In international jurisprudence, judges have been cautious to explicitly recognise the doctrine of servitude as a general principle.[195] This principle, however, has at a minimum played an important role in guiding the development of transit regulation in treaties. For example, the Barcelona Conference Enquiry Commission addressed it in the commentary to the Barcelona Convention as follows:

> Just as, *under existing legislation in most countries*, a person who has to cross his neighbour's property in order to leave his house and reach the thoroughfare *enjoys a right of way over the property*, in the same way every State whose external trade is absolutely or virtually forced to pass across neighbouring territory ought likewise to enjoy a guaranteed right of freedom of transit across that territory.[196]

Moreover, as a treaty norm, it was directly mentioned in the Treaty of Versailles, Article 358, as a right possessed by France vis-à-vis Germany.[197] If one were to accept the promotion of the doctrine of servitude to a general international law principle, it would be somewhat illogical to deny its corollary and close analogy – the principle of freedom of transit.

[194] See Buckland, W. W. and McNair, Arnold D., *Roman Law & Common Law: A Comparison in Outline*, 2nd ed., F. H. Lawson (ed.) (Cambridge: Cambridge University Press, 1952) at 127; Zimmermann, Reinhard, Visser, Daniel and Reid, Kenneth, *Mixed Legal Systems in Comparative Perspective: Property and Obligations in Scotland and South Africa* (New York: Oxford University Press, 2004) at 735; and Gordley, James and von Mehren, Arthur T., *An Introduction to the Comparative Study of Private Law: Readings, Cases, Materials* (New York: Cambridge University Press, 2006) at 196.

[195] In *Right of Passage over Indian Territory*, Portugal referred to the doctrine of servitude to substantiate its alleged right of access to Portuguese enclaves in India. The Court did not address this argument. See *Right of Passage over Indian Territory*, above n 8. The doctrine of servitude has been relied on in: *S.S. 'Wimbledon'*, Dissenting Opinion of Judge Schücking, Collection of Judgments, Series A. – No. 1, 17 August 1923, p. 43 at 43; *Case Concerning Right of Passage over Indian Territory*, Separate Opinion of Judge V. K. Wellington Koo, Judgment of 12 April 1960: I.C.J. Reports 1960, p. 6, p. 54 at 66–67; and *Eritrea – Yemen Arbitration*, Permanent Court of Arbitration, 3 October 1996, http://www.pca-cpa.org, accessed 30 September 2011, para. 126. The principle of servitude, as an individual and separate principle of law, was, however, questioned in: *North Atlantic Coast Fisheries (United States v. Great Britain)*, Permanent Court of Arbitration, 7 September 1910, http://www.pca-cpa.org, accessed 30 September 2011 at 9–10; *S.S. 'Wimbledon'*, above n 8 at 24; and *Case Concerning Right of Passage over Indian Territory*, Dissenting Opinion of Judge Moreno Quintana (translation), Judgment of 12 April 1960: I.C.J. Reports 1960, pp. 6, 88 at 90.

[196] League of Nations, 'Freedom of Transit', above n 45 at 283, emphasis added.

[197] Treaty of Versailles, above n 49.

Another instrument supporting the existence of a general principle of freedom of transit in international law is the doctrine mentioned earlier of 'dedication', which is closely related to servitudes. It is based on the concepts of 'real rights',[198] 'objective regimes'[199] and a legal maxim that 'no one can transfer to another a greater right than he himself possesses' ('*nemo plus juris transferre quam ipse habet*').[200]

The doctrine of 'dedication' has been especially prominent in the context of international canals and rivers as important highways for international communication. Yet there is nothing preventing its application to roads or land if the right to passage had been established on the basis of a permanent grant or long usage.[201] Recall that freedom of transit has been pronounced in numerous treaties, starting with the Treaties of Westphalia and Versailles, and modern treaties discussed in this chapter. The LOSC, for example, established this principle not only in relation to its formal parties, but in relation to the wider international community.[202] What then follows from the doctrine of 'dedication' is that the right of transit could be considered a 'real' or 'objective' right, which, although it was initially created by a commitment by states to their

[198] See this concept explained in this chapter, Section II. The existence of real rights in international law has been explicitly recognised by some international and national courts. See Vali's discussion of the PCIJ's judgment in *Free Zone of Upper Savoy and the District of Gex* in Váli, above n 16 at 171; *Union of India and Others v. Sukumar Sengupta and Others*, Supreme Court of India (3 May 1990), in E. Lauterpacht (ed.), 92 *International Law Reports* (1993), 554 at 574; and Schindler, Dietrich, 'The Administration of Justice in the Swiss Federal Court in Intercantonal Disputes', 15 *American Journal of International Law* (1921) 149 at 164–165 and 174.

[199] According to McNair, a group of great powers or a large number of states can create an 'objective regime' for the benefit of the entire international community by a multilateral treaty, such as dedication of a new international facility (river or a canal route). Due to its importance, such a regime can acquire an objective existence and durability extending beyond the limits of the actual contracting parties. See *International Status of South-West Africa*, Separate Opinion by Sir Arnold McNair, Advisory Opinion of 11 July 1950: I.C.J. Reports 1950, p. 128, p. 146 at 153; and McNair, 'So-Called State Servitudes', above n 38 at 122. See, however, opinions contesting this doctrine expressed during the ILC's work on the codification of the law of treaties, discussed in Rosenne, Shabtai, *Developments in the Law of Treaties 1945–1986* (New York: Cambridge University Press, 1989) at 73–76.

[200] Brownlie, above n 17 at 377–378.

[201] For example, in *Right of Passage over Indian Territory*, the ICJ recognised Portugal's right of passage over Indian territory on the basis of the crystallised local custom (usage). *Right of Passage over Indian Territory*, above n 8 at 38.

[202] In Part X, the LOSC refers to land-locked and transit states in general, as opposed to its 'Contracting Parties'. In addition, it was mentioned earlier that the preamble of the LOSC declares a broad task of the Convention, inter alia, to establish a legal order for the seas and oceans, to codify and progressively develop the law of the sea. United Nations Convention on the Law of the Sea, above n 34.

neighbours or the international community in treaties, may be argued to exist independently from its original source.[203]

The principle of freedom of transit to the sea for land-locked countries has been recognised in a range of international resolutions and declarations. Among examples of these instruments are the Almaty and Vienna Declarations, discussed earlier in this chapter; moreover, the GA Resolution 63/210 notes that 'stable, efficient and reliable energy transportation ... is in the interest of the entire international community'.[204]

As regards international jurisprudence, it was mentioned earlier that the PCIJ has recognised freedom of navigation over international rivers as a matter of general international law already in the beginning of the twentieth century, in *River Oder*.[205] Furthermore, in *Corfu Channel*, the ICJ recognised the existence of the right of 'innocent passage' through international straits located in the territorial seas of transit states. In this case, the Court had to rule, inter alia, on: (i) whether the fleet of Great Britain (Complainant) violated Albania's sovereignty by exercising the alleged general right of innocent passage when crossing the North Corfu Channel (situated in Albania's territorial waters) without prior authorisation by Albania; and (ii) whether Albania (Respondent) violated that right by failing to inform the British fleet of the minefield laid in the Channel, which resulted in damages to property and the death and injury of British navy personnel.[206] The Court stated:

> It is, in the opinion of the Court, generally recognized and in accordance with international custom that States in time of peace have a right to send their warships through straits used for international navigation between two parts of the high seas without the previous authorization of a coastal

[203] The key difference between, on one hand, a river or a canal dedicated to international transit and, on the other hand, a general right of transit over land is that the latter is a non-localised right (not attached to any particular outlet). A question then arises as to which route should be chosen for its implementation. Yet, in private law, servitudes are also not necessarily localised rights. In many jurisdictions, if a landowner whose land is bound by a servitude refuses to appoint a reasonable corridor for the implementation of this right, the determination of the servitude route can be made by a court. See Mommsen, T. et al. (eds.), *The Digest of Justinian*, vol. 1 (Pennsylvania: University of Pennsylvania Press, 1985) at 250, 261; Rheinstein, above n 170 at 721. The potential disagreements over particular terms of passage thus do not undermine the existence of servitudes, nor should they undermine the existence of a right to pass via foreign territory.

[204] See this chapter, Section III.C.3. For details, see also Uprety, above n 14 at 130.

[205] See *River Oder*, above n 26 at 26, discussed in this chapter, Section III.A.1.

[206] See *Corfu Channel Case*, above n 8 at 9–12.

> State, provided that the passage is *innocent*. Unless otherwise prescribed in an international convention, there is no right for a coastal State to prohibit such passage through straits in time of peace.[207]

It was mentioned earlier that, although the Court discussed the right of innocent passage in relation to the passage of warships, this decision is generally considered to have a broader application, including in the context of international commerce.[208] The ICJ has subsequently confirmed the general right of 'innocent passage' through international straits in *Maritime Delimitation and Territorial Questions*.[209]

While the ICJ has referred directly to the right of 'innocent passage' through *international straits*, at least one learned commentator did not find any compelling reason for refusing to apply the Court's reasoning to the issue of passage over *the territory* of a state.[210] Indeed, in modern realities, the rights of passage through territorial sea and land appear to serve the same purpose of ensuring international communication through the territory under the sovereign control of a transit nation and which, in both cases, can be critical to economic development and trade. Notably, in the view of the ICJ, the right of 'innocent passage' did not depend on a particular status of the international strait at issue, such as an 'international highway'. The ICJ held that even if a strait was merely an 'alternative route, it was still a useful route for international maritime traffic'.[211] This confirms that, in the Court's view, the right of 'innocent passage' is not a *sui generis* right but is rather a general right derived from economic necessity, and freedom of communication and transit, recognised by international law.[212]

In *Right of Passage over Indian Territory*, Judge Koo stated in his Separate Opinion in relation to the right of access to enclaves, which is analogous to freedom of transit for land-locked countries: 'it is inconceivable in international law that one sovereignty exists only by the will or caprice of another sovereignty'.[213] He added:

> [T]he fact that an enclaved land in municipal law and an enclaved territory in the international domain has always been able to enjoy passage through the surrounding land of another owner or the surrounding

[207] Ibid. at 28, emphasis original. [208] See this chapter, Section III.A.2.
[209] See *Maritime Delimitation and Territorial Questions*, above n 35, para. 223.
[210] Lauterpacht, 'Freedom of Transit', above n 1 at 333.
[211] *Corfu Channel Case*, above n 8 at 28–29.
[212] However, see the opposite view in Azaria, Danae, *Treaties on Transit of Energy via Pipelines and Countermeasures* (Oxford: Oxford University Press, 2015) at 58.
[213] Judge Koo, above n 195 at 66.

> territory of another State, is based upon reason and the elementary principle of justice. For such land or territory this transit is a necessity and it is reasonable to provide for this necessity both in municipal law and in customary international law.[214]

2. Judicial decisions rejecting freedom of transit as a general principle

Despite these iterations of the principle of freedom of transit, there also are judicial opinions that appear to have either explicitly or implicitly rejected the existence of this principle in general international law. One such opinion is from the *Faber* case. In this arbitration, the Umpire was called on to decide whether the right of transit to the sea existed under general international law. The claim of a violation of this alleged right was brought by Germany on behalf of German investors engaged in trade with Colombia via Venezuelan rivers and Lake Maracaibo, flowing to the Atlantic Ocean. Venezuela (a transit state), on the grounds of national security, imposed a prohibition of commercial traffic coming from Colombia and crossing Venezuelan territory, which allegedly amounted to a commercial blockade with respect to one Colombian region bordering with Venezuela and caused injuries to the German investors.[215] The Umpire rejected the claim, upholding the right of Venezuela to regulate transit and, if necessary for the peace, safety and convenience of its citizens, to prohibit navigation on Venezuelan rivers temporarily.[216] The important factor that appears to have influenced the decision of the Umpire was that the measure at issue was a 'temporal' restriction aiming to defend the national security of Venezuela at a time of existing hostilities between Venezuela and Colombia.[217] Given different opinions among public international lawyers on whether the right to transit exists under general international law, the decision of the Umpire did not escape the criticism of the proponents of this right. For example, Hersch Lauterpacht criticised the decision of the Umpire as 'a very Austinian language on the nature of international law'.[218]

In a few other cases, the adjudicators were also not supportive of a general right to transit existing outside treaty law. This seems to be the case in the *Arbitration Regarding the Iron Rhine Railway (Iron Rhine Arbitration)*, in which the Arbitral Tribunal appears to have agreed with

[214] Ibid. at 67. [215] *Faber Case*, above n 13 at 603–608. [216] Ibid. at 626.

[217] Ibid. at 623–624, 625, 626.

[218] Lauterpacht, Hersch, *The Function of Law in the International Community* (USA: Archon Books, 1966) at 290.

the proposition put forward by one of the parties that the right to transit can only arise as a matter of specific agreement. This view of the Tribunal was, however, expressed in passing, without clarifying whether it concerned the right per se (in the sense of an imperfect right to pass through foreign territory) or the application of this right without obtaining the consent of a transit state.[219]

In *the Right of Passage over Indian Territory*, Judge Chagla was rather antagonistic towards a general right to transit. In his Dissenting Opinion, he argued that a sovereign state must have a complete, absolute and unrestricted right to regulate the passage of goods, men and traffic via its territory, including implementing a complete prohibition.[220]

More recently, the existence of freedom of navigation under general international law was rejected by Judge Guillaume in *Dispute Regarding Navigational and Related Rights (Costa Rica v. Nicaragua)* (*Navigational and Related Rights*). In this case, the ICJ was called on to determine the corresponding scopes of Costa Rica's right of free navigation on the Nicaraguan section of the San Juan River and Nicaragua's sovereign right to regulate the application of Costa Rica's alleged right. According to Costa Rica, its right of free navigation was established by the 1858 Treaty of Limits, and the rules of general international law governing navigation over 'international rivers'. The Court ultimately did not opine on the applicability of general international law to this matter.[221] However, in his Declaration, Judge Guillaume stated that 'customary international law offers no definition of "international rivers" and no regime governing navigation on such rivers'.[222]

C. *Final Remarks on the Status of the Principle of Freedom of Transit in International Law*

The foregoing analysis of doctrinal and judicial opinions reveals different views among scholars and adjudicators as to whether the principle of

[219] See *Belgium v Netherlands (Iron Rhine Arbitration)*, Permanent Court of Arbitration, 24 May 2005, www.pca-cpa.org, accessed 30 March 2012, para. 50. This case is also discussed in Chapter V, Section II.B.3(ii).

[220] *Case Concerning Right of Passage over Indian Territory*, Dissenting Opinion of Judge Chagla, Judgment of 12 April 1960: I.C.J. Reports 1960, p. 6, p. 116 at 117.

[221] *Dispute Regarding Navigational and Related Rights (Costa Rica v. Nicaragua)*, Judgment of 13 July 2009: I.C.J. Reports 2009, p. 213, paras. 34–37.

[222] See *Dispute Regarding Navigational and Related Rights (Costa Rica v. Nicaragua)*, Declaration of Judge Gilbert Guillaume, Judgment of 13 July 2009: I.C.J. Reports 2009, p. 213, p. 290, paras. 2–3.

freedom of transit has crystallised as a principle of general international law. On one hand, some sceptics disagree with this proposition. On the other hand, some authorities believe that, in a time of peace, freedom of transit cannot be prohibited by a transit state without a just cause. Most of the champions of the latter view, however, agree that the principle of freedom of transit is not a self-executing right – on the contrary, it is an imperfect right that must be operationalised by entering into an implementation arrangement with a transit state stipulating the terms and modalities of transit.

On balance, the views advocating the existence of the principle of freedom of transit as an imperfect right, independent from treaties, appear to be better supported by the evidence of state practice in different (if not all) areas of international law, and *opinio juris*. The state practice in question is discernible in a great number of multilateral, regional and bilateral treaties recognising freedom of transit, including the LOSC, the ECT and the GATT 1994. These treaties have also recognised other principles relating to freedom of transit, such as the prohibition on imposing customs duties on traffic in transit as well as charges and regulations that are unreasonable and discriminatory. Furthermore, the wide recognition of the principle of freedom of transit as a principle of general international law can be derived from a number of non-binding declarations and resolutions, such as the GA Resolution 63/210, the Almaty and Vienna Declarations discussed in this chapter. Like treaties, these instruments constitute the evidence of state practice in the area of transit.[223]

This chapter also discussed the evidence of *opinio juris*, suggesting that states have followed the aforementioned state practice 'as a matter of law'.[224] For example, it was demonstrated that various regimes of international rivers in Europe and North America were developed based on the conviction of some states (namely Great Britain, France, Spain and the United States) that a general right of navigation existed independently of treaties.[225] The fact that the principle of freedom of transit has been incorporated into a consistent line of multilateral, regional and bilateral treaties, including the 'codification' conventions regulating transit, such as the 1919 Versailles Treaty, the 1921

[223] The influence of treaties and non-binding international resolutions and declarations on the development of customary international law was explained in Chapter I Section III.A.

[224] See the discussion of this concept in ibid.

[225] See Lauterpacht, 'Freedom of Transit in International Law', above n 1 at 321, 327; Uprety, above n 14 at 39. See Section III.A.1 of this chapter.

Barcelona Convention and the LOSC, in itself indicates that this principle has been perceived by the parties to these treaties as a principle of general international law.[226] The intention to 'codify' this principle can, for example, be derived from the negotiating history of the 1919 Versailles Treaty and the 1921 Barcelona Convention.[227] While, during the process of the codification of the law of the sea, a number of transit states contested the existence of a general right to transit in international law, the UNCLOS negotiations led to the ultimate endorsement of the principle of freedom of transit as an 'imperfect right' in the LOSC.[228] The preamble of the LOSC provides that the convention achieves the 'codification and progressive development of the law of the sea'.[229] Notably, when setting out the basic principles relating to transit, including freedom of transit as an 'imperfect right', LOSC Articles 125, 127, 129, 130 and 131 refer broadly to 'land-locked States' and 'transit States', as opposed to 'States Parties', referred to in some other provisions.[230] This difference in terminology implies that the drafters of the Convention had no doubts that those principles were applicable outside the LOSC, and confirms the 'codification' function of the Convention with respect to transit principles.[231] In the context of trade in energy, the provisions of the ECT and some treaties governing specific gas transit projects, such as the inter-governmental agreement between Georgia and Azerbaijan on the SCP system, refer to the 'principle of freedom of transit', which implies that, in the view of the

[226] On the role of 'codification' conventions and conventions setting out identical or similar obligations in the creation of customary international law, see Chapter I, Section III.A.

[227] See UN Secretariat, 'Question of Free Access to the Sea of Land-Locked Countries', above n 40 at 321–322; League of Nations, 'Freedom of Transit', above n 45 at 286. The 'codification' function of the 1921 Barcelona Convention has been recognised in the academic literature by de Visscher, above n 20 at 12; and McNair, above n 167 at 105–106, 116. See also this chapter, Section III.B.1.

[228] See this chapter, Section III.B.2(i).

[229] United Nations Convention on the Law of the Sea, above n 34.

[230] See inter alia Article 132, ibid.

[231] As noted in section IV.A.1 of this chapter, in 1958, during the UNCLOS negotiations, the UN Secretariat issued a report stating that the practice of states had evolved a number of principles relevant to transit, including the principle of freedom of transit, which apply to traffic by whatever means of communication are chosen. UN Secretariat, 'Question of Free Access to the Sea of Land-Locked Countries', above n 40 at 327–328. The drafters of the LOSC thus were aware of the 'codification' function of the Convention. If they wished to exclude Part X of the LOSC (governing right of access of land-locked countries to/from the sea and freedom of transit) from the scope of the 'codification' process, they would have explicitly stated so in the text of the Convention.

drafters of those treaties, the principle exists as an independent principle of general international law.[232]

The promotion of the principle of freedom of transit to a principle of general international law is also supported by a number of legal instruments linked to transit, such as the principle of 'servitude', discussed in this chapter.[233] The principle of 'servitude' (or its corollary 'easement' in common law), is a close analogy to 'freedom of transit' in international law, and, as some commentators suggest, is recognised in almost all legal systems of the world.[234] Given this wide recognition, the principle of 'servitude' likely qualifies as a 'general principle of law recognised by civilised nations' within the meaning of Article 38(1)(c) of the ICJ Statute. If one were to accept the promotion of the doctrine of servitude to a principle of general international law, it would be somewhat illogical to deny this status to its corollary and close analogy – the principle of freedom of transit.[235]

Naturally, the questions of whether this state practice has crystallised into general international law, and, if so, what exactly the principle of freedom of transit means in public international law can only be answered definitively by a competent and authoritative judicial authority, such as the ICJ. So far, there have been scant judicial pronouncements on this matter; the most direct references were made by either ad hoc tribunals or individual judges in their separate opinions. Having said this, the World Court has already recognised the right to transit on international rivers and the right of innocent passage in the territorial sea of transit states. The strong similarity between these rights and the imperfect right to transit via foreign territory indicates that the latter exists as a matter of general international law, independent from treaties.

V. Summary and Concluding Remarks

This chapter provided a broad overview of the development of the international regulation of transit. The principle of freedom of transit is recognised in different areas of international law, including network-bound transit of energy. Although the status of this principle in public

[232] See Article 7(1) of the Energy Charter Treaty, above n 93; and the preamble of the intergovernmental agreement between Georgia and Azerbaijan, in British Petroleum, *Legal Agreements*, above n 146. See also Sections III.C.2(ii) and III.E.

[233] See this chapter, Section IV.B.1.

[234] See, inter alia, Rheinstein, above n 170 at 714.

[235] For additional arguments supporting the crystallisation of the principle of freedom of transit as a principle of general international law, see Melgar, above n 1 at 311–315.

international law has not been resolved definitively, there is compelling evidence suggesting that this principle exists as an imperfect right in general international law. This means that, while the right to transit per se cannot be denied without a just reason, the detailed conditions and modalities of transit must be agreed upon at a bilateral or regional level, inter alia, in an implementation agreement. This approach to effecting freedom of transit is widely accepted in the context of gas transit.

What follows from this conclusion is that, pursuant to the principle of systemic integration, WTO rules regulating transit must be interpreted in the light of what is canvassed in this study as an imperfect right to transit. The next chapter discusses the WTO's regulation of gas transit, including third-party access and capacity establishment rights.

IV

Pipeline Gas Transit under WTO Law: Assessment of Third-Party Access and Capacity Establishment Rights

I. Introduction

This chapter analyses the question of how – if at all – WTO rules regulate third-party access and capacity establishment rights. It addresses this question by providing an overview of the following WTO obligations relevant to these rights:

- the principle of freedom of transit under GATT Article V:2 (first sentence);
- the principle of non-discrimination in GATT Article V; and
- the non-violation complaint under GATT Article XXIII:1(b).[1]

Throughout this study, the new Agreement on Trade Facilitation is discussed mainly as context, to the extent that its provisions inform the interpretation of GATT Article V.[2] The transit obligations under this Agreement deal primarily with the regulatory aspects of transit, transparency and cooperation between relevant authorities of WTO Members and do not directly address the principles of 'freedom of

[1] Some literature sources discussing WTO transit rules also mention other GATT obligations, namely Articles I (General Most-Favoured-Nation Treatment), III (National Treatment) and XVII (State-Trading Enterprises). These provisions, however, do not regulate transit and are not discussed in this study. See GATT 1994, *WTO Legal Texts*, http://www.wto.org/english/docs_e/legal_e/legal_e.htm, accessed 5 March 2014.

[2] It is hardly controversial that the Agreement on Trade Facilitation, as soon as it enters into force, will constitute the relevant context of GATT Article V within the meaning of Article 31(1) of the VCLT, considering that it is intended to be part of Annex 1A of the WTO Agreement, and its purpose, stated in the preamble, is, inter alia, to 'clarify and improve relevant aspects of Article[] V'. See World Trade Organization, Ministerial Conference, Agreement on Trade Facilitation, Ministerial Decision of 7 December 2013, WT/MIN(13)/36, WT/L/911 (11 December 2013)/ Preparatory Committee on Trade Facilitation, Agreement on Trade Facilitation, WT/L/931 (15 July 2014).

transit' or 'non-discrimination' relevant to our subject.[3] Moreover, as noted earlier, at the time of writing, it has not yet entered into force.[4]

In addition, this chapter examines whether the GATS is also relevant to the WTO's regulation of third-party access and capacity establishment and briefly reviews other rules under WTO covered agreements regulating conditions of transit or establishing exceptions to the freedom of transit obligation.

It will be demonstrated in this chapter that WTO law contains broad rules regulating transit, which may in principle encompass ancillary rights integrally related to the obligation to provide freedom of transit, or to accord non-discriminatory treatment under GATT Article V, such as third-party access and capacity establishment rights.

II. Applicability of WTO Law to Gas Transit: GATT or GATS?

The first question that must be addressed before analysing WTO obligations relevant to gas transit is whether WTO law applies to trade in gas in the first place. Although some commentators argued that trade in energy is not covered by the multilateral agreements on trade in goods either because of its strategic importance or the intention of the drafters,[5] it is demonstrated in this chapter that this opinion no longer fits with reality.

Trade in gas falls squarely within the area of 'trade in goods' regulated by the WTO Multilateral Agreements on Trade in Goods, listed in Annex 1A to the WTO Agreement, including the GATT. The ordinary meaning of the term 'goods' is broad and encompasses any '[m]erchandise or possessions' and '[t]hings to be transported, as distinct from passengers'.[6] The Organisation for Economic Co-operation and Development (OECD) defines 'goods' as 'physical objects for which a demand exists, over which

[3] See paragraphs 1(1)(c) and (f) of Article 1 (Publication), and paragraphs 1–3, 5–6, 8 and 16 of Article 11 (Freedom of Transit) of the Agreement on Trade Facilitation, ibid.

[4] See Chapter II, Section I.

[5] See Aidelojie and Makuch cited in Azaria, Danae, 'Energy Transit under the Energy Charter Treaty and the General Agreement on Tariffs and Trade', 27 *Journal of Energy & Natural Resources Law* (2009) 559 at 565. According to Ehring and Selivanova, there is a misperception in certain countries in the Middle East as to whether trade in oil is regulated by the GATT rooted in the fact that some WTO Members (such as presumably the USA) did not make a respective tariff commitment. Ehring, Lothar and Selivanova, Yulia, 'Energy Transit', in Y. Selivanova (ed.), *Regulation of Energy in International Trade Law: WTO, NAFTA and Energy Charter* (The Netherlands: Kluwer Law International, 2011) 49 footnote 35.

[6] See Oxford Dictionaries, http://oxforddictionaries.com, accessed 30 November 2015.

ownership rights can be established and whose ownership can be transferred from one institutional unit to another by engaging in transactions on markets'.[7] Natural gas falls within these broad definitions, as it has physical composition (a hydrocarbon gas mixture), it is bought and sold in the market like any other merchandise and it is transportable via pipelines and by seagoing vessels.[8]

Trade in gas falls within the regime of 'trade in goods' in a number of regional trade agreements. For example, the ECT classifies gas as an 'energy material and product' (the term synonymous with goods).[9] The EU's rules regulating the free movement of goods apply to trade in gas.[10] In the Eurasian Economic Union, the term 'goods' is defined in Article 4(35) of the Customs Code as 'any movable property, transported through the customs border, including . . . electricity and other kinds of energy'.[11]

Moreover, the Harmonized Commodity Description and Coding System developed by the World Customs Organization contains headings for LNG (code 2711.11) and natural gas in its gaseous state (code 2711.21).[12] These headings have been enshrined in most WTO Members' Schedules of Concessions on *goods* annexed to the GATT by virtue of Article II:7 thereof. The Appellate Body has clarified that 'the Harmonized System is context [that is, an agreement that was made in connection with the conclusion of the WTO Agreement within the meaning of VCLT Article 31(2)(a)] for purposes of interpreting the covered agreements'.[13] The Harmonized System can, therefore, be used to interpret the meaning of the term 'goods' in the GATT, and it indicates unequivocally that the term covers natural gas.

[7] Organisation for Economic Co-operation and Development (OECD), 'Glossary of Statistical Terms', https://stats.oecd.org/glossary, accessed 15 September 2015.

[8] See a similar view in Melgar, Beatriz H., *The Transit of Goods in Public International Law* (Leiden: Brill Nijhoff, 2015) at 7–10; and Ehring and Selivanova, above n 5 at 57–58.

[9] See ECT Article 1(4) and Annex EM-I in Energy Charter Secretariat, 'The Energy Charter Treaty and Related Documents: A Legal Framework for International Energy Co-operation' (Brussels: Energy Charter Secretariat, 2004).

[10] *Case 159/94 Commission v France* [1997] ECR I-5819, paras. 31–42.

[11] See the English version of the Customs Code of the Customs Union (Eurasian Economic Union), http://www.tsouz.ru/Docs/kodeks/Documents/TRANSLATION%20CUC.pdf, accessed 30 November 2015.

[12] See the Harmonized Commodity Description and Coding System, http://www.wcoomd.org/en/topics/nomenclature/instrument-and-tools/hs_nomenclature_2012/hs_nomenclature_table_2012.aspx, accessed 30 March 2014.

[13] See Appellate Body Report, *China – Auto Parts*, para. 149; Appellate Body Report, *EC – Chicken Cuts*, paras. 194–199.

Last but not least, there have already been a number of GATT and WTO cases addressing the regulation of trade in energy and even network-bound renewable electricity under the GATT/WTO agreements on trade in goods. In fact, the very first dispute adjudicated by the Appellate Body – *US – Gasoline* – addressed the consistency of measures regulating the composition and emission effects of gasoline with a view to preventing air pollution with the GATT.[14] In these disputes, no party contested the regulation of trade in energy and energy products under the GATT and other applicable Agreements on Trade in Goods in Annex 1A to the WTO Agreement.

In light of this, the GATT applies to measures affecting trade in gas as a good. Notably, paragraph 1 of Article V (Freedom of Transit) defines the key term 'traffic in transit' as '*goods*' in transit, namely 'goods . . . transit[ing] across the territory of a contracting party when the passage across such territory . . . is only a portion of a complete journey beginning and terminating beyond the frontier of the contracting party across whose territory the traffic passes'.[15] Thus, transit rules set out in Article V apply to the traffic of gas in transit.

In addition, as discussed further, some aspects of gas transportation required for transit may also be regulated by the GATS, such as the transport services via pipeline of natural gas.[16] The overlap between the GATT and the GATS in the context of gas transit could have practical implications for the interpretation of WTO Members' transit-related obligations under these two Agreements. The key difference between the GATT and the GATS lies in their respective negative (top-down) and positive (bottom-up) list approaches to market access, national treatment and other commitments. Whereas under the GATT those commitments are only subject to limited exceptions (inter alia, General Exceptions under GATT Article XX),[17] under the GATS each Member decides whether to undertake these commitments in relation to each

[14] See the *US – Superfund* and *US – Gasoline* disputes discussed in Desta, Melaku G., 'GATT/WTO Jurisprudence in the Energy Sector and Movements in the Marketplace', *Oil, Gas & Energy Law Intelligence* (2003), www.ogel.org, accessed 15 February 2014; and Desta, Melaku G., 'The GATT/WTO System and International Trade in Petroleum: An Overview', 21 *Journal of Energy and Natural Resources Law* (2003) 385, at 391. The most recent case dealing with renewable energy adjudicated in the WTO is *Canada – Renewable Energy /Canada – Feed-in Tariff Program.*

[15] GATT 1994, *WTO Legal Texts*, above n 1, emphasis added.

[16] See the detailed discussion of the regulation of third-party access and capacity establishment under the GATS in this chapter, Section IV.

[17] This provision is discussed in this chapter, Section V.A.2.

specific services sector and mode of supply in its Schedule of Specific Commitments.[18] Consequently, the absence of specific energy-related commitments under the GATS could create an impression of a lack of a relevant transit obligation under the GATT.

There has been some debate among scholars and WTO Members' delegates on which particular aspects of gas transit are covered by the GATT, dealing with trade in goods, and the GATS, dealing with trade in services. For example, on one hand, during the accession negotiations of Russia to the WTO, when responding to a question relating to the alleged restrictions in Russia on third-party access to oil and gas pipelines for energy exportation purposes, the Russian delegate stated that 'access to pipelines was considered by the Russian Federation to be an issue covered by the provisions of the GATS'.[19] On the other hand, the note the WTO Secretariat issued on energy services highlights a general difficulty in identifying energy services regulated by the GATS, since the energy industry has traditionally not distinguished between goods and services.[20] In the course of the WTO Doha negotiations on trade facilitation, the EU invited WTO Members to consider the inter-relationship between the GATT and the GATS, referring particularly to transit by road, but also addressing gas transit through pipelines.[21]

However, the room for a clash between commitments under the GATT and the GATS has been significantly reduced in the WTO jurisprudence. In *EC – Bananas III*, the Appellate Body clarified that the GATT and the GATS are not mutually exclusive agreements.[22] This means that they can apply to the same measure simultaneously, although covering its different aspects.[23] More specifically, Cossy states that: 'the GATS protects suppliers of services, in this case suppliers of pipelines transportation

[18] See this chapter, Section IV.

[19] World Trade Organization, Working Party on the Accession of the Russian Federation, 'Draft Report of the Working Party on the Accession of the Russian Federation to the World Trade Organization', WT/ACC/SPEC/RUS/25/Rev.2 (27 May 2003), para. 301.

[20] World Trade Organization, Council for Trade in Services, 'Energy Services', Background Note by the Secretariat, S/C/W/52 (9 September 1998), para. 7.

[21] World Trade Organization, Council for Trade in Goods, 'WTO Trade Facilitation – Strengthening WTO Rules on GATT Article V on Freedom of Transit', Communication from the European Communities, G/C/W/422 (30 September 2002) at 5.

[22] See Appellate Body Report, *EC – Bananas III*, para. 221. See also Panel Report, *China – Publications and Audiovisual Products*, paras. 7.127–128.

[23] Gaffney, John P., 'The GATT and the GATS: Should They Be Mutually Exclusive Agreements?', 12 *Leiden Journal of International Law* (1999) 135 at 148–149.

services' (not suppliers of goods).[24] The WTO Secretariat's note on energy services further clarifies that the GATS concerns the transportation and distribution of energy services only if they are provided independently.[25] By contrast, the GATT establishes disciplines regulating trade in goods, whereby gas sellers either deliver gas by using their own pipeline networks or, in general, are indifferent as to which particular pipeline operators or gas shippers (domestic or foreign) will ship their gas to the final destination, including through transit states.

Based on this demarcation between the scope of the GATT and the scope of the GATS, it is unlikely that there could be an inherent contradiction between these agreements in the context of pipeline gas transit.[26] Moreover, as will be explained later, while the GATS could potentially play an important role in facilitating the creation of an international gas market and the expansion of cross-border pipeline infrastructure, current GATS rules regulate gas transit only to a very limited extent. In the following sections, the obligations under the GATT and the GATS relevant to gas transit and in particular third-party access and capacity establishment rights are discussed separately.

III. Third-Party Access and Capacity Establishment under the GATT

A. *Third-Party Access and Capacity Establishment under GATT Article V:2 (First Sentence)*

1. GATT Article V:2 in scholarly debates

Article V:2 (first sentence) incorporates the principle of freedom of transit into the GATT by providing: 'There shall be freedom of transit through the territory of each contracting party, via the routes most convenient for international transit, for traffic in transit to or from the territory of other contracting parties.'[27] It was explained earlier that the

[24] Cossy, Mireille, 'Energy Trade and WTO Rules: Reflexions on Sovereignty over Natural Resources, Export Restrictions and Freedom of Transit', in Herrmann, C. and Terhechte, J. P. (eds.), 3 *European Yearbook of International Economic Law* (2012) 281 at 301.

[25] World Trade Organization, 'Energy Services', above n 20, para. 9.

[26] This is not to say that the obligations under the GATT and the GATS can never overlap and clash. See Pauwelyn, Joost, *Conflict of Norms in Public International Law: How WTO Law Relates to Other Rules of International Law* (New York: Cambridge University Press, 2003) at 201 at 399–405. For the purposes of this study, however, these hypothetical scenarios need not be considered.

[27] GATT 1994, *WTO Legal Texts*, above n 1.

term 'traffic in transit' covers natural gas. However, the provision on its face does not address either a third-party access or capacity establishment right, which brought about disagreements among scholars on whether these rights are regulated by this provision.[28]

For example, in relation to *capacity establishment*, Friedrich states that 'there is no indication whatsoever in [GATT] that the parties intended to commit themselves to such a far reaching obligation'.[29] Ehring and Selivanova also noted: 'There is nothing explicit in Article V of GATT 1994 obliging WTO Members to expand existing transit capacity or to allow infrastructure construction, [footnote omitted] for instance, where constraints exist.'[30] Azaria, Clark, Cossy, Konoplyanik and Rakhmanin expressed similar views in relation to both *capacity establishment* and *third-party access*, denying the existence of these rights under the GATT altogether.[31]

Notably, in her recent monograph, Azaria adopted a contrary view with respect to *third-party access*. Azaria now contends that 'mandatory third-party access arguably falls within the scope of these wide terms [namely "there shall be freedom of transit" under GATT Article V:2]' and that it can be enforced even against 'private companies [operating pipeline systems], whose conduct is not attributed to the

[28] It follows also from the draft consolidated negotiating texts submitted during the negotiations on trade facilitation that there is indeed uncertainty among some WTO Members as to whether GATT Article V:2 implies third-party access and capacity establishment. In this respect, the relevant part of one document begins: 'For greater certainty . . . '. World Trade Organization, Negotiating Group on Trade Facilitation, 'Draft Consolidated Negotiating Text', TN/TF/W/165/Rev.11 (7 October 2011) at 21.

[29] Friedrich, H. M., 'Legal Aspects of Transit Carriage in Gas Transmission Systems: State Obligations and Private Ownership' (thesis on file at the Centre for Energy, Petroleum and Mineral Law and Policy, 1989) at 36.

[30] An argument of these authors holds: 'Article V:2, first sentence ("via the routes most convenient") seems to suggest that the freedom of transit exists only within existing capacities: only existing "routes" can be "convenient" and only one of several existing routes may be the "most convenient".' Ehring and Selivanova, above n 5 at 70.

[31] See Azaria, 'Energy Transit under the Energy Charter Treaty and the General Agreement on Tariffs and Trade', above n 5 at 572; Clark, Bryan, 'Transit and the Energy Charter Treaty: Rhetoric and Reality', 5 *Web Journal of Current Legal Issues* (1998) at 3; Cossy, 'Energy Trade and WTO Rules', above n 24 at 298–299; Konoplyanik, Andrei A., 'Russia-EU Summit: WTO, the Energy Charter Treaty and the Issue of Energy Transit', 2 *International Energy Law and Taxation Review* (2005) 30 at 32; and Rakhmanin, Vladimir, 'Transportation and Transit of Energy and Multilateral Trade Rules: WTO and Energy Charter', in J. Pauwelyn (ed.), *Global Challenges at the Intersection of Trade, Energy and the Environment* (Geneva: The Graduate Institute, Centre for Trade and Economic Integration, 2010) 123 at 124.

state'.[32] Azaria further states that 'the terms "unnecessary restrictions" and "unreasonable regulations" allow the transit state to refuse transit access [when there are capacity constraints]'.[33]

Along the same lines, Ehring and Selivanova do not share the views of Clark, Cossy, Konoplyanik and Rakhmanin concerning the regulation of third-party access under Article V:2. These authors contend that when the issue of a pipeline capacity constraint arises, the third-party access right can follow from freedom of transit as such and the obligation of non-discrimination in GATT Article V:2 (second sentence).[34] A similar opinion was expressed by Kurmanov.[35] Grewlich considers that the third-party access right follows from GATT Article V, based on the competition law doctrine of 'essential facilities'.[36]

The argument that GATT Article V:2 implies a mandatory third-party access right is addressed in greater detail in other chapters of this monograph.[37] Nevertheless, at this stage, it must be mentioned that it has far-reaching implications and is highly problematic for a number of legal and practical reasons. The proponents of this argument essentially suggest that a transit Member can be held liable for a restriction imposed by a private company on third-party access to its privately owned pipeline, even if the conduct at issue is not compelled by the Member, and the company does not perform any 'governmental function' in the Member's jurisdiction; in other words, the conduct is not *attributed* to the state within the meaning of the general international law principles on state responsibility for internationally wrongful acts.[38] In this way, they read into Article V:2 a competition law–type obligation of transit states to ensure the third-party access to gas pipeline facilities operated by private entities. However, despite certain failed attempts of GATT and WTO

[32] Azaria, Danae, *Treaties on Transit of Energy via Pipelines and Countermeasures* (Oxford: Oxford University Press, 2015) at 64, 67.

[33] Ibid. at 64.

[34] Ehring and Selivanova, above n 5 at 70–71. The principle of non-discrimination in GATT Article V:2 is addressed in this chapter, Section III.B.

[35] Kurmanov, Baurzhan, 'Transit of Energy Resources under GATT Article V', *Transnational Dispute Management* (provisional issue, January 2013) at 24–25.

[36] Grewlich, Klaus W., 'International Regulatory Governance of the Caspian Pipeline Policy Game', 29 *Journal of Energy & Natural Resources Law* (2011) 87, footnote 91.

[37] See this chapter, Section III.A.2(i); and Chapter V, Section III.B.

[38] Azaria states explicitly that 'WTO members are obliged to achieve non-discriminatory freedom of transit . . . even when private companies, whose conduct is *not attributed* to the state, operate pipeline systems.' Azaria, *Treaties on Transit of Energy via Pipelines and Countermeasures*, above n 32 at 67, emphasis added.

Members to create GATT/WTO competition rules, such rules do not exist under the GATT, including Article V.[39]

Grewlich's assertion that the third-party access right can be enforced in the WTO based on the doctrine of 'essential facilities' does not appear to correspond well with international law realities. The concept of 'essential facilities' is not harmonised even between the most developed national systems of competition law, let alone in jurisdictions where there is no competition regulation at all. In other words, it is unlikely that the doctrine of 'essential facilities' has crystallised into a principle of general international law applicable to all states and can be used to interpret WTO law.[40]

Finally, Azaria's suggestion that the regulatory autonomy of transit states is preserved by their right to adopt 'necessary restrictions' and 'reasonable regulations' under Article V is also not convincing. It follows from this argument that a Member that cannot enforce a third-party access right against a private pipeline operator through administrative or judicial means would bear the burden of demonstrating to a WTO panel that the restriction on third-party access satisfies the requirements of 'reasonableness' or 'necessity'.[41] As part of this exercise, the transit Member will likely have to establish that the restriction contributes to the achievement of a 'legitimate objective' protected by the GATT. If it fails to do so, the restriction will be found to be inconsistent with Article V:2. Unfortunately, Azaria does not explain which particular objectives recognised by WTO law such measures would address.

[39] It was mentioned earlier that there are scant rules in WTO law addressing competition. See Chapter II, Section III.C.1. The only exception here is the WTO Telecom Reference Paper (incorporated into some WTO Members' GATS Schedules of Specific Commitments). See World Trade Organization, Reference Paper on Telecommunications Services (Telecom Reference Paper), www.wto.org/english/tratop_e/serv_e/telecom_e/tel23_e.htm, accessed 30 March 2012. This legal instrument is discussed, inter alia, in Chapter VI, Section III.B.3.

[40] On the doctrine of 'essential facilities', see Talus, Kim, 'Just What Is the Scope of the Essential Facilities Doctrine in the Energy Sector?: Third Party Access-Friendly Interpretation in the EU v. Contractual Freedom in the US', 48 *Common Market Law Review* (2011) 1571; and Stratakis, Alexandros, 'Comparative Analysis of the US and EU Approach and Enforcement of the Essential Facilities Doctrine', 27 *European Competition Law Review* (2006) 434. At the international level, the doctrine of 'essential facilities' is recognised to a very limited extent, inter alia, in the context of the international telecommunications services. See WTO Telecom Reference Paper, above n 39.

[41] On the principles of the burden of proof applied in the WTO dispute settlement, see Appellate Body Report, *US – Wool Shirts and Blouses* at 12–17. The terms 'unnecessary restrictions' and 'reasonable regulations' referred to in paragraphs 3 and 4 of Article V, respectively, are discussed in this chapter, Sections III.A.3 and V.A.1.

From a practical perspective, this argument is even more problematic, as it assumes that even in monopolised gas markets, such as Russia or Saudi Arabia, there exists the right of access to pipeline networks similar to the right of third-party access established in liberalised gas markets, such as the EU. It thus fails to acknowledge important differences between regulatory approaches adopted by WTO Members in their respective gas markets. Finally, according to this argument, to ensure freedom of transit, some transit states would be required to expropriate pipeline capacities owned by private companies, which may lead to costly investment disputes and serious financial consequences for those states.[42]

As regards other scholarly views on the meaning of the obligation under GATT Article V:2 (first sentence), Roggenkamp states that this provision could be interpreted as incorporating *the right to establish additional pipeline capacity*, as by adopting the language 'routes *most convenient*' it implies a deeper commitment than that previously assumed under the Barcelona Convention on Freedom of Transit referring merely to 'routes *in use*'.[43] Kurmanov shared this view.[44]

These views, although providing interesting arguments both against and in favour of the regulation of third-party access and capacity establishment rights under the GATT, can generally be characterised by one common limitation. While placing strong emphasis on the text of the GATT, they stop short of engaging in its contextual and systematic analysis, grounded in general principles of public international law in the way required by the principles of treaty interpretation, embodied in VCLT Article 31.[45] This study fills this essential gap. The following sections examine whether Article V:2 (first sentence) can in principle

[42] For example, as noted in Chapter III, Section III.E, under the agreement between Georgia and Azerbaijan on transit through the SCP System, parties explicitly agreed that the pipeline must not be regulated as a public utility, will not involve the provision of services to the public at large, and is not intended or required to operate in the service of the public benefit or interest. This language makes clear that the SCP system is a private facility, the access to which by third parties is restricted. See Article II(8) of the 2001 Agreement between Georgia and Azerbaijan relating to Transit, Transportation and Sale of Natural Gas in and beyond the Territories of Georgia and Azerbaijan through the SCP System (Mutual Representations, Warranties and Covenants), in British Petroleum, *Legal Agreements*, http://www.bp.com/sectiongenericarticle.do?categoryId=9029334&contentId=7053632, accessed 10 December 2012.

[43] See Roggenkamp, Martha M., 'Implications of GATT and EEC on Network-Bound Energy Trade in Europe', 12 *Journal of Energy & Natural Resources Law* (1994) 59 at 72–73; Convention and Statute on Freedom of Transit of 1921, 8 *American Journal of International Law, Supplement* (1924), 118.

[44] Kurmanov, above n 35 at 28. [45] See Chapter I, Section III.B.

imply third-party access and capacity establishment rights by analysing the key terms in this provision – namely 'freedom of transit' and 'route most convenient' – as informed by the context of Article V and other interpretative elements, including the principles of freedom of transit and evolutionary interpretation. Chapter V analyses a number of additional principles of general international law that further inform the meaning of the obligations under GATT Article V:2.

2. The meaning of the key terms in Article V:2 (first sentence)

(i) **'Freedom of transit'** The meaning of the term 'freedom of transit' in Article V:2 (first sentence) is of key importance for this study. Indeed, one of the objectives of this study is to determine whether, in the context of pipeline gas transit, freedom of transit implies third-party access and/or capacity establishment rights. The Panel in *Colombia – Ports of Entry* has clarified this term. In this case, the Panel defined 'freedom' as 'the *unrestricted* use of something', which was based on this term's ordinary meaning taken from 'The New Oxford Dictionary of English'.[46] The Panel thus appears to interpret the obligation under GATT Article V:2 broadly. In its view, *any* restriction on 'freedom of transit' could potentially violate this provision. This, at a first glance, may suggest that the obligation to provide 'freedom of transit' is broad enough to cover any ancillary rights (such as third-party access and capacity establishment rights) related to this primary obligation.

Nevertheless, it is important to note that, in *Colombia – Ports of Entry*, the term 'freedom' was defined in a particular technical context. The Panel aimed to see whether Colombia's requirement of trans-shipment, imposed on 'traffic in transit' as a prerequisite for an exception from this state's regulation restricting ports of entry, was consistent with Article V:2.[47]

[46] Panel Report, *Colombia – Ports of Entry*, paras. 7.399 and 7.414, emphasis added. The term 'freedom' is used in some other provisions of WTO-covered agreements, such as GATT Article XXIV:4; Article 40.3 of the Agreement on Trade-Related Aspects of Intellectual Property Rights (TRIPS Agreement); para. 7 of the Decision on Negotiations on Maritime Transport Services; and preamble of the Agreement on Trade in Civil Aircraft, but was not interpreted before the *Colombia – Ports of Entry* case. See these legal instruments in *WTO Legal Texts*, above n 1. The word 'restriction' (i.e. the corollary of the word 'unrestricted') can be found, inter alia, in GATT Article XI:1 and has been defined in that context as 'a limitation on action, a limiting condition or regulation'. Panel Report, *China – Raw Materials*, footnote 1521.

[47] It is noteworthy that Article V:2, as complemented by the definition of 'traffic in transit' in Article V:1, clearly establishes the obligation to allow transit without requiring 'trans-shipment'. Panel Report, *Colombia – Ports of Entry*, para. 7.423.

Perhaps for that narrow purpose, the ordinary meaning of the term 'freedom', taken from the dictionary definition, was sufficient. However, it is not clear at all whether such a broad definition could be extended to other disputes.

The interpretation of the term 'freedom' in GATT Article V can be informed by reference to analogies of this term in legal theory, namely to the fundamental legal conceptions of 'freedoms', 'liberties' and 'privileges'.[48] The classical exponent of legal theory, Hohfeld explained that 'freedoms', 'liberties' and 'privileges' must be contrasted with 'rights' and 'claims'. While a jural corelative of X's 'right' to something is a corresponding 'positive duty' of Y to guarantee this right, the corelative of X's 'freedom' to do something (inter alia, a constitutional 'freedom of speech' or 'freedom of movement') is a mere 'no-right' of Y to interfere in X's exercising of this freedom. However, this freedom alone does not create a duty for Y to help X enjoy it, if this, for example, would be physically or naturally impossible.[49]

If Hohfeld's distinction between 'freedoms' and 'rights' were transposed to public international law, this distinction would mean that a broad principle of freedom of transit under GATT Article V:2 must be contemplated as an obligation imposed on a transit state to abstain from interfering in the enjoyment of this freedom, provided that there are no physical or natural obstacles to do so. Arguably, the example of such an interference would be a positive measure of a transit state restricting transit. This restriction can be applied directly through the prohibition of gas transit via the territory of a transit state or indirectly by restricting inherent ancillary rights, including third-party access or capacity establishment. Both types of measures can in principle violate WTO rules.[50]

That said, in light of Hohfeld's theory, freedom of transit alone would not compel a transit state to take positive measures to guarantee this freedom. The examples of such positive measures could be the construction of new pipeline facilities or the enforcement of a third-party access right against a private pipeline operator, independent from the state. It would be unreasonable to argue that the broad principle of

[48] The terms 'freedom', 'liberty' and 'privilege' are often used interchangeably in legal theory. Hohfeld, Wesley N., 'Some Fundamental Legal Conceptions as Applied in Judicial Reasoning', 23 *Yale Law Journal* (1913–1914), 16 at 30; Campbell, David, and Thomas, Philip (eds.), *Fundamental Legal Conceptions as Applied in Judicial Reasoning by Wesley Newcomb Hohfeld* (Aldershot: Ashgate Darmouth, 2001).

[49] Hohfeld, above n 48 at 30; and Campbell and Thomas, above n 48 at 12–14.

[50] See Chapter VI, Sections II.A and II.B.

freedom of transit under GATT Article V:2 could imply such far-reaching obligations.

The ECT transit rules establish the relationship between freedom of transit and ancillary rights integrally related to this freedom in a similar manner. As noted earlier, the ECT imposes on transit states an obligation to operationalise freedom of transit through either a negotiated third-party access right or by not placing obstacles in the way of new pipeline capacity being established, in accordance with the principle of freedom of transit. Notably, the ECT does not require a transit state to take positive measures aiming to ensure transit through its territory, with the exception of the obligations to facilitate energy transit and to encourage entities, such as domestic pipeline operators, to cooperate on matters relevant to energy transit.[51]

This, however, does not suggest that freedom of transit can never mean more than a claim against interference. As another renowned legal theorist, Alexy, pointed out, in national legal orders, a broad 'freedom' can be 'enhanced' and turned into a precise and positive 'right to something' by putting in place additional laws and regulations, thus protecting it.[52] Consequently, in light of Alexy's understanding of the concept of 'freedom', a positive duty to ensure transit can result from 'rights' established by additional legal instruments. In the context of the WTO's regulation of freedom of gas transit, these instruments can take the form of an implementation agreement stipulating the detailed conditions of transit.

The meaning of the term 'freedom of transit' can be further informed by the principle of freedom of transit in general international law through the principle of systemic integration. As explained in the previous chapter, the negotiating history of the GATT 1947 and the GATT 1994 indicates that the text of Article V was based on a similar (and at times even identical) wording of the 1921 Barcelona Convention. This Convention, in turn, intended to codify the basic principles and state practice that had evolved in the area of transit, including the principle of 'freedom of transit' itself.[53] This principle was defined earlier as an imperfect right to transit, which, in complex cases, is normally operationalised through

[51] See Chapter III, Section III.C.2(ii). See the detailed analysis of the relationship between third-party access and capacity establishment rights under the ECT in Chapter VI, Section III.A.2.

[52] See the so-called concept of 'protected liberty' in the context of national constitutional laws in Alexy, Robert, *A Theory of Constitutional Rights*, J. Rivers (trans.) (New York: Oxford University Press, 2010) at 148–149.

[53] See Chapter III, Sections III.B.1 and III.B.2(ii).

implementation agreements setting out the terms and modalities of transit.[54] In the context of pipeline gas, these terms and modalities must necessarily include third-party access and/or capacity establishment rights.

This analysis of the term 'freedom of transit' reveals several approaches to its interpretation. First, the *Colombia – Ports of Entry* Panel suggested that this term imposes a broad obligation on a transit state to provide 'unrestricted' access to its territory for the purposes of international transit. Second, in light of the analogy with the fundamental legal conception of 'freedom', the term implies a negative right against interference, which can further be enhanced by additional legal instruments creating positive duties for a transit state. Third, the interpretation of this term in conjunction with the principle of freedom of transit leads to a conclusion that this term must establish an imperfect right to transit, which has to be operationalised through an implementation agreement.

The first approach interprets 'freedom of transit' in too broad a manner, leaving no regulatory autonomy for a transit state. This approach will, therefore, unlikely be suitable for each situation. The context of Article V:2, analysed in the following section, confirms that freedom of transit cannot be understood as such an absolute (unrestricted) right.[55]

The last two approaches do not appear to be considerably different, since in essence they describe freedom of transit as an incomplete right, the practical implementation of which must be executed through an additional legal arrangement between a transit state and a transit-dependent state. In the case of establishing a new flow of gas via a transit pipeline, this legal arrangement will have to take the form of a regional or bilateral treaty stipulating conditions of third-party access and/or capacity establishment. Given that under both the GATT and general international law, freedom of transit is extended to network-bound energy (including pipeline gas), the refusal of a transit state to give effect to freedom of transit by negotiating such a treaty would deprive this freedom of any practical meaning and would, therefore, appear to be not in line with the basic principle of freedom of transit. As explained in the

[54] See Chapter III, Sections IV.C.

[55] It has to be recalled that VCLT Article 31(1) requires the contextual reading of treaty provisions – that is, interpretation with reference to the text as a whole. See Vienna Convention on the Law of Treaties of 1969, 1155 *United Nations Treaty Series* (1980) 331; and Gardiner, Richard, *Treaty Interpretation* (New York: Oxford University Press, 2008) at 177.

remainder of this study, this proposition is supported by various sources of WTO law and public international law.

(ii) 'Route most convenient' The meaning of the term 'route most convenient' is also of key relevance for interpreting GATT Article V:2 (first sentence), since, pursuant to this provision, a transit state must provide freedom of transit in its territory 'via the routes most convenient for international transit'. However, in the context of gas transit, the meaning of this term is not clear in the following respects. First of all, does the word 'route' cover both fixed infrastructure, namely a pipeline, and a land corridor, where a pipeline will have to be constructed for the purposes of international transit? Second, the term is not clear as to who is the party determining the factor of 'convenience' – a transit state, a transit-dependent state or their pipeline operators? Both questions are important to understand the scope of the obligation under Article V:2 and the regulatory autonomy of a transit state in determining the 'route'.

A starting point in answering the first question can be a dictionary definition of the term 'route'.[56] This term is literally defined as 'a way or course taken in getting from a starting point to a destination . . . the line of a road, path, railway, etc.'.[57] This definition indicates that the term is broad enough to cover both fixed infrastructure (for it refers to 'railways')[58] and a land corridor ('a way or course', 'path', etc.). In other words, the ordinary meaning of the term 'route' envisages the possibility that gas could access the territory of another state either through an existing pipeline or through land where the pipeline does not exist as yet. Certainly, this meaning alone would be a weak basis for asserting the existence of ancillary construction rights in the territory of a transit state. These rights must be derived from other legal instruments, such as in our case the obligation to provide freedom of transit.

Interestingly, the negotiating history of the Barcelona Convention on Freedom of Transit may suggest that the drafters of this treaty, upon which GATT Article V was closely modelled, had not intended the term 'route' to cover additional or new routes. In this respect, a report on the

[56] See the obligation to interpret the treaty provisions in accordance with the ordinary meaning to be given to the terms of the treaty in Article 31(1) of the Vienna Convention on the Law of Treaties, above n 55.

[57] See Oxford Dictionaries, http://oxforddictionaries.com, accessed 30 March 2012.

[58] This is also supported by the fact that the obligation in Article V:2 is established in relation to 'traffic in transit', which covers gas, transportable mainly via pipelines.

draft Convention prepared by the Barcelona Conference Enquiry Commission reads:

> The words *by the routes most convenient for international transit* [in Article 2 of the Statute on Freedom of Transit] ... do not in any way signify that, in virtue of this Convention, a State can demand from another State the bringing into traffic of new lines of greater utility for transit, or the carrying out of alterations to routes already existing, but refer to the routes most convenient at the actual time of transit.[59]

Contrary to this suggestion, however, it must be mentioned that while, during the course of the negotiations, the final wording of Article 2 of the Statute on Freedom of Transit (perhaps inspired by the previously described intention) was changed into 'routes in use', the text of GATT Article V reinvigorated the formulation 'routes most convenient'.[60] As stated earlier, in Roggenkamp's view, this linguistic amendment could well mean that the intentions of the drafters of the GATT were more ambitious than those of the drafters of the Barcelona Convention and that they could intend GATT Article V to cover certain aspects of capacity establishment.[61] In the same connection, when interpreting GATT Article V, one should also take into account the vast difference in time, circumstances and technologies surrounding the creation of the GATT (both 1947 and 1994) and the 1921 Barcelona Convention. Unlike the latter treaty, concluded after WWI, the major post-WWII transit-related treaties have addressed the issue of infrastructure constraints, imposing in this respect, cooperation, equitable sharing or other obligations on their Contracting Parties.[62]

Furthermore, it must be noted that the aforementioned report of the Enquiry Commission proposed to exclude from the scope of Article 2 of the Statute on Freedom of Transit a *positive* obligation of a transit state to

[59] League of Nations, 'Verbatim Reports and Texts Relating to the Convention on Freedom of Transit', Barcelona Conference (Geneva, 1921), emphasis original.

[60] Compare the current version of the Barcelona Convention on Freedom of Transit, above n 43 with GATT 1994, *WTO Legal Texts*, above n 1.

[61] See Roggenkamp, above n 43 at 72–73.

[62] See Article 33(6) of the International Trade Organization Charter, http://www.wto.org/english/docs_e/legal_e/havana_e.pdf, accessed 10 January 2014; Article 129 of the United Nations Convention on the Law of the Sea, http://www.un.org/depts/los/convention_agreements/texts/unclos/unclos_e.pdf accessed 30 December 2012; and ECT Article 7(4) in Energy Charter Secretariat, 'The Energy Charter Treaty and Related Documents', above n 9. As explained in this chapter, Section V.C, the new Agreement on Trade Facilitation also regulates some aspects of cooperation between WTO Members relevant to the use of transit infrastructure.

'bring into traffic new lines of greater utility' or to 'carry out alterations to routes already existing'. Such an obligation would have required a transit state to construct a new route or to modify/repair the existing ones. It is obviously far-reaching and does not have analogies even in the legal frameworks providing explicitly for capacity establishment rights, such as the ECT. However, the report of the Enquiry Commission does not address the possibility that new transit facilities or routes are developed by other interested parties, such as a transit-dependent state, private investors or international organisations sponsoring trade facilitation and the improvement of cross-border infrastructure. In other words, the report does not exclude the existence of a *negative* obligation of a transit state not to create obstacles to the establishment of new transit routes by third parties.

The second question that must be answered in interpreting the term 'route most convenient' is who determines 'convenience'. In *Colombia – Ports of Entry*, the Panel notes: 'a Member is not required to guarantee transport on necessarily any or all routes in its territory, but only on the ones "most convenient" for transport through its territory'.[63] Accordingly, the Panel appears to regard the term 'convenient' as a limitation on the obligation to provide freedom of transit.[64] The Panel, however, did not address the question of who determines 'convenience'.

On one hand, it is hardly possible to conceive of a situation where a transit-dependent state, at its will, would pick and choose any transit 'route'. On the other hand, it would not seem reasonable to suggest that the choice of that route must depend solely on the discretion of the transit state.[65] Indeed, this could defeat the very purpose of establishing freedom of transit under GATT Article V:2. Even though a transit state may naturally be in a better position to decide on the route, the choice of convenience should not be discretionary and must instead entail a certain balance of interests struck in an objective manner.[66] Similarly, WTO

[63] Panel Report, *Colombia – Ports of Entry*, para. 7.401. [64] Ibid., para. 7.400.

[65] In this regard, Ehring and Selivanova note the difference between the linguistic structures of GATT Articles XXI (Security Exceptions) and Article V:2. While the former provision uses the formula 'which it *considers* necessary for . . . ', clearly giving the discretion to the territorial power, GATT Article V:2 does not provide for such discretion. Ehring and Selivanova, above n 5 at 72. See a contrary opinion in Azaria, 'Energy Transit under the Energy Charter Treaty and the General Agreement on Tariffs and Trade', above n 5 at 571.

[66] Along the same lines, the preparatory work of the Barcelona Convention reads: 'it is for the States across which transit takes place to regulate the conditions of transit, to select the routes to be employed.' The preparatory work, however, acknowledges: 'The point is to

adjudicators have taken an objective approach to interpreting other broad standards in WTO law, such as 'necessary' and 'reasonable'.[67]

The text of Article V:2 (first sentence) itself indicates that the element of 'convenience' is linked to the phrase 'international transit', as opposed to a particular party determining the route.[68] This suggests that the 'route most convenient' must be determined on the basis of its ability to satisfy the needs of 'international transit'. In circumstances in which international transit depends inherently on an ancillary right to construct a new pipeline, this pipeline will be the most convenient – if not the only – route for the international transit of gas. In this situation, the choice of convenient routes would appear rather narrow, although the transit state will retain its ability to negotiate on the particular terms and modalities of a given pipeline construction project.

In light of the foregoing, the term 'route most convenient' appears broad enough to cover transit via both existing infrastructure, such as pipelines, and transit through a land corridor, where a pipeline has not yet been constructed. The term in and of itself does not provide explicitly for the right to construct new transit facilities. It merely indicates that the choice of the 'route most convenient' entails the balancing of interests of a transit state and a transit-dependent state and must be linked to the interests of international transit. In certain circumstances, due to particular facts, this choice can naturally be limited. However, should the right to create new transit facilities be derived from other legal sources, such as the principle of freedom of transit under both GATT Article V:2 and general public international law, this right would not contradict the ordinary meaning of the term 'route most convenient'.

3. Relevant context

WTO law contains a number of provisions, including GATT Article V itself, that can be considered in interpreting the obligation to provide freedom of transit under Article V:2 (first sentence). In *Colombia – Ports of Entry*, the Panel observed that the obligation in Article V:2 (to

obtain *the best possible terms for transit traffic* by selecting *the least expensive route*.' League of Nations, 'Freedom of Transit', above n 59 at 284 and 9, emphasis added.

[67] See this chapter, Section V.A.1.

[68] It reads as follows: 'There shall be freedom of transit . . . via the routes most convenient *for international transit*.' GATT 1994, *WTO Legal Texts*, above n 1, emphasis added. See also Valles, Cherise, 'Article V Freedom of Transit', in R. Wolfrum et al. (eds.), *WTO – Trade in Goods* (Leiden: Martinus Nijhoff Publishers, 2011) 183 at 188; Willems, Arnoud R. and Li, Qing, 'Using WTO Rules to Enforce Energy Transit and Influence the Transit Fee', 4 *European Energy Journal* (2014) 34 at 38; and Ehring and Selivanova, above n 5 at 71.

guarantee freedom of transit) is informed by the term of art 'traffic in transit' (that is a subject of this obligation) as defined in the first paragraph of the same provision.[69] To recall, this term refers to *'goods'* whose passage across a transit territory is only a portion of a complete journey beginning and terminating beyond the frontier of a transit state.[70] GATT Article V:1 thus confirms that Article V:2 creates obligations relevant to the transit of natural gas, which is a 'good'.[71] This should logically mean that, to have a practical application in the context of gas transit, Article V:2 must regulate rights integrally related to the transit of this commodity, namely third-party access and capacity establishment rights.

One could argue that the constituent elements of the term 'traffic in transit' in GATT Article V:1 – i.e. the 'passage' or 'journey' of 'goods' – are associated with the 'movement' of somebody or something.[72] Thus, based on the literal meaning of these terms, the obligation to provide freedom of transit has immediate relevance to goods in passage, not to traffic that has to be established in the first place (such as through capacity establishment). This argument, however, if upheld, would mean that, in the context of network-bound transit, GATT Article V would have no meaning unless there is an existing infrastructure in the territory of a transit state or even a technical connection to the already existing infrastructure. Such an interpretation would significantly reduce the utility of Article V in this context.

Furthermore, two provisions in the Agreement on Trade Facilitation provide guidance on how Members should operationalise their obligations under GATT Article V:2 in the areas where freedom of transit can face technical difficulties, such as pipeline gas. In particular, Article 11(5) of the Agreement contains the best-endeavour commitment of Members to 'make available, where practicable, physically separate infrastructure (such as lanes, berths and similar) for traffic in transit'.[73] Article 11(16) requires Members to 'cooperate and coordinate with one another with a view to enhanc[ing] freedom of transit'.[74] In light of these provisions, Members must facilitate freedom of transit in such areas, through, inter alia, good faith cooperation/coordination and, where practicable, by

[69] Panel Report, *Colombia – Ports of Entry*, para. 7.400 and footnote 680.
[70] GATT 1994, *WTO Legal Texts*, above n 1. [71] See Section II of this chapter.
[72] GATT 1994, *WTO Legal Texts*, above n 1. See the definitions of the terms 'journey' ('an act of travelling from one place to another'), and 'passage' ('the action or process of moving') in 'Oxford Dictionaries', above n 57.
[73] World Trade Organization, Agreement on Trade Facilitation, above n 2. [74] Ibid.

allocating physically separate infrastructure to transit, arguably including pipelines.

The context of Article V:2 (first sentence) also sets the outer limits of the obligation to provide freedom of transit by defining matters not covered by this provision and establishing exemptions from this obligation. Starting from the former, Article V establishes a comprehensive standard of non-discrimination in paragraphs 2 (second sentence), 5 and 6.[75] In *Colombia – Ports of Entry*, the Panel clarified that Article V:2 (second sentence) '*complements and expands* upon the obligation to extend freedom of transit', prohibiting, in addition, distinctions based on the place of origin, departure, entry, exit or destination, or on any circumstances relating to the ownership of goods, or of other means of transport.[76] It can be inferred from this clarification that the principle of freedom of transit as such was not intended to deal with discriminatory treatment, which is left to other provisions in GATT Article V and the Agreement on Trade Facilitation. The role of this principle is rather to establish a general obligation requiring the absence of 'restrictions' on traffic in transit.

Following the same logic, it can be argued that 'freedom of transit' does not per se aim at addressing measures levying customs duties and other charges on traffic in transit (including transportation charges). Such measures are covered specifically by Articles V:3 and V:4 of the GATT as well as Article 11(2) of the Agreement on Trade Facilitation. These provisions allow the imposition of only: (i) charges for transportation; (ii) charges commensurate with administrative expenses entailed by transit; and (iii) charges corresponding to the cost of services rendered.[77]

A number of provisions in GATT Article V and the Agreement on Trade Facilitation define the scope of the obligation to provide freedom of transit by establishing exemptions from this obligation. In particular, Article V:3 of the GATT implies that GATT Article V will not tolerate only delays or restrictions that are 'unnecessary'; and Article V:4 of the GATT rules out only 'unreasonable' regulations.[78] It follows from these provisions that GATT Article V allows certain limitations on

[75] See these provisions discussed in this chapter, Sections III.B and V.A.1.

[76] Panel Report, *Colombia – Ports of Entry*, para. 7.397, emphasis added; and GATT 1994, *WTO Legal Texts*, above n 1.

[77] GATT Article V:4 stipulates that these charges must be reasonable. See GATT 1994, *WTO Legal Texts*, above n 1; and World Trade Organization, Agreement on Trade Facilitation, above n 2.

[78] GATT 1994, *WTO Legal Texts*, above n 1. Valles, above n 68 at 189.

freedom of transit, such as those imposed via 'necessary' and 'reasonable' regulations. These provisions are further elaborated on by paragraphs 1 and 6 of Article 11 of the Agreement on Trade Facilitation.[79]

The context of GATT Article V:2 (first sentence) confirms that this provision regulates ancillary rights integrally related to freedom of gas transit. In addition, it provides guidance on how the scope of the obligation to provide freedom of transit must be delimited in the light of other obligations in Article V. Some of these obligations address regulatory concerns of transit states by allowing these states to impose 'necessary' and 'reasonable' restrictions on transit.

4. GATT Article V:2 from the perspective of evolutionary interpretation

It was mentioned earlier that, according to some scholars, the initial purpose of the GATT was not to specifically regulate pipeline gas transit.[80] Yet, when interpreting an international agreement, one has to keep in mind that international law obligations are rarely meant to be static or self-contained in the sense of a regime 'closed to public international law'.[81] The fact that DSU Article 3(2) mandates the interpretation of WTO covered agreements in accordance with customary rules of interpretation of public international law, including the principle of 'systemic integration', is a clear indication that the WTO Agreement does not establish a self-contained, static regime. This, in turn, suggests that certain broad provisions in WTO agreements, especially if they contain generic terms such as the term 'freedom' of transit under GATT Article V:2, can undergo evolution. Consequently, such provisions must be interpreted in the light of the contemporary concerns of WTO Members, which include sustainable development, energy trade and energy security, and which are highly dependent on the creation of cross-border pipelines.[82] This approach to treaty interpretation is supported by the well-recognised principle of 'evolutionary interpretation', based on the idea that 'no legal relationship can remain unaffected by

[79] See World Trade Organization, Agreement on Trade Facilitation, above n 2, and this chapter, Section V.C.

[80] See Chapter III, Section III.C.1

[81] Dispute Settlement Understanding, *WTO Legal Texts*, above n 1; International Law Commission, 'Fragmentation of International Law: Difficulties Arising from the Diversification and Expansion of International Law', A/CN.4/L.682 (13 April 2006) at 71–72.

[82] Chapter II, Section III.

time'.[83] However, how can one discern the evolutionary meaning of a treaty term?

If a WTO law term is interpreted in harmony with the broader environment of public international law, as the principle of systemic integration requires,[84] this interpretation must take into account relevant extraneous legal sources. Not only do the latter contain legal obligations binding on WTO Members, but they can also be indicative of the intention of the international community to deal with a complex matter (such as gas transit) in a particular way. These sources may, therefore, point to the direction of the evolution of a given treaty term and will support a particular method of evolutionary interpretation.[85] Other sources of guidance have traditionally been the object and purpose of an interpreted treaty, which in essence set out the long-term goals of the contracting parties. The object and purpose are often derived from the treaty preamble.[86] Finally, certain principles developed within the WTO legal system may also shed light on how a treaty term must be interpreted in an evolutionary manner.

The same approach to reading WTO obligations in an evolutionary manner appears to be adopted by the Appellate Body and a number of panels. For example, in *US – Shrimp*, the Appellate Body assessed the consistency of US measures regulating the harvesting of shrimp both inside and outside US territorial waters with the requirements of general exceptions set out in GATT Article XX(g).[87] When interpreting the generic term 'natural resources' in this provision, the Appellate Body deduced the evolutionary feature of this term from the preamble of the

[83] International Law Commission, 'Fragmentation of International Law', above n 81 at 241.

[84] See Appellate Body Report, *EC and certain member States – Large Civil Aircraft*, para. 845; Chapter I, Section III.B.1.

[85] In this regard, Koskenniemi suggests that principles of international law may provide convenient arguments for a Court to justify its decision and make it seem in line with the values and goals of the international legal order. Koskenniemi, Martti, 'General Principles: Reflexions on Constructivist Thinking in International Law', in M. Koskenniemi (ed.), *Sources of International Law* (Aldershot: Ashgate Darmouth, 2000) 360 at 381. The main criticism of the evolutionary interpretation is based on the fact that, apart from using 'generic meanings', judges rarely support their reference to this method with other evidence. Dawidowicz, Martin, 'The Effect of the Passage of Time on the Interpretation of Treaties: Some Reflections on Costa Rica v. Nicaragua', 24 *Leiden Journal of International Law* (2011) 201 at 218–219 and 221.

[86] See Panel Report, *EC – Asbestos*, para. 8.47. See also Gardiner, above n 55 at 192.

[87] These measures included the import ban on shrimp and shrimp products that were not certified by US authorities as those that had been produced by using turtle-excluding devices. Appellate Body Report, *US – Shrimp*, paras. 2–6. See GATT Article XX discussed in Section V.A.2 of this chapter.

WTO Agreement, which declares 'sustainable development' one of the goals of the WTO.[88] Then, after consulting relevant sources clarifying the term 'natural resources' (including those outside WTO law, such as the LOSC and a number of soft-law instruments), the Appellate Body held that the term covers both 'living and non-living resources'. In effect, this interpretation expanded the scope of the obligation as it was initially understood.[89]

In *China – Publications and Audiovisual Products*, the United States (Complainant) argued that China's (Respondent) specific commitments on 'Sound recording distribution services' undertaken in China's GATS Schedule encompass the distribution of these services in both physical and electronic forms, that is, music embedded in CDs and that distributed through the Internet, respectively. China contended that its Schedule covered the distribution of services in a physical form only.[90] After examining the text and the context of China's specific commitments in question, as well as the object and purpose of the GATS, the Appellate Body upheld the US claim. It noted that the term 'sound recording' is broad and 'can be used to refer to "recorded content", irrespective of how it is distributed'.[91] Moreover, in line with the evolutionary interpretation of WTO law in *US – Shrimp*, the Appellate Body stated that 'the terms used in China's GATS Schedule ("sound recording" and "distribution") *are sufficiently generic that what they apply to may change over time*'.[92]

Along the same lines, in *EC – IT Products*, the Panel suggested that new commitments could evolve with technological development, especially when the wording of an obligation is overly broad, as it appears to be in GATT transit provisions. In *EC – IT Products*, one of the main questions was whether the EU's duty-free treatment of certain information technology (IT) goods, granted pursuant to the Information Technology Agreement (ITA), incorporated into some WTO Members' Schedules of Concessions and subject to GATT Article II:1(a) and (b) (Schedules of Concessions), was extended to new, multifunctional products. The EU (Respondent) argued that new technological functions put such products outside its original concession and caused them to be covered by other dutiable headings of its Schedule.[93] Nevertheless, the Panel did not find

[88] Appellate Body Report, *US – Shrimp*, paras. 129–131. [89] Ibid., paras. 130–131.

[90] Appellate Body Report, *China – Publications and Audiovisual Products*, para. 349. See a brief overview of the structure of the GATS specific commitments in this chapter, Section IV.A.

[91] Ibid., para. 395. [92] Ibid., para. 396, emphasis added.

[93] Panel Report, *EC – IT Products*, paras. 7.704 and 714.

anything in the broad language of the Schedule that would exclude the evolved new technologies from the concession and duty-free treatment.[94] One of the arguments supporting the Panel's decision was that the object and purpose of the WTO Agreement and the ITA is: 'to provide security and predictability in the reciprocal and mutually advantageous concessions negotiated by parties for the *reduction of tariffs* and *other barriers to trade*'.[95]

The evolutionary interpretation of WTO law is supported by the principle of 'technological neutrality', largely shared among WTO Members.[96] This principle appears to be relevant to gas transit, the regulation of which has to be adapted to changes in gas transportation technologies, although, in WTO jurisprudence, this principle has so far been used to interpret GATS commitments. In *US – Gambling*, the Panel explained the principle of 'technological neutrality' as follows: 'a market access commitment for mode 1 [in the GATS] implies the right for other Members' suppliers to supply a [cross-border] service through *all means of delivery*, whether by mail, telephone, Internet etc., *unless otherwise specified in a Member's Schedule*'.[97] In other words, the principle of 'technological neutrality' means that, unless a Member makes an explicit reservation in its Schedule that its market access commitment refers to a particular means of delivery of a cross-border service, this commitment must be read in an evolutionary manner as encompassing *all* means.

In *China – Electronic Payment Services*, the Panel referred to another important principle of WTO law – the principle of an 'integrated service'. This principle appears to be closely linked to the principle of 'effective or integrated rights' in general public international law discussed in the following chapter. The principle of integrated service supports the interpretation of GATT Article V:2 in a manner that gives effect to the obligation to provide freedom of transit in the area of trade in pipeline gas, even if this requires the recognition of certain ancillary rights derived from this freedom.

In *China – Electronic Payment Services*, the United States (Complainant) claimed that a series of restrictions and requirements imposed by China (Respondent) on electronic payment services and service suppliers involved in payment card transactions were inconsistent with China's obligations under Articles XVI (Market Access)

[94] Ibid., paras. 7.730 and 734. [95] Ibid., para. 7.683, emphasis added.
[96] See Panel Report, *US – Gambling*, para. 6.285. [97] Ibid., emphasis added.

and XVII (National Treatment) of the GATS.[98] To make a substantive ruling on US claims, the Panel had to examine whether the electronic payment services that otherwise were not explicitly listed in the relevant sector of China's GATS Schedule, such as authentication, authorisation, clearing and settlement services, nonetheless fell under this sector.[99]

The United States argued that these electronic payment services (EPS), namely, authentication, authorisation, clearing and settlement services, must be treated as one 'integrated service', covered by China's Schedule. In particular, the United States asserted that 'without the entire system supplied by the EPS supplier, no issuer would be able individually to offer a card that is as widely accepted by merchants, and no acquirer could offer merchants a service that can deliver such a large number of card holders'.[100] China's view was that these services were distinct services, which, in several countries, are supplied by different service suppliers, and the United States failed to prove that they fell within China's Schedule.[101]

The Panel noted that, although EPS consist of different services that may be individually identified, all elements of the system, together, are necessary for the payment card transaction to materialise.[102] In the Panel's view, the completion of a payment card transaction would include, at a minimum, what the Panel referred to as 'front-end processing' (i.e. authentication and authorisation of transactions) and 'back-end processing' (i.e. clearing and settlement of the transaction).[103] On this basis, the Panel upheld the US argument that the services at issue constituted an 'integrated service'; it further found that these services fell within the scope of China's GATS Schedule.[104]

It is noteworthy that, in reaching its decision, the Panel relied, inter alia, on the object and purpose of the GATS to expand trade in services under conditions of transparency and progressive liberalisation and the fact that the language of the relevant section of China's Schedule was broad and did not include limitations on particular services or service

[98] See the list of US claims in Panel Report, *China – Electronic Payment Services*, paras. 2.4, 3.1.

[99] In particular, sector 7.B(d) (Banking services) of China's Schedule read as follows: 'All payment and money transmission services, including credit, charge and debit cards, travellers cheques and bankers drafts (including import and export settlement).' Ibid., para. 7.518. For a detailed description of services involved in a payment card transaction see ibid., paras. 7.12–7.24.

[100] Ibid., para. 7.55 and para. 7.63. [101] Ibid., paras. 7.56, 7.60, 7.64.

[102] Ibid., paras. 7.58–7.59. [103] Ibid., para. 7.180. [104] Ibid., paras. 7.62, 7.204.

suppliers.[105] In this respect, the interpretative approach taken by the Panel in *China – Electronic Payment Services* is similar to those adopted by the Appellate Body in *US – Shrimp* and *China – Publications and Audiovisual Products* as well as the Panel in *EC – IT Products*.

A certain common trend can be discerned from these interpretation approaches adopted by the Appellate Body and panels. In all four cases, WTO provisions were interpreted broadly, which in effect expanded the scope of the obligation as it was understood initially. This interpretation, however, was in line with the object and purpose of the WTO Agreement and was supported by the principles of 'evolutionary interpretation', 'technological neutrality' or 'integrated service'.[106] In *US – Shrimp*, in making its decision, the Appellate Body consulted various non-WTO sources, which supported its interpretative method.

The WTO case law and legal principles discussed in this section provide support for the interpretation of GATT Article V:2 in an evolutionary manner, in the light of contemporary concerns of the WTO, which cover trade in gas. Whereas, originally, GATT drafters, perhaps, did not attach any particular importance to this aspect of trade, in the preamble of the WTO Agreement they have declared that the *expansion* of the regime of 'trade in goods' is one of the objectives of the WTO.[107] This objective bestows on the generic term 'freedom' of transit under GATT Article V:2 a feature of evolution. The process of evolution may, in turn, expand the scope of rights and obligations initially assumed by GATT/WTO Members under GATT Article V, encompassing additional rights integrally related to freedom of transit, such as third-party access and capacity establishment rights. The broad language of this provision does not create obstacles to such an evolution.

Moreover, the evolutionary interpretation of GATT Article V:2 is also supported by a number of non-WTO sources that recognise the application of the principle of freedom of transit in the context of network-bound

[105] Ibid., paras. 7.129, 7.179, and 7.196.

[106] While some of these principles have not yet been applied in the GATT context, there is no compelling reason why they cannot be transposed to the latter Agreement. After all, both the GATT and the GATS are the integral parts of one and the same treaty – the WTO Agreement.

[107] The preamble of the WTO Agreement states: '[t]he Parties to this Agreement, [r]ecognizing that their relations in the field of trade and economic endeavour should be conducted with a view to . . . *expanding* the production of and trade in goods and services, while allowing for the optimal use of the world's resources in accordance with the objective of sustainable development . . . '. WTO Agreement, *WTO Legal Texts*, above n 1, emphasis added.

energy trade, such as the ECT and various soft law instruments.[108] These sources confirm that, in the context of network-bound energy trade, freedom of transit must imply the ability to obtain access to pipeline infrastructure. This can be achieved by giving effect to third-party access and/or capacity establishment rights.

5. Freedom of gas transit in light of '*in dubio mitius*'

The last question that must be addressed in interpreting GATT Article V:2 (first sentence) is whether a panel could rely on the concept of '*in dubio mitius*' (i.e. 'more leniently in case of doubt') to justify its narrow reading of the obligation to provide freedom of transit in the context of trade in pipeline gas. This reading could, for example, restrict the application of GATT Article V:2 to cases in which there is already a pipeline connection between the WTO Members concerned. At the outset, however, it must be noted that this concept is controversial in public international law and the conditions of its application are rather strict.

In *EC – Hormones*, the Appellate Body referred to this concept as a 'supplementary means of interpretation'. It held: '[w]e cannot lightly assume that sovereign states intended to impose upon themselves the more onerous, rather than the less burdensome, obligation.'[109]

The concept of '*in dubio mitius*' has, however, been harshly criticised by a number of authoritative commentators as an outdated concept that secures the sovereign rights of one party to a treaty at the expense of the others. In the view of these commentators, the concept must, therefore, be used with extreme caution, namely when the treaty terms are ambiguous and cannot be clarified by reference to the regular interpretation principles set out in Articles 31–33 of the VCLT, including the principle of systemic integration.[110] The Appellate Body itself appears to have realised that it may not have fully considered the implications of its use

[108] See Chapter III, Section III.C.

[109] Appellate Body Report, *EC – Hormones*, para. 165.

[110] See Jackson, John H., 'The WTO Dispute Settlement System after Ten Years: the First Decade's Promises and Challenges', in Taniguchi, Y., Yanovich, A. and Bohanes, J., *The WTO in the Twenty-First Century: Dispute Settlement, Negotiations, and Regionalism in Asia* (New York: Cambridge University Press, 2007) at 33; Wälde, Thomas W., 'Interpreting Investment Treaties: Experiences and Examples', in C. Binder et al. (eds.), *International Investment Law for the 21st Century: Essays in Honour of Christoph Schreuer* (New York: Oxford University Press, 2009) 724 at 735; and Lauterpacht, Hersch, 'Restrictive Interpretation and the Principle of Effectiveness in the Interpretation of Treaties', 26 *British Year Book of International Law* (1949) 48 at 59.

of the concept of '*in dubio mitius*' in *EC – Hormones*. In *China – Publications and Audiovisual Products*, it acknowledged that this concept, even if applicable in WTO dispute settlement, would be relevant only when the interpretation under VCLT Articles 31 and 32 leads to inconclusive results.[111] It is clear that, in the case at hand, this condition has not been met.

6. Summary and concluding remarks on GATT Article V:2 (first sentence)

So far, it has been demonstrated that the obligation to provide freedom of transit under GATT Article V:2 (first sentence) may encompass ancillary rights integrally related to this freedom. This does not mean that a transit state has a positive obligation to construct pipeline facilities in its territory or to enforce a compulsory third-party access right against a private pipeline operator. In fact, it will be demonstrated in the following chapter of this study that neither of these measures is required.

However, freedom of transit must have a practical meaning in the area of pipeline gas, especially given that gas falls within the definition of traffic in transit in GATT Article V:1. As demonstrated in Chapter III, freedom of transit in this area is normally operationalised through implementation agreements, stipulating the detailed terms and modalities of transit. In such agreements, freedom of gas transit will necessarily have to be established through either third-party access or capacity establishment. In this regard, a transit state may have certain discretion, provided that the ultimate choice of implementation does not contradict the principle of freedom of transit.

[111] Appellate Body Report, *China – Publications and Audiovisual Products*, paras. 410–411. Along the same lines, in *River Oder*, the PCIJ stated: 'Nor can the Court, on the other hand, accept the Polish Government's contention that, the text being doubtful, the solution should be adopted which imposes the least restriction on the freedom of States. This argument, though sound in itself, must be employed only with the greatest caution. To rely upon it, it is not sufficient that the purely grammatical analysis of a text should not lead to definite results; there are many other methods of interpretation, in particular, reference is properly had to the principles underlying the matter to which the text refers; it will be only when, in spite of all pertinent considerations, the intention of the Parties still remains doubtful, that that interpretation should be adopted which is most favourable to the freedom of States.' *Territorial Jurisdiction of the International Commission of the River Oder*, Collection of Judgments, Series A. – No. 23, 10 September 1929 at 26. See a similar reasoning of the PCIJ in *S.S. 'Wimbledon'*, Collection of Judgments, Series A. – No. 1, 17 August 1923 at 24–25; and *Polish Postal Service in Danzig*, Collection of Advisory Opinions, Series B. – No. 11, 16 May 1925 at 39.

Precisely because in the context of pipeline gas the practical operation of the right to transit depends on an additional implementation agreement, it has been referred to in this study as an 'imperfect right'. The imperfect nature of this right, however, does not mean that a transit state can disregard its obligation to provide freedom of transit by, inter alia, refusing to negotiate on an implementation agreement with a transit-dependent state. Since it is generally recognised that freedom of gas transit cannot be effected without access to pipeline infrastructure, such a refusal would appear to be inconsistent with the principle of freedom of transit established by both GATT Article V:2 and general public international law.

This interpretation of GATT Article V:2 (first sentence) is supported by the principle of freedom of transit in general international law, the context of this provision, the object and purpose of the WTO Agreement and the principle of evolutionary treaty interpretation. It will be demonstrated in Chapter V that this interpretation of GATT Article V:2 is also in line with other principles of public international law.

This reading of GATT Article V:2 raises a number of practical questions. For example, can public international law compel a transit state to negotiate in good faith on the terms and modalities of gas transit? Does a transit state have an obligation to open a particular transit route or to allow investments in its territory in order to give effect to a capacity establishment right? These questions are also addressed in Chapter V.

B. Third-Party Access and Capacity Establishment in Light of the Non-discrimination Obligations in GATT Article V

GATT Article V establishes a comprehensive standard of non-discrimination, including:

- prohibition of any distinctions based on the place of origin, departure, entry, exit or destination of traffic in transit, as well as on any circumstances relating to the ownership of goods, vessels or other means of transport (Article V:2, second sentence);
- the most-favoured-nation (MFN) obligation with respect to all charges, regulations and formalities in connection with transit (Article V:5); and
- a non-discrimination obligation with respect to products based on whether or not they went through transit (Article V:6).

The first two obligations are relevant to third-party access and capacity establishment rights. Even though they do not directly require a transit state to operationalise its freedom of transit obligation through these inherent ancillary rights, should a transit state grant any of these rights to a third party (such as a gas shipper of one Member), it may be compelled to extend the same treatment to gas shippers of other Members. These obligations are, therefore, discussed later from this perspective. Article V:6 falls outside our subject.[112]

1. Prohibition of 'distinctions' in Article V:2 (second sentence)

Article V:2 (second sentence) establishes a somewhat atypical non-discrimination obligation, broadly prohibiting 'distinctions' based on the origin or ownership of goods.[113] Like in the case of other provisions in Article V, these 'distinctions' must relate to 'traffic in transit' – the term clarified in the previous sections. As explained earlier, the fact that this term is associated with the movement of goods does not automatically preclude Article V from being applicable to measures restricting ancillary rights linked inherently to effective transit, such as third-party access and capacity establishment rights. However, before addressing the relationship between Article V:2 (second sentence) and third-party access/capacity establishment, it is important to examine the nature of the non-discrimination obligation in this provision.

The key question that must be discussed in this regard is which particular standard of non-discrimination Article V:2 establishes:

[112] This provision reads: 'Each contracting party shall accord to products which have been in transit through the territory of any other contracting party treatment no less favourable than that which would have been accorded to such products had they been transported from their place of origin to their destination without going through the territory of such other contracting party.' GATT 1994, *WTO Legal Texts*, above n 1. Its slightly modified version can also be found in Article 11(4) of the Agreement on Trade Facilitation. The latter provision provides as follows: 'Each Member shall accord to products which *will be* in transit through the territory of any other Member treatment no less favourable than that which would be accorded to such products if they were being transported from their place of origin to their destination without going through the territory of such other Member.' World Trade Organization, Agreement on Trade Facilitation, above n 2, emphasis added.

[113] The precise language of this provision is as follows: '*No distinction* shall be made which is based on the flag of vessels, *the place of origin*, departure, entry, exit or destination, or on *any circumstances relating to the ownership of goods*, of vessels or of other means of transport.' GATT 1994, *WTO Legal Texts*, above n 1, emphasis added.

- the prohibition of distinctions between the traffic in transit of foreign goods of different origins (i.e. an MFN-type of an obligation); and/or
- the prohibition of *distinctions* between the traffic of goods in transit and that originating in the territory of the transit state and/or destined for that territory, such as internal traffic, export or import (i.e. a national treatment-type of an obligation).

In the context of third-party access to a transit pipeline, the answer to this question is relevant, for example, to determine whether a transit state B is obliged: (i) to provide non-discriminatory access to its pipeline network for gas originating in transit-dependent states A and D; or (ii) to split its pipeline capacity among gas shipments originating in A, D and its domestic market.

The same question can be asked in relation to capacity establishment. Does the fact that B provides a licence to its national company to build a pipeline from a domestic gas source to a third-country market but denies a licence to foreign gas suppliers to build a transit pipeline through its territory to the same market on similar conditions result in a violation of Article V:2 (see Figure 4)?

(i) The meaning of the non-discrimination standard under Article V:2 (second sentence) The question whether GATT Article V:2 covers an MFN or a national treatment standard or both has not been clarified sufficiently in WTO jurisprudence,[114] but it has generated debates in the academic literature. On one hand, a commentary written by the Energy Charter Secretariat as well as Nychay and Shemelin contend that only the MFN standard falls under Article V:2.[115] On the other hand, Cossy argues that Article V:2 entails a form of national treatment obligation, but, at the same time, she asserts that 'there is no requirement to treat goods in transit like goods imported to, or exported from, the domestic

[114] In *Colombia – Ports of Entry*, the Panel merely stated that 'Article V:2, second sentence requires that goods from *all* Members must be ensured *an identical level of access and equal conditions* when proceeding in international transit', without explaining whether this obligation relates only to MFN treatment or also to national treatment. Panel Report, *Colombia – Ports of Entry*, para. 7.402, emphasis added.

[115] Energy Charter Secretariat, 'Trade in Energy: WTO Rules Applying under the Energy Charter Treaty' (Brussels: Energy Charter Secretariat, 2001), paras. 99–100 and 108; Nychay, Nadiya and Shemelin, Dmitry, 'Interpretation of Article 7(3) of the Energy Charter Treaty', the Graduate Institute of International and Development Studies, Trade and Investment Law Clinic Papers (Geneva 2012), http://graduateinstitute.ch/home/research/centresandprogrammes/ctei/projects-1/trade-law-clinic.html, accessed 12 September 2013.

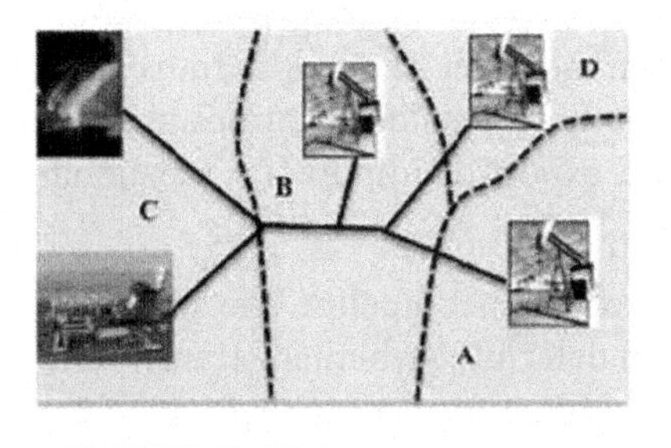

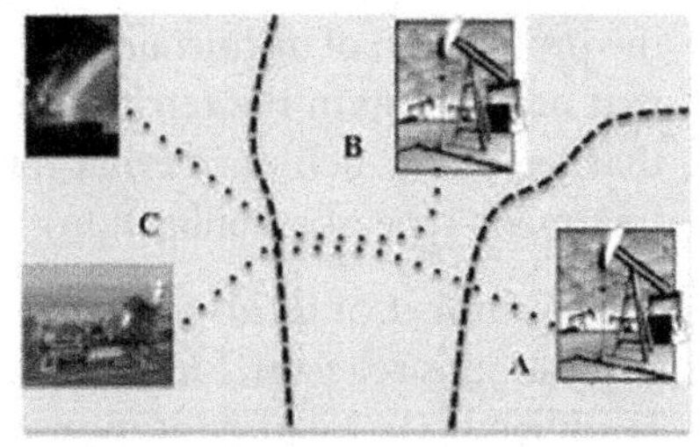

Solid line: existing pipeline
Round dot line: planned pipeline

Figure 4: The Scope of a Non-discrimination Obligation under GATT Article V:2

market, or like goods transported within the domestic market'.[116] Ehring and Selivanova assert that Article V:2 covers both MFN and 'at least certain elements of national treatment'.[117]

The opinion that GATT Article V:2 (second sentence) covers national treatment is strongly supported by the text of this provision, which broadly reads that 'no distinction shall be made' based on, among other things, 'the place of origin' or 'any circumstances relating to the ownership of goods'.[118] This provision does not indicate that those terms only refer to foreign, not domestic, places or goods. Put differently, there is no indication in GATT Article V that it allows a transit state to utilise its geographical location in its own favour, upsetting the competitive opportunities of the goods in transit that originate in other WTO Members.

The context of Article V:2 also supports the view that this provision must refer to both MFN and national treatment. In particular, pursuant

[116] Cossy, 'Energy Trade and WTO Rules', above n 24 at 297–298.

[117] Ehring and Selivanova, above n 5 at 65. These authors analyse two opposite approaches to interpreting GATT Article V:2, terming them as 'textual' and 'contextual' approaches. On one hand, according to their 'textual' approach, there is nothing in Article V:2 limiting the scope of its non-discrimination obligation only to instances of transit. In addition, as these authors argue in a different part of their work, the context of Article V:2 – Article V:6 – suggests that Article V does not only deal with restrictions directly imposed on *'traffic in transit'*, but it also addresses discrimination against *imports* that went through transit. Ibid. at 63; see also Panel Report, *Colombia – Ports of Entry*, paras. 7.477–78. On the other hand, the 'contextual' approach that they analyse is based on the assumption that Article V was intended to discipline measures directly affecting 'traffic in transit'. As Ehring and Selivanova admit, taking the latter approach would read words into Article V:2 that are not there. Ibid. at 65.

[118] GATT 1994, *WTO Legal Texts*, above n 1.

to the chapeau of GATT Article XX, a measure inconsistent with the GATT can be justified if it is not applied in a manner which would constitute a means of arbitrary or unjustifiable discrimination between countries where the same conditions prevail.[119] Like in the case of Article V:2 (second sentence), Article XX does not specify which standard of non-discrimination it implies. The Appellate Body, however, clarified that this standard encompasses the prohibition of discrimination 'not only between different exporting Members, but also between exporting Members and the importing Member concerned'.[120]

Furthermore, this view is in line with the object and purpose of Article V, which promotes freedom of transit. Clearly, by allocating transit infrastructure or pipeline capacity only to traffic of domestic origin or internal traffic, a transit state could circumvent its obligations under Article V and make this provision inutile in circumstances where the pipeline infrastructure is claimed to be used for the purposes of such traffic. Note that, in practice, pipeline owners or operators are rarely willing to disclose information regarding the available capacity in their pipeline networks. Even in the EU, where there are strong competition rules and energy legislation promoting the creation of a single gas market, the EU Commission has systematically discovered instances of capacity hoarding by private pipeline operators.[121]

Finally, the view that GATT Article V:2 covers both types of non-discrimination appears to be endorsed by the negotiating history of GATT Article V, traced back to the negotiations on the Barcelona Convention on Freedom of Transit.[122] The negotiating history shows

[119] Ibid. See this provision discussed in this chapter, Section V.A.2.

[120] Appellate Body Report, *US – Shrimp*, para. 150 (citing Appellate Body Report, *US – Gasoline* at 23–24).

[121] See the EU Commission's investigations against the infringement of EU competition rules by GDF, E.ON, RWE and Eni in Slotboom, Marco, 'Recent Developments of Competition Law and the Impact of the Sector Inquiry', in M. M. Roggenkamp and U. Hammer (eds.), *European Energy Law Report VII* (Antwerp: Intersentia, 2010) 97 at 106–108.

[122] Article V:2 (second sentence) (as well as then Article 10 of the ITO Charter) was borrowed almost verbatim from Article 2 of the Statute on Freedom of Transit, annexed to the Convention, on the request of Belgium-Luxembourg, France, and the Netherlands. Economic and Social Council, Preparatory Committee of the International Conference on Trade and Employment (Committee II), 'Report of the Technical Sub-committee', E/PC/T/C.II/54 (16 November 1946) at 8. The only textual differences between GATT Article V:2 and Article 2 is that the latter also referred to the distinction based on the nationality of persons and ownership of coaching or goods stock – issues left outside the GATT. See Barcelona Convention on Freedom of Transit, above n 43.

that the standard of non-discrimination, contemplated under Article 2 of the Statute on Freedom of Transit, was the standard of 'equality' between *all* parties. The drafters, however, were concerned that transit states could defeat the principle of 'equality' by imposing, inter alia, unfavourable transit conditions on *their own nationals*, which, in their view, warranted the adoption of additional standards protecting freedom of transit.[123] In this respect, the very distinction between the treatment of 'foreign' vis-à-vis 'own' attests to the intention of the drafters to establish the national treatment standard under the Convention.

More specifically, the question of which standard of non-discrimination (that is MFN or national treatment) should be established by the Barcelona Convention was discussed in the context of the negotiations on Article 4 thereof. This provision regulates the imposition of transportation charges on traffic in transit and contains wording similar to that in Article 2. The consensus among the drafters was that, whereas commercial differentiation between charges imposed on traffic of domestic and foreign origin could be tolerated, provided that this differentiation did not create obstacles to transit, political discrimination was a priori excluded.[124]

It is noteworthy that some negotiating drafts of the Agreement on Trade Facilitation contained language that aimed to establish an explicit national treatment obligation with respect to regulations and formalities imposed on or in connection with traffic in transit.[125] One could argue that the fact that this language was removed from the final text of the Agreement in the course of the negotiations may suggest that GATT Article V:2 (second sentence) had not intended to cover national treatment. Such an argument, however, would not fully appreciate the political realities of the negotiations at the Bali Ministerial Conference, where the negotiations on the Agreement on Trade Facilitation were concluded under extreme time pressure and were regarded by many stakeholders as a test of the institutional relevance of the WTO in global

[123] League of Nations, 'Freedom of Transit', above n 59 at 285–286.

[124] Ibid. at 290. See the texts of these provisions in Barcelona Convention on Freedom of Transit, above n 43 and Appendix 1.

[125] Article 11(5) of the Draft Consolidated Negotiating Text of 7 October 2011 read as follows: 'With respect to all regulations and formalities imposed on or in connection with traffic in transit, including charges for transportation, traffic regulations, safety regulations and environmental regulations, Members shall accord to traffic in transit treatment no less favourable than that accorded to [export or import traffic/domestic traffic/traffic which is not in transit]. This principle refers to like products being transported on the same route under like condition.' See World Trade Organization, 'Draft Consolidated Negotiating Text', above n 28 at 23.

trade.[126] The proposed language was particularly supported by Cuba and reportedly targeted the US trade embargo against Cuba, which has been in place for more than fifty years.[127] Except in the case of a special license from the US Treasury Department, this embargo imposes a prohibition on any vessels carrying goods to or from Cuba, or carrying goods in which Cuba or a Cuban national has any interest, to enter US ports.[128] The Cuban proposal was removed on the last days of the Bali talks, which, understandably, caused a strong reaction from Cuba and other WTO Members developing countries.[129] As a compromise, WTO Members agreed to include an explicit statement in the Bali Ministerial Declaration to the effect that 'the non-discrimination principle of Article V of GATT 1994 *remains valid*'.[130] This statement indicates that, should GATT Article V regulate national treatment, the status quo with respect to this treatment must be maintained.

In light of these considerations, an argument can be made that Article V:2 (second sentence) covers both MFN and national treatment standards. The next question that must be addressed is what this standard should mean in practice when applied in the context of pipeline gas transit. For example, does the prohibition of 'distinctions' mean the equal sharing of an existing pipeline capacity among all gas shippers, or does it imply identical conditions for capacity establishment?

(ii) Article V:2 (second sentence) and pipeline gas transit In *Colombia – Ports of Entry*, the Panel held that 'Article V:2, second sentence requires that goods from all Members must be ensured *an*

[126] Larry, Elliott, 'Bali summit invigorated World Trade Organisation, says Roberto Azevêdo', *The Guardian*, Wednesday 18 December 2013, http://www.theguardian.com/business/2013/dec/18/roberto-azevedo-wto-bali-global-trading, accessed 15 January 2014.

[127] See International Centre for Trade and Sustainable Development, 'Historic Bali Deal to Spring WTO, Global Economy Ahead' (Bridges, 7 December 2013).

[128] See, inter alia, Section 6005(b) of the Cuban Democracy Act of 1992 at http://www.treasury.gov/resource-center/sanctions/Documents/cda.pdf, accessed 20 December 2013. The US embargo was the subject of a WTO dispute between the United States and the EU, which, however, did not reach the panel stage. See World Trade Organization, 'United States – The Cuban Liberty and Democratic Solidarity Act', http://www.wto.org/english/tratop_e/dispu_e/cases_e/ds38_e.htm, accessed 1 February 2014.

[129] World Trade Organization, Ministerial Conference, 'Statement by Bolivia, Cuba, Nicaragua and Venezuela at the Ninth WTO Ministerial Conference', WT/MIN(13)/29 (7 December 2013).

[130] World Trade Organization, Ministerial Conference, Bali Ministerial Declaration, WT/MIN(13)/DEC (11 December 2013), emphasis added.

identical level of access and equal conditions when proceeding in international transit.'[131] In the case of pipeline gas transit, the meaning of 'an identical level of access and equal conditions' requires qualification, since each gas transit project is unique in terms of its stakeholders and financial arrangements. As explained in other chapters, pipelines may not always be under state control or contain idle capacity that can be equally divided among all interested parties. Furthermore, even if the pipeline capacity is physically available, the requirement of non-discrimination cannot be imposed in a just manner by disregarding the distinct financial contributions made by various entities (domestic and foreign) to build the pipeline in question, especially if the pipeline or its element were dedicated to a particular gas field or a long-term contractual arrangement. Pipeline projects often require billions in investments, which are normally recouped from the project output – gas sales.[132] In other words, a transit state may not always be able to ensure that no formal differences exist between the treatment of various gas shipments transported through or within its territory. These complicating factors triggered an academic debate on how the prohibition of 'distinctions' under GATT Article V:2 should be operationalised in the context of gas transit.

For example, Azaria notes that 'there are limits to the duty to provide a non-discriminatory (and physically identical) level of access for energy goods via fixed infrastructure under GATT Article V'.[133] In her view, in the context of gas transit, the obligation to provide 'an *identical* level of access' should be understood as the requirement for a transit state to 'establish a procedure to allow the owners of goods *identical possibilities* to access the infrastructure [idea underpinned by the principle of transparency]'.[134] In addition, Azaria argues that a transit state would be required to allow identical possibilities of access to the pipeline on equal conditions (inter alia, with respect to transportation charges) only to the extent that the transit routes are equal. She derives this caveat from Ad Note to Article V:5 (discussed later). However, Azaria notes that, in reality, the transportation charges cannot be monetarily equal, since they depend on different factors, such as geography, pipeline parameters, financing costs, ownership of the pipeline and whether the

[131] Panel Report, *Colombia – Ports of Entry*, para. 7.402, emphasis added
[132] See Chapter II, Section III.A.3, and Chapter V, Section III.B.1.
[133] Azaria, 'Energy Transit under the Energy Charter Treaty and the General Agreement on Tariffs and Trade', above n 5 at 572.
[134] Ibid. at 572–573, emphasis added.

pipeline is used purely for transit or also domestic transport.[135] Cossy expressed a similar opinion.[136]

In contrast to these views, Ehring and Selivanova assert that when infrastructure constraints (bottlenecks) arise, GATT Article V:2 (second sentence) can oblige a transit State to allocate scarce transit capacity in such a way that transit is possible. In the view of these authors, while Article V:2 does not provide clear guidance on how this allocation between foreign and domestic traffic must be executed, it is clear that a distinction based on the ownership of the goods would be inconsistent with this provision.[137]

This study posits that, in the case of pipeline gas transit, Article V:2 (second sentence) does not – and cannot – require a transit state to provide formally identical treatment to gas originating in different Members (including its own gas). Such an application of the non-discrimination obligation under Article V:2 would appear to contradict the text of this provision, its context, and would likely defeat the object and purpose of the GATT to protect 'competitive opportunities' of WTO Members vis-à-vis each other.[138] Recall that GATT Article V:2 (second sentence) prohibits distinctions that are '*based on* ... the place of origin ... or on any circumstances relating to the ownership of goods'.[139] It follows from this wording that distinctions that stem from factors other than those explicitly mentioned in GATT Article V:2 (for example, purely commercial considerations) should not fall afoul of this provision.

The same approach is adopted in GATT Article III:4 (National Treatment on Internal Taxation and Regulation), which prohibits discrimination between imported and domestic like products, including regulations and requirements affecting their internal sale and transportation. It states, however, that: '[t]he provisions of this paragraph shall not prevent the application of *differential internal transportation charges* which *are based exclusively on the economic operation of the means of transport* and *not on the nationality of the product*.'[140] In addition, as will

[135] Ibid. [136] Cossy, 'Energy Trade and WTO Rules', above n 24 at 298–299.

[137] Ehring and Selivanova, above n 5 at 70–71.

[138] In *Colombia – Ports of Entry*, the panel stated that 'Article XI:1, like Articles I, II and III of the GATT 1994 [provisions dealing with market access and non-discrimination], protects *competitive opportunities* of imported products not trade flows'. Panel Report, *Colombia – Ports of Entry*, para. 7.236, and footnote 463, emphasis added. The notion of competitive opportunities was similarly referred to in connection with the non-discrimination obligation in the TBT Agreement. See Appellate Body Report, *US – COOL*, para. 18.

[139] GATT 1994, *WTO Legal Texts*, above n 1, emphasis added.

[140] Ibid, emphasis added.

be explained later, with respect to the MFN obligation under GATT Article V:5, the explanatory AD Note to this provision clarifies that this obligation applies to transportation charges that are levied on products being transported *on the same route under like conditions.*[141] This clarification acknowledges that, in commercial realities, charges may differ depending on the conditions of transportation.

Along the same lines, in *EC – Tariff Preferences*, the Appellate Body defined the ordinary meaning of the term 'non-discriminatory' in footnote 3 to paragraph 2(a) of the WTO Enabling Clause[142] as a distinction 'among *similarly-situated* beneficiaries'.[143] In *US – COOL*, while interpreting the non-discrimination obligation under the TBT Agreement, the Appellate Body held that, to the extent that there are adverse effects on imported products, the focus of the analysis of this obligation is, inter alia, on 'whether these effects are attributable to the measure *itself* or are based on *external non-origin related factors*, such as pre-existing market conditions and the independent actions of private market actors'.[144]

Under the GATS, the national treatment provision – Article XVII – states explicitly that '[a] Member may meet [the standard of non-discrimination] by according to services and service suppliers of any other Member, either formally identical treatment or formally *different* treatment to that it accords to its own like services and service suppliers.' This provision further clarifies that the treatment will be considered less favourable (i.e. inconsistent with the standard of national treatment) 'if it modifies the conditions of competition'.[145] In *China – Publications and Audiovisual Products*, the Panel confirmed that the 'less favourable treatment' under Article XVII cannot be presumed lightly and must be demonstrated 'through careful analysis of the measure and the market'.[146]

In light of these examples, it is not a novel proposition that the analysis of the consistency of any alleged 'distinction' with Article V:2 must take into account the competitive positions of market participants (such as

[141] Ibid.

[142] The Enabling Clause is an exception to the MFN treatment obligation under the GATT granted to developing and least-developed countries under the 1979 GATT Decision on Differential and More Favourable Treatment, Reciprocity and Fuller Participation of Developing Countries. On this exception, see Van den Bossche, Peter and Zdouc, Werner, *The Law and Policy of the World Trade Organization*, 3rd ed. (New York: Cambridge University Press, 2013) at 330.

[143] Appellate Body Report, *EC – Tariff Preferences*, para. 153, emphasis added.

[144] Appellate Body Report, *US – COOL*, para. 18, emphasis in original.

[145] See Article XVII(2) and (3) of the GATS in *WTO Legal Texts*, above n 1, emphasis added.

[146] Panel Report, *China – Publications and Audiovisual Products*, para. 7.1131.

foreign and domestic gas shippers) against each other. This means, for example, that, in the absence of free capacity in a pipeline that was built by a domestic gas producer and seller at its expense, a licence to build another pipeline, along a similar or the same route and subject to equivalent regulatory conditions, should not be considered as a measure amounting to a 'distinction' in the sense of Article V:2.[147] Put differently, if equal opportunities are offered to all market actors to utilise the most convenient route in the transit state's territory (either for transit, export/import or domestic transportation purposes), the obligation under GATT Article V:2 should be satisfied. Among the examples of measures inconsistent with Article V:2 (second sentence) would be:

- a restriction imposed by a transit state (inter alia, through a legislative or administrative act, including those regulating transit charges) on the transportation of gas of foreign origin via domestic pipelines, while similar restrictions do not apply to gas of domestic origin; or
- a discriminatory application of a licensing regime governing pipeline construction in relation to foreign pipeline operators, especially when there are no other pipeline facilities capable of transporting foreign gas through the territory of the transit state.

2. MFN obligation in GATT Article V:5

GATT Article V:5 is the last provision in Article V establishing a non-discrimination obligation relevant to third-party access and capacity establishment rights. It requires an MFN treatment with respect to all charges (applied to like products being transported via the same route under like conditions), as well as regulations and formalities. This provision (including Ad Note thereto) reads as follows:

> With respect to *all charges, regulations* and *formalities* in connection with transit, each contracting party shall accord to *traffic in transit* to or from the territory of any other contracting party treatment no less favourable than the treatment accorded to traffic in transit to or from any third country.
>
> ...
>
> With regard to *transportation charges*, the principle laid down in paragraph 5 refers to like products being transported on *the same route under like conditions*.[148]

[147] As explained in Chapter VI, Section III.A.2, this is a legitimate option for operationalising the obligation to provide freedom of transit in the ECT.

[148] See GATT Article V:5 and Ad Note thereto, GATT 1994, *WTO Legal Texts*, above n 1, emphasis added.

In the context of third-party access and capacity establishment rights, the key question is whether this provision compels a WTO Member providing favourable treatment to gas originating in one Member (in the form of, inter alia, a pipeline transportation charge or a license to build a pipeline) to extend this treatment to other WTO Members. This question overlaps with the questions examined earlier with respect to Article V:2 (second sentence) and, therefore, does not warrant a separate detailed discussion. The particular terms in Article V:5 (namely, 'charges', 'regulations' and 'route') in principle do not preclude the applicability of this provision to measures affecting third-party access and capacity establishment.[149]

Note, however, that, in contrast to GATT Article V:2, the non-discrimination obligation under Article V:5 has essential limitations, somewhat reducing its practical relevance to pipeline gas transit: (i) it disciplines a particular group of measures (namely, charges, regulations and formalities); and (ii) it covers only the MFN standard.

C. Non-violation Complaint under the GATT

Another provision relevant to third-party access and capacity establishment is GATT Article XXIII:1(b), or the so-called non-violation complaint. Under Article XXIII:1(b), WTO Members can raise claims regarding the application by a Member of any measure if they consider that any benefit accruing to them directly or indirectly under the GATT is being nullified or impaired or the attainment of any objective of the GATT is being impeded as the result of the application of the measure.[150] This provision can in principle be invoked regardless of whether there is a violation of the GATT,[151] although it has been used primarily in the latter case. So far, there have only been a few cases in which Article XXIII:1(b) was applied

[149] Ibid. In this regard, the references to '*all* charges' in Article V:5 and the term 'transportation charges' in its Add Note indicate that this provision encompasses transportation charges for pipeline transit. The term 'regulation' is broad and can be defined as a rule, requirement or condition having a binding or quasi-binding effect. See Panel Report, *China – Publications and Audiovisual Products*, para 7.1448. It was mentioned earlier that the term 'route' refers to both fixed infrastructure (a pipeline) and a land corridor. See this chapter, Section III.A.2(ii).

[150] See GATT 1994, *WTO Legal Texts*, above n 1 and the text of this provision in Appendix 1.

[151] In this respect, Article XXIII:1(b) applies to 'any measure, whether or not it conflicts with the provisions of [the GATT]'. Ibid.

successfully.[152] Nevertheless, it is worth examining whether the 'non-violation complaint' could serve as a complementary or alternative cause of action for claims concerning restrictions on third-party access and capacity establishment, especially considering that the relevance of Article XXIII:1(b) to gas transit does not appear to have been addressed by previous researchers.

1. Legal requirements of Article XXIII:1(b)

In *Japan – Film*, the Panel summarised the following elements that a complaining party must demonstrate in support of its non-violation complaint under Article XXIII:1(b): (i) application of a measure by a WTO Member; (ii) a benefit accruing under the relevant agreement; and (iii) nullification or impairment of the benefit as the result of the application of the measure.[153] In addition, a consistent line of GATT/WTO panel reports has developed a test of 'reasonable (legitimate) expectations', explaining which particular kinds of benefits a Member could reasonably expect to obtain from the GATT.[154] This test sets an additional condition of 'reasonableness' on otherwise unlimited expectations. This section examines whether a challenge of the transit measure affecting third-party access or capacity establishment rights could satisfy the aforementioned three basic requirements. The next section explores how the challenge should be assessed in light of the 'reasonable expectations' test.

The first element of the legal test under Article XXIII:1(b) is application of a measure by a WTO Member. The 'measure' has to be a binding, governmental plan or course of action.[155] If the measure on the face of it has a non-binding nature, such as a mere guidance or recommendation, the complaining party must prove that it, in fact, creates certain incentives or disincentives inducing private parties to act in a particular manner causing the nullification or impairment of benefits.[156] The

[152] The Appellate Body has only partially ruled on this provision, without any substantive finding, in *EC – Asbestos*. At the WTO panel level, the non-violation complaint has been relied on in: *EC – Asbestos, Korea – Procurement*, and *Japan – Film*. Among adopted GATT panel reports, this complaint was raised in: *US – Sugar Waiver, EEC – Oilseeds, Japan – Semi-conductors, Germany – Sardines* and *Australian Subsidy on Ammonium Sulphate*. The complainant only succeeded in three GATT cases: *Australian Subsidy on Ammonium Sulphate, Germany – Sardines*, and *EEC – Oilseeds*.

[153] Panel Report, *Japan – Film*, para. 10.41.

[154] See ibid., para. 10.72.

[155] Ibid., para. 10.43–56. The implications of relevant non-governmental measures are discussed in Chapter V, Section III.B in the context of third-party access to pipelines.

[156] For example, in *US – Corrosion-Resistant Steel Sunset Review*, the Appellate Body did not find any compelling reason for concluding that, in principle, *a non-binding measure*

particular form of a measure can differ.[157] The choice of domestic policies or objectives promoted by the measure as well as the regulatory area of its application can also vary, since the text of Article XXIII:1(b) broadly refers to *any* measure. This means that a measure restricting the access to gas-transit infrastructure by rejecting third-party access or capacity establishment, could, in principle, be considered 'a measure' within the meaning of Article XXIII:1(b).[158]

The important qualifying condition imposed on a 'measure' under Article XXIII:1(b) is that it must be applied currently, as the text of this provision is written in the present tense. This condition envisages a higher evidentiary burden for the complainant, since the measure may not merely be introduced or adopted in the past, without having a continuing effect.[159] In light of this requirement, the question arises whether the denial of third-party access or capacity establishment can be considered a 'continuing' measure, as opposed to a one-off restriction. It can be argued that such a measure would indeed have a continuing effect in the sense of Article XXIII:1(b) until either the barrier is lifted or the expected benefit expires.

could not be challenged as such under WTO law. Appellate Body Report, *US – Corrosion-Resistant Steel Sunset Review*, para. 88. See also Panel Report, *Japan – Film*, para. 10.49–50.

[157] In *US – Corrosion-Resistant Steel Sunset Review*, the Appellate Body noted that a broad range of 'measures', whether legislative acts or not, can be submitted, as such, to WTO dispute settlement. Appellate Body Report, *US – Corrosion-Resistant Steel Sunset Review*, para. 85.

[158] A rather broad reading of the term 'measure' was, for example, adopted in Appellate Body Report, *EC – Asbestos*, para. 188; and Panel Report, *Japan – Film*, para. 10.38. This broad reading of the term is also supported by the negotiating history of Article XXIII:1(b), which, in Durling and Lester's opinion, takes roots in the discussions of the so-called equitable treatment ('indirect protection') clause at the League of Nations' economic conferences, held in the 1930s, for the purposes of its subsequent inclusion in bilateral commercial treaties between the League's Members. These discussions show that, despite the initial intention of the drafters to devise a complete list of examples of indirect protection, they later realised that this task was impractical and instead decided to draw up a rather broad text of the clause. Interestingly, the League of Nations' report on the Geneva Conference, held on 4–23 May 1927, among the examples of indirect protection, mentions 'discrimination arising from the *conditions of transport*'. Durling, James P., and Lester, Simon N., 'Original Meanings and the Film Dispute: The Drafting History, Textual Evolution, and Application of the Non-violation Nullification or Impairment Remedy', 32 *George Washington Journal of International Law and Economics* (1999) 211 at 216–226, emphasis added.

[159] See Durling and Lester, above n 158 at 244–245. See also Panel Report, *Japan – Film*, paras. 10.57–59.

The second element of the legal test under Article XXIII:1(b) is 'any benefit' accruing under the GATT. In most GATT cases interpreting this provision, the expected benefit was linked to the improved market access opportunities arising from tariff concessions under GATT Article II.[160] Reportedly, in one such case, the respondent even blocked the adoption of the GATT panel report mainly because the Panel had extended the application of the non-violation provision beyond the issue of tariff concessions.[161]

Nevertheless, despite this GATT practice, there is nothing in the text of Article XXIII:1(b) that narrows its scope to particular benefits. On the contrary, this provision refers to '*any* benefit' accruing potentially under any provision of the GATT, including Article V. Along the same lines, Durling and Lester suggest that if the negotiators of the GATT intended to tie up the non-violation complaint with the issue of tariff concessions, they would use a more specific language, such as that adopted in a trade agreement between the United Kingdom and Lithuania: 'full benefits of the tariff concessions'.[162] The broad understanding of the notion of 'any benefit' is also supported by the fact that, under Article XXIII:1(b), the non-violation complaint can be raised even if, more generally, 'the attainment of any objective of the GATT is being impeded'. This provision, however, has not been used effectively as an independent cause of action.[163]

The third and last element is nullification or impairment of the benefit 'as the result of' the application of the measure. This means that, to fall under Article XXIII:1(b), a measure must weaken or invalidate a benefit established under the GATT.[164] Notably, the term 'as the result of' creates a requirement of a causal link between the nullified or impaired benefit, on one hand, and the measure, on the other. Unlike in the case of a violation complaint in which there is a

[160] Panel Report, *Japan – Film*, para. 10.61.

[161] See the discussion of *EC – Citrus* in Chua, A. 'Reasonable Expectations and Non-violation Complaints in GATT/WTO Jurisprudence', 32 *Journal of World Trade* (1998) 27 at 40.

[162] Durling and Lester, above n 158 at 247.

[163] Cottier, Thomas and Nadakavukaren Schefer, K., 'Non-violation Complaints in WTO/GATT Dispute Settlement: Past, Present and Future', in E-U. Petersmann (ed.), *International Trade Law and the GATT/WTO Dispute Settlement System* (London: Kluwer Law International, 1997) 143 at 170.

[164] See the definition of the terms 'nullification' and 'impair' in 'Oxford Dictionaries', above n 57.

rebuttable presumption of an existing adverse effect on Members, in a non-violation complaint the burden of proof is on the complaining party to show that a GATT-consistent measure created the adverse effect.[165] Yet, in the context of a gas transit restriction, considering that a denial of the third-party access or capacity establishment rights could preclude a WTO Member from trading with the outside world, it should be possible for an aggrieved party to demonstrate that the third element of an Article XXIII:1(b) claim is fulfilled and the nullification or impairment of the benefit is the result of the application of the measure.

2. The sources of 'reasonable expectations' and their relevance to third-party access and capacity establishment

The notion of 'reasonable expectations' was developed in early GATT cases interpreting Article XXIII:1(b). For example, in *Australian Subsidy on Ammonium Sulphate*, the Report of a Working Party found that the withdrawal by Australia of a subsidy to fertilisers imported from Chile, while maintaining a subsidy on domestic fertilisers, nullified or impaired benefits accruing to Chile under the GATT. The Working Party held that '[i]n the case under consideration, the inequality created and the treatment that Chile could *reasonably* have expected at the time of the negotiation . . . were important elements in the working party's conclusion'.[166] Likewise, in *Germany – Sardines*, the Panel concluded that a range of measures applied by Germany to the imports of biologically distinct but commercially competitive kinds of sardines from different Members, such as tariff and tax rates as well as quantitative restrictions, had reduced the value of concessions obtained by Norway. The Panel explained that Norway had '*reason* to assume' that, after the Torquay round of negotiations, the type of sardines Norway exported would not be treated less favourably than those coming from other Members.[167]

In light of these cases, it appears that the mere expectation that no benefits under the GATT could ever be affected by a measure is not sufficient to challenge the measure under Article XXIII:1(b). The benefit/expectation should fall within the scope of 'reasonable expectations'. While the existence of this additional qualifying element is somewhat logical, the pertinent questions are: (i) what could be the sources of

[165] See DSU Article 3.8. Dispute Settlement Understanding, *WTO Legal Texts*, above n 1.
[166] GATT Panel Report, *Australia – Ammonium Sulphate*, para. 12, emphasis added.
[167] GATT Panel Report, *Germany – Sardines*, paras. 16–18, emphasis added.

'reasonable expectations' in general; and (ii) those specifically relating to third-party access and capacity establishment rights? These questions are especially important given that the initial burden to justify why such expectations were indeed created rests on the complaining party.[168]

In *Japan – Film*, the Panel states that '[a]n obvious starting point for determining whether a measure was reasonably anticipated is to consider whether the measure was adopted before or after the conclusion of the relevant round of tariff negotiations'.[169] The purpose of this condition is to see whether the complainant could be aware of the application of a challenged measure during the most relevant negotiations. Notably, however, in WTO jurisprudence, panels have narrowed the scope of this assessment to measures that: (i) were applied in the market of the responding Member, as opposed to other possible markets where similar measures could be in force; and (ii) were consistent with a specific governmental policy, as opposed to an abstract or general policy.[170]

This set of conditions is relevant for examining whether a particular measure relating to third-party access or capacity establishment could be reasonably anticipated during the most recent negotiations on the GATT transit rules – that is the Uruguay round or the accession negotiations of each new Member following that round. For example, if prior to the conclusion of relevant negotiations the government officials or authorities of a Member were consistently denying the existence of third-party access and capacity establishment obligations under the GATT, it would be challenging for other adversely affected Members to prove that they could not reasonably expect the application of those measures after the conclusion of these negotiations.[171]

[168] In addition to DSU Article 3.8, see also Article 26.1(a) of the same agreement requiring a detailed justification in support of any complaint relating to a measure that does not conflict with the relevant covered agreement. Dispute Settlement Understanding, *WTO Legal Texts*, above n 1. Importantly, the extent of this burden is not clearly resolved in the GATT/WTO jurisprudence. For example, in *Japan – Film*, the Panel appears to take the position that, '*failing evidence to the contrary*', it would not be reasonable for a complainant to expect that a measure of another Member would impair or nullify its benefits under the GATT. See, Panel Report, *Japan – Film*, para. 10.74, emphasis added (quoting 'Working Party Report on Other Barriers to Trade', adopted on 3 March 1955, BISD 3S/222, 224, para. 13). By contrast, in *EC – Asbestos*, the Panel states that *it is for the complaining party* to present detailed evidence showing why it could reasonably expect the concessions not to be affected by a legitimate health-protection measure (that is a ban on products containing asbestos in that case). Panel Report, *EC – Asbestos*, para. 8.292.

[169] Panel Report, *Japan – Film*, para. 10.78. [170] Ibid., para. 10.79.

[171] In this respect, Chua discusses the *Australian Subsidy on Ammonium Sulphate* case, where a long-standing practice of the Australian government to grant a subsidy on imports

That said, it is also possible to conceive of a situation in which, during relevant negotiations, Members were indeed cognisant of certain infrastructure-related transit restrictions imposed by a transit Member. Yet, arguably, they could expect that such restrictions would lapse due to: (i) promises or assurances provided by that transit Member (either in an oral or written form); or (ii) the crystallisation or existence of certain principles and standards commonly accepted either within the WTO or a broader system of public international law. The question can, therefore, be asked whether these two additional sources of 'reasonable expectations' are legitimate.

Concerning promises and assurances, for example, during the accession negotiations of Russia and Ukraine to the WTO, in 2012 and 2008 respectively, WTO Members knew about the interruptions of gas transit between Ukraine, Russia and the EU, as well as about the denial by Russian Gazprom of third-party access to its pipeline network for the imports of Central Asian gas into Western Europe through Russia.[172] Consequently, both Russia and Ukraine were requested to undertake an explicit commitment in their accession documents that they would apply laws, regulations and other measures governing the transit of goods (*including energy*) in conformity with relevant WTO rules.[173] While this commitment per se does not address third-party access and capacity establishment, would such a commitment not give rise to WTO Members' reasonable expectations that freedom of gas transit would not be nullified or impaired by the denial of rights inherently linked to this freedom? While the answer to this question may vary depending on particular factual circumstances, in general, in the GATT practice, there were instances where certain assurances and promises, made during negotiations, created reasonable expectations which were enforced through a non-violation complaint.[174]

resulted in a reasonable expectation that this subsidy would not be withdrawn after the negotiations on tariff concessions were concluded. Although in that case the presumption was created in favour of the complainant, arguably, this situation could develop in the opposite direction in favour of the responding party. Chua, above n 161 at 35.

[172] See Chapter II, Section III.C.3.

[173] See World Trade Organization, Working Party on the Accession of Ukraine, 'Report of the Working Party on the Accession of Ukraine to the World Trade Organization', WT/ACC/UKR/152 (25 January 2008), para. 367; World Trade Organization, Working Party on the Accession of the Russian Federation, 'Report of the Working Party on the Accession of the Russian Federation to the World Trade Organization', WT/ACC/RUS/70 (17 November 2011), para. 1161.

[174] See the discussion of *Germany – Sardines* and *Germany – Starch Duties* in Chua, above n 161 at 32.

As regards international principles and standards, in *EC – Asbestos*, the Panel rejected the argument Canada made that no recent scientific development could have made the respondent's measure restricting the importation of asbestos foreseeable. The Panel based this decision, inter alia, on the relevant guidelines of the World Health Organization and the International Labour Organization's Convention 162 (on asbestos), known to WTO Members during the relevant GATT negotiations.[175] These instruments recommend the replacement of asbestos and asbestos-based products with alternatives.[176] In addition, Roessler, while explaining the reasons why the non-violation complaint was successful in each particular case, states that, for instance, in *EEC – Oilseeds*, the Panel's finding of impairment was based on the normative guidance adopted by the GATT Contracting Parties by consensus in the form of the 1955 Decision. This Decision explicitly provided that the Contracting Parties could have reasonable expectations that the value of their concessions would not subsequently be impaired by the introduction or increase of a subsidy.[177] In other words, it can be argued that, in this case, the key source of reasonable expectations was a principle (or a normative guidance), which the GATT community recognised by consensus, although it was not then a part of formal GATT law.

In light of the foregoing, this section concludes that, in principle, GATT Article XXIII:1(b) can cover WTO Members' expectations that freedom of gas transit would not be nullified or impaired by measures restricting third-party access and capacity establishment. These expectations, however, could be better substantiated and justified by reference to relevant principles or practices that evolved in the international regulation of gas transit, as well as promises or assurances of particular Members given in the course of relevant WTO negotiations. In WTO law, among such principles are the principles of 'technological neutrality' and 'integrated service'. In this respect, if a non-violation case could be brought against a transit restriction, it would serve as a complement or alternative to a claim on the violation of GATT Article V. Considering that Article XXIII:1(b) envisages a higher burden of proof for the complainant than the latter provision, it is unlikely to be used as a primary cause of action.

[175] Panel Report, *EC – Asbestos*, para. 8.295. [176] Ibid.

[177] Roessler, Frieder, 'Should Principles of Competition Policy Be Incorporated into WTO Law through Non-violation Complaints?', 2 *Journal of International Economic Law* (1999) 413 at 419–420.

IV. Third-Party Access and Capacity Establishment under the GATS

This chapter has already looked at the relationship between the GATT and the GATS in the context of gas transit. It has been explained that these two agreements cover different aspects of this matter – trade in goods and trade in services. So far, this chapter has only analysed the former.

From the services perspective, the GATS directly regulates trade in energy. It contains commitments with respect to transport services of natural gas via pipeline (pipeline transportation of fuels), bulk storage services of gases, general construction services of long distance pipelines, as well as other services having an indirect relation to third-party access and capacity establishment, such as engineering-related scientific and technical consulting services, including surface surveying and map-making services.[178]

Yet the regulation of energy transit and, in particular third-party access and capacity establishment rights, under the GATS has significant limitations. This section discusses some uncertainties in this Agreement concerning the proper classification of energy services and the scope of energy-related specific commitments. More importantly, it demonstrates that the GATS covers only some aspects of energy transit, which reduces its relevance to transit proper.

A. *Overview of Relevant GATS Obligations*

The scope of the GATS is broadly defined in Article I by using the 'effects' approach. This provision establishes that the GATS applies to measures

[178] These commitments are inscribed in WTO Members' Schedules of Specific Commitments, based on the document 'W/120', prepared by the GATT Secretariat during the Uruguay round of negotiations, which, in turn, was drawn up based on the UN Provisional Central Product Classification (CPC). The CPC codes covering the aforementioned energy services are as follows: 7131, 7422, 5134 and 8675, respectively. In addition, the Explanatory Note to one of the revised versions of the CPC (version 1.1) clarifies that 'services incidental to mining' (code 86210) cover 'liquefaction and regasification of natural gas for transportation'. See World Trade Organization, 'Services Classification List', MTN.GNS/W/120 (10 July 1991); United Nations Statistics Division, Provisional Central Product Classification, http://unstats.un.org/unsd/cr/registry/regcst.asp?cl=9&lg=1, accessed 30 September 2012. For a more detailed discussion of energy-related services, see Meggiolaro, Francesco and Vergano, Paolo R., 'Energy Services in the Current Round of WTO Negotiations', 11 *International Trade Law & Regulation* (2005) 97; and Nartova, Olga, *Energy Services and Competition Policies under WTO Law* (Moscow: Infra-M, 2010).

by Members *affecting* trade in services.[179] Article I defines 'trade in services' as a 'supply of a service' via four modes, including cross-border trade in services (mode 1) and trade in services through commercial presence (mode 3).[180] The term 'supply of a service' is clarified in Article XXVIII as the production, distribution, marketing, sale and delivery of a service.[181] Finally, the term 'service' itself is defined in Article I:3 as 'any service in any sector', except services supplied in the exercise of governmental authority – that is neither on a commercial basis nor in competition with one or more service suppliers.[182]

The GATS contains both general obligations and specific commitments, which are undertaken in Members' Schedules of Specific Commitments on services with respect to particular services sectors and the aforementioned modes of supply, by using a positive list approach. The latter commitments are:

- market access commitments, scheduled pursuant to Article XVI, and prohibiting terms, limitations, and conditions listed in Article XVI:2;[183]
- national treatment commitments, scheduled pursuant to Article XVII;[184] and
- additional commitments (i.e. positive undertakings), not falling under the scope of the market access and national treatment obligations but applicable only to the extent the latter are also undertaken, scheduled pursuant to Article XVIII.[185]

In regard to additional commitments, Bigos states that these commitments should cover measures that, even though they restrict trade in services, are neither discriminatory nor captured by the list in Article XVI:2.[186]

179 GATS, *WTO Legal Texts*, above n 1.

180 Ibid. These are the most relevant modes. Evans, Peter C., 'Strengthening WTO Member Commitments in Energy Services: Problems and Prospects', in P. Sauvé and A. Mattoo (eds.), *Domestic Regulation and Services Trade Liberalization* (Washington, DC: World Bank Publications 2003) 167 at 172.

181 GATS, *WTO Legal Texts*, above n 1. 182 Ibid.

183 They cover limitations on the number of service suppliers and service operations, as well as measures either restricting or requiring specific types of legal entity for the supply of a service. Ibid.

184 See a detailed discussion on the distinction and interaction between the market access and national treatment commitments under the GATS in Panel Report, *China – Electronic Payment Services*, paras. 7.655–65.

185 GATS, *WTO Legal Texts*, above n 1; Bigos, Bradley J., 'Contemplating GATS Article XVIII on Additional Commitments', 42 *Journal of World Trade* (2008) 723 at 734.

186 They should also not repeat general obligations under the GATS, such as those under Article II (MFN) and VI (Domestic Regulation). Bigos, above n 185 at 731–732.

Table 1: *Schedule of Specific Commitments of Ukraine*[187]

7. Pipeline Transport			
(a) Transportation of fuels (CPC 7131)	(1) None (2) None (3) None (4) Unbound, except as indicated in the horizontal section	(1) None (2) None (3) None (4) Unbound, except as indicated in the horizontal section	Ukraine commits itself to provide full transparency in the formulation, adoption and application of measures affecting access to and trade in services of pipeline transportation. Ukraine undertakes to ensure adherence to the principles of non-discriminatory treatment in access to and use of pipeline networks under its jurisdiction, within the technical capacities of these networks, with regard to the origin, destination or ownership of product transported, without imposing any unjustified delays, restrictions or charges, as well as without discriminatory pricing based on the differences in origin, destination or ownership.

Among the examples of Members' specific commitments in the area of energy services, those of Ukraine are most relevant to this subject:

As Table 1 illustrates, by inscribing the term 'None' in the second and third columns, Ukraine undertook respectively the market access and national treatment commitments with regard to the transportation of

[187] See World Trade Organization, 'Schedule of Specific Commitments of Ukraine', GATS/SC/144 (10 March 2008) at 34.

fuels via three out of four modes (excluding the supply of a service through presence of natural persons). In addition, in column 4, Ukraine made quite extensive additional commitments relevant to third-party access.

Apart from specific commitments, the GATS establishes general obligations indirectly relevant to third-party access and capacity establishment, such as MFN (Article II), Transparency (Articles III and III*bis*), Domestic Regulation (Article VI), Monopolies and Exclusive Service Suppliers (Article VIII), Business Practices (Article IX), as well as general and security exceptions (Articles XIV and XIV*bis*). For example, under Article III, Members are required to publish promptly all relevant measures of general application which pertain to or affect the operation of the GATS. As the provision indicates, Members, however, are not required to publish measures addressing particular companies or operations, such as high transportation charges applied to a foreign gas supplier as a condition of third-party access. Moreover, Article III*bis* makes clear that the GATS does not require any Member to provide confidential information, the disclosure of which would, inter alia, 'prejudice legitimate commercial interests of particular enterprises, public or private'. In light of these stipulations, the relevance of Article III to the regulation of third-party access and capacity establishment rights under the GATS is significantly diminished.[188]

B. *Problems Relating to the Classification of Specific Commitments*

A number of scholars and practitioners suggest that the current services classification system does not reflect the commercial reality of the energy industry and that it inadequately explains which service sectors cover which particular energy-related activities.[189] Indeed, as demonstrated in this section, there is some mismatch and overlap between the scopes of specific commitments under the GATS. This factor creates difficulties for understanding the exact relevance of the GATS to third-party access and

[188] See these provisions in GATS, *WTO Legal Texts*, above n 1. See similar transparency obligations applicable to trade in goods discussed in this chapter, Section V.C.

[189] See Musselli, Irene and Zarrilli, Simonetta, 'Oil and Gas Services: Market Liberalization and the Ongoing GATS Negotiations', 8 *Journal of International Economic Law* (2005) 551 at 559–562; Meggiolaro and Vergano, above n 178 at 103; Poretti, Pietro and Rios-Herran, Roberto, 'A Reference Paper on Energy Services: The Best Way Forward?', 4 *Oil, Gas & Energy Law Intelligence* (2006), www.ogel.org, accessed 15 February 2014, Section II.B; and Evans, above n 180 at 74–77.

capacity establishment. In this study, it is explained in relation to pipeline transportation services.

The problem of services classification becomes pertinent when relevant services that interact and are mutually dependent can arguably belong to different energy sectors containing conflicting commitments. For example, a meaningful commitment on market access with respect to the pipeline transportation services supplied through commercial presence may depend on the pipeline construction services, pipeline management, surface surveying and map-making services. The effective cross-border supply of the pipeline transportation services may require access to bulk storage services of gases.

In certain cases, where a Member's Schedule and the Services Classification System do not explicitly address a particular auxiliary service, the problem can be resolved by integrating that service into the relevant primary service existing in the Schedule.[190] However, if integrally related services are explicitly scheduled in different parts of the Schedule, the principle of 'mutual exclusivity' of scheduled commitments would not allow them to be merged.[191] This means that the lack of access to one service can undermine the effectiveness of the commitments made with respect to the other.

Another problem concerns the uncertainty as to which specific commitments cover third-party access and capacity establishment in the context of pipeline transportation services. On one hand, Musselli and Zarrilli argue that the *third-party access right* is not subject to market access and national treatment commitments regulated by GATS Articles XVI and XVII.[192] The Schedule of Ukraine, discussed in the previous section, also indicates that third-party access belongs to additional commitments, scheduled pursuant to Article XVIII. Similarly, the Telecom Reference Paper, incorporated into the GATS as an additional commitment, contains obligations similar to third-party access, such as interconnection (Article 2).[193]

[190] To recall, in *China – Electronic Payment Services*, the Panel clarified that a service can be considered an 'integrated service' even when it is made up of several services, and if all the elements of such service, together, are necessary for the transaction to materialise. Panel Report, *China – Electronic Payment Services*, paras. 7.55–62. See this case discussed in this chapter, Section III.A.4.

[191] This principle indicates that a specific service cannot fall within two different sectors or subsectors. Appellate Body Report, *US – Gambling*, para. 180.

[192] Musselli and Zarrilli, above n 189 at 578.

[193] WTO Telecom Reference Paper, above n 39. The same can be said of some additional commitments drafted by the WTO Secretariat to facilitate the Members' negotiations on

On the other hand, the Panel's ruling in *Mexico – Telecoms* suggests that at least some aspects of third-party access can indeed be covered by the market access and national treatment commitments. In this case, the Panel defined the 'cross-border supply' of a service under GATS Article I:2(a) (such as the pipeline transportation service in our case) as implying, when necessary, 'interconnection' between different service suppliers. In other words, under this mode, the service supplier of the exporting Member does not have to be located in (or to move physically to) the territory of the importing Member where a given service terminates – for example, it can merely subcontract the domestic supplier of the importing Member to terminate the cross-border supply of the service.[194] By the same token, a gas shipper may decide to operationalise its third-party access right through 'interconnection', by subcontracting the domestic supplier of pipeline transportation services in the importing Member to deliver its gas to a consumer in that Member, which must also qualify as the 'cross-border supply' of a service. Consequently, measures affecting this type of the supply of a service (inter alia, by imposing market access limitations set out in Article XVI:2 or according discriminatory treatment in favour of domestic services or service suppliers within the meaning of Article XVII)[195] would appear to fall within the scope of these two provisions.

The question thus arises as to which aspects of the third-party access right implied in the cross-border supply of a pipeline transportation service are regulated by Articles XVI or XVII and which ones are regulated by Article XVIII. As explained in the previous section, to the extent that a measure falls under market access/national treatment commitments, it cannot be covered by additional commitments and vice versa.

While it is true that these specific commitments cannot overlap, they can regulate different aspects of a third-party access right. Articles XVI and XVII appear to impose limitations on the ability of *Members* to interfere with 'interconnection' arrangements between different private service suppliers. In contrast, commitments under Article XVIII

maritime transport services. Reportedly, such additional commitments would include access to a range of transport services for the purposes of providing multimodal transportation – in essence the analogy of a third-party access right.

[194] See Panel Report, *Mexico – Telecoms*, paras. 7.28–45.

[195] For example, Articles XVI:2(a) and XVI:2(c) discipline limitations imposed by Members on the number of service suppliers or service operations. GATS, *WTO Legal Texts*, above n 1, emphasis added.

appear to be deeper and may take the form of a positive anti-trust type obligation to ensure that a foreign service provider can obtain the third-party access permission from *domestic private pipeline operators*. For example, the additional commitments in the Schedule of Ukraine read as follows: 'Ukraine *undertakes to ensure* adherence to the principles of non-discriminatory treatment in access to and use of pipeline networks under its jurisdiction.'[196] Similarly, Article 2.2 of the Telecom Reference Paper establishes that '[i]nterconnection with a major supplier *will be ensured* at any technically feasible point in the network.'[197] The major difference between these two groups of obligations (under Articles XVI/XVII and XVIII respectively) is that the latter appears to capture not only governmental measures, but also private measures affecting third-party access. In addition, commitments regulated by Article XVIII may concern particular standards and conditions on the basis of which a third-party access right must be guaranteed (such as the prohibition of unjustified delays, restrictions, charges and discriminatory pricing; or non-discriminatory terms, conditions and rates for access).[198] The demarcation between measures affecting third-party access rights that fall under Articles XVI and XVIII of the GATS is illustrated in Figure 5.

In relation to *capacity establishment*, Cossy suggests that the market access and national treatment commitments under the GATS with respect to mode 3 could cover rights of way for building transportation network facilities, such as pipelines.[199] She supports this proposition by referring to the example of the US negotiating offer on the pipeline transportation via commercial presence, establishing certain limitations

[196] World Trade Organization, 'Schedule of Specific Commitments of Ukraine', above n 187, emphasis added. In respect of this obligation, Cossy states: 'The second sentence breaks new grounds as it aims at introducing a non-discriminatory third-party access obligation ("non-discriminatory treatment in access to and use of pipeline networks").' Cossy, Mireille, 'Energy Services under the General Agreement on Trade in Services', in Y. Selivanova (ed.), *Regulation of Energy in International Trade Law: WTO, NAFTA and Energy Charter* (The Netherlands: Kluwer Law International, 2011) 149 at 160.

[197] WTO Telecom Reference Paper, above n 39, emphasis added. Along the same lines, Mathew interprets the meaning of the obligation under Section 2.2 of the Telecom Reference Paper as follows: 'The introductory sentence of this very Paragraph is written as an absolute obligation, meaning that not only do measures have to be maintained, but they must also be carried out by the WTO Member. Thus, failure to enforce them would be cause for dispute settlement.' Mathew, Bobjoseph, *The WTO Agreements on Telecommunications* (Bern: Peter Lang, 2003) at 160–161.

[198] See WTO Telecom Reference Paper, above n 39; World Trade Organization, 'Schedule of Specific Commitments of Ukraine', above n 187.

[199] Cossy, 'Energy Services under the GATS', above n 196 at 165.

Article XVI measure: restriction imposed by Member B on the transportation by the domestic pipeline operator Y of natural gas of foreign origin, including gas shipped by the foreign service supplier X.

Article XVIII measure: violation of a positive commitment undertaken by B that Y will provide non-discriminatory third party access to its pipeline network.

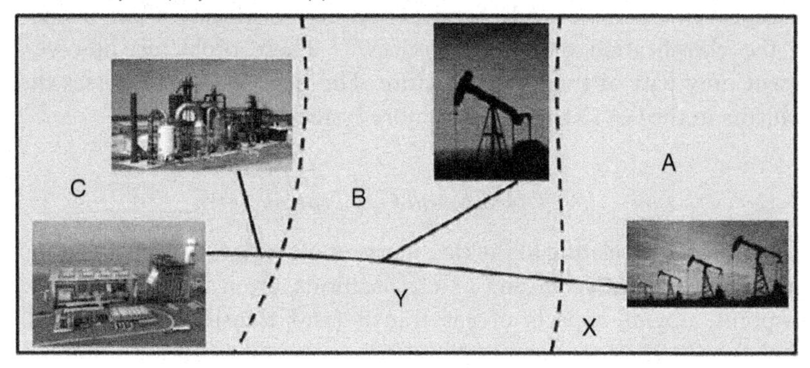

Solid line: pipeline connection between Members A, B and C

Figure 5: Distinction between Measures Regulated by Articles XVI and XVIII

on the right of way supposedly derived from this commitment.[200] More generally, Gomez-Palacio and Muchlinski argue that the supply of a service via commercial presence can create 'the right of establishment' of investment (the term used in investment law and relevant to capacity establishment).[201]

Despite this view, Cossy also notes that the fact that 'rights of way' are mentioned in the Telecom Reference Paper (Article 6) and that, pursuant to the model schedule for basic telecommunications, it is a type of measure on which additional commitments could be undertaken creates uncertainty as to the proper placement of this measure either within market access or additional commitments.[202] Furthermore, it was mentioned earlier that, according to the current Services Sectoral Classification List, capacity establishment, even though relevant to pipeline transportation services, appears to belong to a completely different services sector – general construction services of long-distance pipelines.

This discussion reveals a range of uncertainties and structural problems in the GATS Specific Commitments relevant to third-party access and

[200] Ibid.

[201] Gómez-Palacio, Ignacio and Muchlinski, Peter, 'Admission and Establishment', in P. Muchlinski et al. (eds.), *The Oxford Handbook of International Investment Law* (New York: Oxford University Press, 2008) 227 at 245–246.

[202] Cossy, 'Energy Services under the GATS', above n 196 at 165–166.

capacity establishment. Although they can adversely affect the current regulation of these rights under the GATS, note that energy services are subject to on-going negotiations of WTO Members, which may result in the improvement of the GATS Services Sectoral Classification List, including the classification of energy services.[203] These problems, however, present only part of the general picture. The next section addresses the problems in the GATS that are of a more systemic nature.

C. GATS and Gas Transit

While challenges relating to the classification of services can be overcome through interpretative means or clarifications, from a more systemic viewpoint, not all aspects of gas transit (and transit in general) fall under the GATS. For example, there appears to be a common opinion that the GATS does not regulate vertically integrated activities of service suppliers.[204] Thus, when gas is supplied by a vertically integrated gas producer, which, according to a gas sale contract, does not charge separately for the gas shipment services, this activity falls outside the GATS.[205] Of course, this view should not be taken as absolute, and nothing precludes the parties to a contract from including a clause in the contract providing for an independent transportation charge. In this case, a broad definition of 'services' and 'supply of a service' in the GATS could, arguably, cover services supplied in conjunction with the sale of goods. In this regard, in *Canada – Autos*, the Appellate Body clarified:

[203] Energy services is one of the subjects of the services negotiations, which began in January 2000. The negotiations are based on a range of proposals submitted by WTO Members. See World Trade Organization, 'Energy services', http://www.wto.org/english/tratop_e/serv_e/energy_e/energy_e.htm, accessed 30 November 2012. See the discussion of Members' proposals in Musselli and Zarrilli, above n 189 at 559–562; and Nartova, above n 178 at 201–218.

[204] In this respect, Musselli and Zarrilli state: 'Energy-related activities are commonly regarded as included within the scope of energy services when they are performed by an independent services supplier and not by a vertically integrated manufacturer [of goods, such as gas].' Musselli and Zarrilli, above n 189 at 562. See also Cossy, 'Energy Services under the GATS', above n 196 at 161; World Trade Organization, 'Energy Services', above n 20, paras. 9, 12.

[205] For example, see Contract between Gazprom and Naftogaz Ukrayini (Ukrainian national gas company) on the Sales of Natural Gas, signed in Moscow, on 19 January 2009, available at Ukrayinska Pravda, 'Gas Deal of Timoshenko-Putin – Full Text' ('Газова угода Тимошенко-Путіна. Повний текст'), http://www.pravda.com.ua/articles/2009/01/22/3686613, accessed 10 October 2012.

> even if a company is vertically-integrated, and even if it performs other functions related to the production, importation, distribution and processing of a product, *to the extent that it is also engaged in providing 'wholesale trade services'* and is therefore *affected in that capacity* by a particular measure of a Member *in its supply of those 'wholesale trade services'*, that company *is a service supplier* within the scope of the GATS.[206]

Yet, for various reasons (including the vertical integration of gas sellers), the price formulae for gas shipments, especially in Eastern Europe, do not appear to mention transportation costs.[207]

However, a more important limitation in the regulation of 'gas transit' under the GATS is that this Agreement does not appear to have been designed to discipline measures restricting the supply of a service *through* the territory of a Member – transit State – to its export destination (that is a situation contemplated under GATT Article V:1). All relevant modes of supply and specific commitments under the GATS focus on restrictions applied to services supplied either across the border to an importing Member or by a foreign-service operator in the territory of that Member. With the exception of additional commitments, the GATS does not appear to regulate restrictions on the exit of services outside the territory of Members, including export of services.[208] However, the regulation of restrictions on the destination of a service is essential to cover all aspects of gas transit. For example, restrictions on the export of services in the field of trade in energy have been applied in practice.[209]

In light of the foregoing, having weak rules on export operations or the supply of services through the territory of transit states, the GATS is only partially relevant to transit. For example, under the GATS modes 1 and 3,

[206] Appellate Body Report, *Canada – Autos*, para. 163, emphasis original.

[207] See Contract between Gazprom and Naftogaz on the Sales of Natural Gas, above n 204.

[208] For example, in *US – Gambling*, while citing the Explanatory Note on Scheduling of Initial Commitments in Trade in Services, the Panel clarified: 'cross-border supply concerns a service "delivered within the territory of the Member, from the territory of another Member"' (that is only involving two – as opposed to three – jurisdictions). Panel Report, *US – Gambling*, para. 6.30. Along the same lines, Neven and Mavroidis are also of the view that mode 1 of the GATS does not regulate export of services. See Neven, Damien J. and Mavroidis Petros, C., 'El Mess in TELMEX: A Comment on Mexico – Measures Affecting Telecommunications Services', 5 *World Trade Review* (2006) 271 at 279–282.

[209] In 1954, Saudi Arabia granted a priority right of oil transportation abroad (export) to a transport monopolist ('Satco'), thereby adversely affecting the trading rights of 'Aramco' (a company that invested in oil production in Saudi Arabia) under the Oil Concession Agreement. This measure was successfully challenged before an investment arbitration tribunal. See *Saudi Arabia v Arabian American Oil Company (Aramco Arbitration)*, in E. Lauterpacht (ed.), 27 *International Law Reports* (1963) 117 at 227–228.

a transit state may undertake a number of specific commitments allowing foreign gas shipping companies to render their gas transportation services in its territory and even build the pipeline network for this purpose. Yet these commitments may only result in the de facto expansion of the pipeline infrastructure *inside* the transit state. De jure, under the GATS, the transit state is not obliged to allow this gas to *exit* its territory, unless it has undertaken specific commitments to that effect. The latter obligation is established under GATT Article V, albeit with respect to trade in goods. Moreover, the existence of a pipeline in the territory of a WTO Member – transit state – does not mean that it is connected to gas networks of other WTO Members.

D. Summary and Concluding Remarks on the GATS

The overview of GATS obligations reveals that this Agreement is only partially relevant to the WTO's regulation of third-party access and capacity establishment. While it indirectly addresses these rights through different specific commitments, its rules do not cover all aspects of transit, such as the supply of pipeline transportation services through the territory of a transit state. This makes the GATS relevant to third-party access and capacity establishment mostly from the perspective of the energy services supplied between two states.

In this respect, it should also be mentioned that currently only a few WTO Members have undertaken relevant specific commitments on energy services. For example, to date, only sixteen Members have undertaken specific commitments on the transport services via pipelines of natural gas, including Ukraine and Montenegro, which have undertaken comprehensive additional commitments.[210] However, even the existing market access and national treatment commitments are not unconditional and do not necessarily cover all the relevant modes of supply.[211]

[210] World Trade Organization, Secretariat, 'Member/Sector Matrix Report: Transport Services', http://tsdb.wto.org, accessed, 10 January 2012; World Trade Organization, 'Schedule of Specific Commitments of Ukraine', above n 187; World Trade Organization, Working Party on the Accession of Montenegro, 'Report of the Working Party on the Accession of Montenegro – Addendum, Part II (Schedule of Specific Commitments on Services)', WT/ACC/CGR/38/Add.2/WT/MIN(11)/7/Add.2 (5 December 2011) at 28.

[211] In particular, Brazil has undertaken market access and national treatment commitments on transport services via pipelines of natural gas only with respect to the third mode (commercial presence). Cambodia, while undertaking commitments on the same

If the GATS can play any role in regulating gas transit from the perspective of trade in services, this role is rather forthcoming, provided that Members agree to undertake extensive commitments relating to the movement of gas through the territory of transit states. These commitments must address third-party access and capacity establishment rights. Chapter VI discusses the examples of such commitments.

V. WTO Rules Regulating Conditions of Transit, Exceptions and Institutional Arrangements

In this section, WTO rules are discussed that, although they do not directly regulate third-party access and capacity establishment, are relevant to determining conditions of transit and establishing exceptions from transit-related obligations. From this perspective, they are indirectly relevant to our subject and, therefore, must be briefly addressed.

A. Relevant GATT Obligations

1. Provisions that delimit and define the scope of the obligation to provide freedom of transit

In this chapter, a number of provisions in GATT Article V (such as Articles V:3 and V:4) that delimit and define the scope of the obligation to provide freedom of transit have already been mentioned. These provisions have to be distinguished from general exceptions in the GATT invoked when a GATT norm is violated, such as GATT Article XX, which will be discussed later. By contrast, if a Member's measure falls within the scope of a legitimate regulatory condition imposed on traffic in transit, in accordance with other provisions in Article V, there is no violation of the GATT and no need to invoke an exception.[212]

services with respect to mode 1 and 3, inscribed in its Schedule the following condition: 'Services must be provided through a contract of concession granted by the State on case-by-case basis.' Nepal has undertaken similar commitments, stipulating in the context of the third mode that the service can be supplied only through incorporation in Nepal and with maximum foreign equity capital of 51 per cent. WTO Secretariat, 'Member/Sector Matrix Report', above n 209.

[212] Similarly, see the relationship between GATT Articles XI:1 and XI:2 as explained in *China – Raw Materials*. In this case, the Appellate Body notes that these provisions must be read together and that the obligation not to impose quantitative restrictions under GATT Article XI:1 is limited by Article XI:2(a). Appellate Body Report, *China – Raw Materials*, paras. 319, 334. Although in this example there is a more explicit textual link

For example, it has been mentioned that GATT Article V:3 prohibits the imposition of customs duties and other charges on transit. Importantly, however, this provision, complemented by Article V:4, allows the imposition of: (i) a reasonable charge for transportation, having regard to the conditions of the traffic; and (ii) charges commensurate with administrative expenses entailed by transit or with the cost of services rendered.[213] Likewise, Articles V:3 and V:4 imply that a WTO Member can apply regulations that are reasonable and do not cause unnecessary delays or restrictions.[214] All these rules are instrumental in understanding what legitimate conditions can be imposed on traffic in transit by a transit state in accordance with Article V.

It is clear that what is a 'reasonable' charge or regulation or a 'necessary' delay or restriction has to be assessed on a case-by-case basis. However, this assessment cannot be an arbitrary exercise. In the WTO, there is extensive jurisprudence on the 'necessity' test applied, inter alia, in the context of GATT Article XX (General Exceptions) and the TBT Agreement.[215] In a range of WTO cases, panels have also interpreted the meaning of 'reasonableness'.[216] Furthermore, the

between Articles XI:1 and Article XI:2(a) established by the phrase 'shall not extend to' in the latter provision, the proposition that the scope of Article V:2 is similarly delimited and defined by other provisions in Article V is supported by the requirement of contextual interpretation in VCLT Article 31(1). Vienna Convention on the Law of Treaties, above n 55. Furthermore, it is noteworthy that the permission of certain exceptional restrictions on freedom of transit, resulting from domestic regulations, can be traced back to the preparatory work of the Barcelona Convention of 1921. The negotiators of the Convention noted: 'On no account is Freedom of Transit, so conceived in the Convention, to imply any infringement of the rights of States exercising sovereignty or authority over the territories crossed, to administer transit routes in their own way, so far as is compatible with the provisions contained in the Convention.' League of Nations, 'Freedom of Transit', above n 59 at 284.

[213] See this chapter, Section III.A.3. [214] Ibid.

[215] See Neumann, Jan and Türk, Elisabeth, 'Necessity Revisited: Proportionality in World Trade Organization Law after *Korea-Beef, EC-Asbestos* and *EC-Sardines*', 37 *Journal of World Trade* (2003) 199; Marceau, Gabrielle, 'The New TBT Jurisprudence in *US – Clove Cigarettes, WTO US – Tuna II* and *US – COOL*', 8 *Asian Journal of WTO & International Health Law and Policy* (2013) 1 at 13–22. For the WTO jurisprudence on this subject, see, inter alia, Appellate Body Report, *Brazil – Retreaded Tyres*, para. 133; Appellate Body Report, *US – Tuna II (Mexico)*, paras. 318–322.

[216] For instance, in *Dominican Republic – Cigarettes*, the Panel clarified the ordinary meaning of the term 'reasonable' in GATT Article X (Publication and Administration of Trade Regulations) as referring to notions such as: '"in accordance with reason, not irrational or absurd", "proportionate", "having sound judgement", "sensible", "not asking for too much" ... '. Panel Report, *Dominican Republic – Import and Sale of Cigarettes*, para. 7.385. See also Panel Report, *China – Raw Materials*, para. 7.741; Panel Report, *US – COOL*, para. 8.850; and Panel Report, *Mexico – Telecoms*, para. 7.182.

preparatory work of the Barcelona Convention explains that the test of 'reasonableness' was initially introduced in the Treaty from the common law system, where it denotes, inter alia, the idea of common sense. When referring to this notion in the context of transit tariffs in Article 4 of the Statute on Freedom of Transit, the Czech delegate described a 'reasonable tariff' as, among other things, 'one which opposes no direct or indirect obstacle to transit traffic'.[217]

In the context of gas transit, the assessment of the 'reasonableness' and 'necessity' of a transit measure could be informed by different international or quasi-international standards, comparative assessments and model instruments, developed by international agencies and bodies specialising in energy-related matters (such as the Energy Charter, OECD and International Energy Agency (IEA)), as well as the industry itself.[218]

Finally, it is worth noting that the criterion of a 'reasonable' regulation has often been interpreted by international tribunals in connection with the principle of the prohibition of abuse of rights (*abus de droit*), which will be discussed in Chapter V.[219] The requirements developed with respect to a 'reasonable' regulation in this context may, therefore, play a useful role in defining the term 'reasonable' in WTO law. WTO panels could use these tests of a 'reasonable' charge or regulation, or a 'necessary' delay or restriction, as balancing tools to interpret the scope of the obligation to provide freedom of transit in particular cases.

2. Exceptions

A transit measure inconsistent with Article V can be subject to a number of exceptions under the GATT: Articles XX (General Exceptions), XXI (Security Exceptions) and XXIV (Customs Unions and Free-trade Areas). GATT Article XX contains a list of exceptions from the GATT rules, which, for example, are: (i) necessary to protect human, animal or

[217] See League of Nations, 'Freedom of Transit', above n 59 at 9.

[218] See, inter alia, Energy Charter Secretariat, 'Bringing Gas to the Market. Gas Transit and Transmission Tariffs in Energy Charter Treaty Countries: Regulatory Aspects and Tariff Methodologies' (Brussels: Energy Charter Secretariat, 2012); and Energy Charter Secretariat, 'Model Intergovernmental and Host Government Agreements for Cross-Border Pipelines', 2nd ed. (Brussels: Energy Charter Secretariat, 2007). See also Association of International Petroleum Negotiators, https://www.aipn.org, accessed 28 July 2013. This study does not delve into the problem of the acceptability of international or quasi-international standards or practices in the WTO, which requires a separate analysis, including that in the context of energy trade.

[219] See Chapter V, Section II.B.3.

plant life or health; (ii) necessary to secure compliance with laws or regulations of a WTO Member that are not in themselves inconsistent with the provisions of the GATT; and (iii) relate to the conservation of exhaustible natural resources.[220] The first and the third exceptions relate to the conservation of environment and preventing various health risks. They can be invoked when, for example, a pipeline construction project in a transit state's territory has the potential to create an environmental hazard. The second exception can arguably address, among other things, laws and regulations governing the secure and stable functioning of the pipeline system of a transit state.[221] These exceptions, however, do not create a blanket permission for a transit state to adopt any measure remotely relevant to these causes. On the contrary, the measure must make a material contribution to the achievement of its legitimate objective and be taken in the absence of a less trade restrictive alternative that is reasonably available.[222] The chapeau of GATT Article XX introduces additional conditions requiring that a measure (subject to a possible exception) not be applied in a manner that constitutes 'a means of arbitrary or unjustifiable discrimination' or 'a disguised restriction on international trade'.[223] The essential purpose of the latter conditions is to prevent the abuse of the exceptions.[224]

Article XXI establishes additional exceptions to GATT rules that relate to the protection of essential security interests. In particular, Article XXI, paragraph (a) allows WTO Members not to furnish any information the disclosure of which they consider contrary to their essential security interests.[225] Article XXI, paragraph (b) states that '[n]othing in this Agreement shall be construed . . . to prevent any contracting party from taking any action *which it considers necessary* for the protection of its *essential security interests*'.[226] The three sub-paragraphs of Article XXI(b) provide an exhaustive list of examples when such actions can be taken,

[220] See GATT Article XX(b), (d) and (g). GATT 1994, *WTO Legal Texts*, above n 1.

[221] Notably, one of the draft consolidated negotiating texts, prepared during the WTO negotiations on trade facilitation to amend GATT Article V, mentions an exception to transit obligations relating to the protection and maintenance of transit infrastructure, which in this draft is textually linked to GATT Articles XX and XXI. World Trade Organization, 'Draft Consolidated Negotiating Text', above n 28 at 28.

[222] See the standard of 'necessity' explained in Appellate Body Report, *Brazil – Retreaded Tyres*, paras. 150 and 156; and Appellate Body Report, *EC – Asbestos*, para. 172.

[223] GATT 1994, *WTO Legal Texts*, above n 1.

[224] Appellate Body Report, *US – Gasoline* at 22. See the detailed analysis of Article XX in Van den Bossche and Zdouc, above n 142 at 543–583.

[225] GATT 1994, *WTO Legal Texts*, above n 1. [226] Ibid., emphasis added.

including 'in time of war or other *emergency in international relations*' (i. e. sub-paragraph XXI(b)(iii)).[227] There has been a debate among GATT/ WTO Members, as well as in the academic literature, as to whether the 'essential security' exceptions are justiciable or self-judging.[228] On one hand, given that the provision uses the terms 'considers necessary' and lacks the Article XX chapeau-type of a balancing test, it clearly grants a wide scope of discretion to WTO Members protecting essential security interests. On the other hand, as Van den Bossche and Zdouc rightly note, 'it is imperative that a certain degree of "judicial review" be maintained; otherwise the provision would be prone to abuse without redress'.[229] Notably, during the Falkland crisis, GATT Contracting Parties adopted a Decision which recognised that '[w]hen action is taken under Article XXI, all contracting parties affected by such action *retain their full rights under the General Agreement*', arguably including the right to challenge it pursuant to GATT Article XXIII.[230] In the WTO, at least two panels have been established to examine trade measures that were allegedly taken to protect the essential security interests, although one of these panels was ultimately suspended by the Complainant and the other one was never composed.[231]

If the Panel were to review a restriction imposed under GATT Article XXI (inter alia, paragraph (b)(iii)) on energy transit, the party invoking this exception would have to establish a link between the restriction and its essential security interests: in particular, it would have to demonstrate that the restriction was imposed in circumstances of 'war' or 'other emergency in international relations'. Although the term 'emergency in international relations' is not defined in the GATT, its ordinary meaning indicates that a mere reluctance to allow transit originating in and destined for a friendly nation, in the ordinary course of trade, because

[227] Ibid., emphasis added.

[228] See this issue discussed in Van den Bossche and Zdouc, above n 142 at 595–599; Hestermeyer, Holger P., 'Article XXI Security Exceptions', in R. Wolfrum et al. (eds.), *WTO – Trade in Goods* (Leiden: Martinus Nijhoff Publishers, 2011) 569; and Alford, Roger P., 'The Self-Judging WTO Security Exception', 3 *Utah Law Review* (2011) 697.

[229] Van den Bossche and Zdouc, above n 142 at 596. See also Hestermeyer, above n 227 at 580; and GATT Panel Report, *US – Nicaraguan Trade*, para. 5.17.

[230] See GATT Council, 'Decision Concerning Article XXI of the General Agreement', L/ 5426 (30 November 1982), emphasis added.

[231] See World Trade Organization, 'United States – The Cuban Liberty and Democratic Solidarity Act', above n 128; Alford, above n 227 at 719–721; World Trade Organization, 'Nicaragua – Measures Affecting Imports from Honduras and Colombia', http://www.wto.org/english/tratop_e/dispu_e/cases_e/ds188_e.htm, accessed 1 February 2014; and World Trade Organization, DSB Minutes, WT/DSB/M/80, 26 June 2000, paras. 28–39.

of some reasons of economic or commercial nature, would hardly pass the test of Article XXI(b)(iii). The restriction must be imposed in a situation which is 'unexpected, ... dangerous ... requiring immediate action'.[232] The terms 'war' and 'essential security' in the same provision further inform the meaning of the term 'emergency'. These terms suggest that the circumstances permitting the application of Article XXI(b)(iii) must be similar to a *war* or concern the security interests that are *essential* (i.e. 'absolutely necessary' or 'extremely important').[233] Given that the UN Members have, in general, promoted the spirit of friendly cooperation among all nations in improving the conditions of energy transit and developing transit infrastructure, a mere request for permission to transit energy materials would unlikely qualify as an '*emergency* in international relations'.[234] In ruling on this matter, the panel may also rely on the principle of the prohibition of abuse of rights (*abus de droit*) discussed in the next chapter, which could inform the nature of the obligation under GATT Article XXI.[235]

Last but not least, GATT Article XXIV allows WTO Members to form customs unions and free trade areas in which the duties and other restrictive regulations of commerce are eliminated on substantially all trade between the constituent territories in products originating in such territories.[236] So far, there has not been any authoritative guidance in WTO jurisprudence on whether this provision allows discrimination, inter alia, with respect to pipeline transportation charges and other transit regulations between WTO Members that are participants of the aforementioned regional trade arrangements and other WTO Members.[237]

B. *The TBT Agreement*

The TBT Agreement has a somewhat curious relationship with WTO transit disciplines. On one hand, Article 11(8) of the Agreement on Trade Facilitation provides that 'Members shall not apply technical

[232] See the term 'emergency' in 'Oxford Dictionaries', above n 57.

[233] See the meaning of the term 'essential' in ibid. This term was also discussed in the context of 'essential products' within the meaning of GATT Article XI:2(a) in Panel Report, *China – Raw Materials*, para. 7.275.

[234] See Chapter III, Section III.C.3. [235] See Chapter V, Section II.B.3.

[236] GATT 1994, *WTO Legal Texts*, above n 1.

[237] In this regard, one of the essential purposes of GATT Article XXIV is to grant an exception to WTO Members from a general MFN obligation under GATT Article I:1. See Van den Bossche and Zdouc, above n 142 at 651.

regulations and conformity assessment procedures within the meaning of the Agreement on Technical Barriers to Trade to goods in transit'. On the other hand, the Agreement on Trade Facilitation indicates, in Article 24(6) (Final Provisions), that 'nothing in this Agreement shall be construed as diminishing the rights and obligations of Members under the Agreement on Technical Barriers to Trade'.[238] It remains to be seen how these two provisions will be interpreted and reconciled in WTO cases. In general, the TBT measures can be regarded as protectionist when imposed on goods in transit, since these goods do not intend to be marketed in the transit states' territory and it would be unreasonable for exporters to comply with the technical regulations, such as labelling standards, of each transit state through which their goods intend to pass. Nevertheless, in the context of pipeline gas, certain technical requirements, such as gas quality specifications, are understandable, since injecting the so-called off-spec gas (with different chemical composition) into the pipeline system may affect other gas supplies. Such requirements, however, are already addressed by WTO agreements other than the TBT Agreement, inter alia, GATT Articles V:3 and V:4, as well as Article 11(1) of the Agreement on Trade Facilitation.

It is pertinent to make a few remarks on the TBT rules that, absent Article 11(8) of the Agreement on Trade Facilitation, would have been relevant to gas transit. The Agreement disciplines the application by Members of specific measures, including the so-called technical regulations. The term 'technical regulation' is defined as a document that lays down the mandatory requirements concerning product characteristics, their related processes and production methods or labelling requirements.[239] Among the examples of substantive provisions of the TBT Agreement regulating these measures, Article 2.2 requires WTO Members to ensure that their technical regulations are not prepared, adopted or applied with a view to, or with the effect of, creating unnecessary obstacles to international trade – they must not be more trade-restrictive than necessary to fulfil a legitimate objective pursued by the Member (e.g. environment protection).[240]

[238] World Trade Organization, Agreement on Trade Facilitation, above n 2.

[239] See Annex 1(1) of the TBT Agreement. TBT Agreement, *WTO Legal Texts*, above n 1.

[240] Ibid. On the detailed analysis of the TBT Agreement, see Van den Bossche and Zdouc, above n 142 at 850.

C. *Agreement on Trade Facilitation*

Some amendments to transit regulation were made by the new Agreement on Trade Facilitation, which was concluded at the Bali Ministerial Conference but has not yet entered into force.[241] Although one of the purposes of the Agreement is 'to clarify and improve relevant aspects of Article[] V', it is an independent agreement, which will be included in Annex 1A of the WTO Agreement.[242] Importantly, the Agreement on Trade Facilitation states explicitly that it does not diminish the obligations of Members under the GATT, including Article V, as well as the TBT Agreement.[243] It further provides that '[a]ll exceptions and exemptions under the General Agreement on Tariffs and Trade 1994 shall apply to the provisions of this Agreement', which arguably include GATT Articles XX and XXI.[244] These provisos are atypical, since normally agreements listed in Annex 1A of the WTO Agreement take precedence over the GATT as *lex specialis*.[245] Furthermore, with certain limited exceptions, such as the TRIMS Agreement, the Annex 1A agreements do not contain express references to GATT exceptions.[246]

Among provisions that are relevant to the regulation of gas transit, Article 11(2) of the Agreement (Freedom of Transit) reiterates the obligation in GATT Article V:3 with respect to charges imposed on traffic in transit. It provides as follows: 'Traffic in transit shall not be conditioned upon collection of any fees or charges imposed in respect of transit, except the charges for transportation or those commensurate with administrative expenses entailed by transit or with the cost of services rendered.'[247] In Article 1(1)(c) (Publication), the Agreement requires that Members promptly publish, in a non-discriminatory and easily accessible manner, inter alia, fees and charges imposed by governmental agencies on or in connection with transit.[248] However, given that this requirement does not extend to charges imposed by private entities

[241] See Chapter II, Section I.

[242] Ibid.; preamble to the Agreement on Trade Facilitation in World Trade Organization, Agreement on Trade Facilitation, above n 2.

[243] See Article 24(6) (Final Provisions) in World Trade Organization, Agreement on Trade Facilitation, above n 2.

[244] See Article 24(6), ibid.

[245] See General Interpretative Note to Annex 1A, *WTO Legal Texts*, above n 1.

[246] This is logical, since most of such agreements (e.g. the TBT Agreement) are the elaborations of GATT general exceptions, such as Article 2.2 of the TBT Agreement borrowing the 'necessity' test from GATT Article XX.

[247] World Trade Organization, Agreement on Trade Facilitation, above n 2.

[248] Ibid.

(such as private or quasi-private pipeline operators), it does not appear to expand the already existing obligation under GATT Article X:1 (Publication and Administration of Trade Regulations) that regulations pertaining to, inter alia, charges or affecting the transportation of products must be published promptly.[249] Like the obligation under GATT Article X:1, Article 1(1)(c) appears to apply only to regulations of 'general application', as opposed to contractual charges agreed on between a government and a particular company, let alone between two companies. However, in reality, charges for pipeline transportation services are rarely set by governments through general regulations. For example, the Treaty on the Eurasian Economic Union between Belarus, Kazakhstan and Russia provides that gas transportation charges must be determined in commercial contracts between the entities of the constituent Members.[250]

In paragraphs 1, 3 and 6 of Article 11, the Agreement imposes additional requirements on regulatory measures. In particular, Article 11(1) provides that:

> Any regulations or formalities in connection with traffic in transit imposed by a Member shall not be:
> a. maintained if the circumstances or objectives giving rise to their adoption no longer exist or if the changed circumstances or objectives can be addressed in a reasonably available less trade-restrictive manner,
> b. applied in a manner that would constitute a disguised restriction on traffic in transit.[251]

In Article 11(3), it prohibits the requirement of a 'voluntary restraint' and other similar measures on traffic in transit.[252] Article 11(6) of the Agreement provides that '[f]ormalities, documentation requirements and customs controls in connection with traffic in transit shall not be more burdensome than necessary to: (a) identify the goods; and (b) ensure fulfilment of transit requirements'.[253]

[249] GATT 1994, *WTO Legal Texts*, above n 1.

[250] See Article 83(4) and Annex 22 of the Treaty on the Eurasian Economic Union of 29 May 2014 (Договор 'О Евразийском Экономическом Союзе'), http://www.economy.gov.ru/wps/wcm/connect/economylib4/mer/about/structure/depsng/agreement-eurasian-economic-union, accessed 4 July 2014.

[251] World Trade Organization, Agreement on Trade Facilitation, above n 2.

[252] Ibid.

[253] See also similar obligations established by Article 10(1) of the Agreement (Formalities connected with Importation and Exportation and Transit). Ibid.

Apart from Article 11(3), establishing an obligation that is novel in the context of transit,[254] all the other provisions set out earlier appear to elaborate on the already existing commitments under GATT Article V. For example, Articles 11(1)(a) and 11(6) are the elaborations of the obligation under GATT Article V:3 that traffic in transit must not be subject to 'unnecessary delays or restrictions'. Article 11(1)(b) refers to a regulation that would hardly be considered 'reasonable' under GATT Article V(4) and would certainly not pass the test under the chapeau of GATT Article XX, which explicitly prohibits 'disguised restrictions on international trade'. Thus, the main function of all of these obligations under the Agreement on Trade Facilitation (with the exception of Article 11(3)) is clarificatory.

Nevertheless, there are some new obligations in the Agreement on Trade Facilitation indirectly relevant to gas transit, albeit of a soft-law nature. These obligations are established in Article 11(5), which contains the best-endeavour commitment of Members to 'make available, where practicable, physically separate infrastructure (such as lanes, berths and similar) for traffic in transit';[255] and Article 11(16), which requires Members to 'cooperate and coordinate with one another with a view to enhancing freedom of transit'.[256] It was argued earlier that these provisions constitute a useful context for the interpretation of GATT Article V:2.

Another contribution that the Agreement on Trade Facilitation makes to the WTO's transit regulation is that it established a political body specialised in trade facilitation matters, namely the Committee on Trade Facilitation.[257] As the Agreement provides, the Committee: may establish such subsidiary bodies as may be required; must develop procedures for sharing by Members of relevant information and best practices as appropriate; must maintain close contact with other international organisations in the field of trade facilitation and, to this end, may invite representatives of such organisations or their subsidiary bodies to attend meetings of the Committee; and has to encourage and facilitate the

[254] A similar obligation exists with respect to 'voluntary export restraints' or similar measures on the export or the import side under Article 11(1)(b) of the Agreement on Safeguards. See Agreement on Safeguards, *WTO Legal Texts*, above n 1. The term 'voluntary export restraint' can be defined as 'actions taken by exporting countries involving a *self-imposed* quantitative restriction on exports'. Van den Bossche and Zdouc, above n 142 at 491, emphasis original.

[255] World Trade Organization, Agreement on Trade Facilitation, above n 2.

[256] Ibid.

[257] See Article 23(1)(1), ibid.

discussion among Members of specific matters related to the implementation of the Agreement.[258]

This contribution is extremely important, especially in the area of network-bound transit, since the latter significantly depends on cooperation between the states involved,[259] and Members may need a political platform to discuss pertinent problems in this area. These problems may include the lack of sufficient infrastructure, the need for funding, differences in regulatory approaches and others. It is noteworthy that the idea to establish such a platform is not new. Just in the area of international trade alone, it was already set forth in the ITO Charter. Articles 33(6) of the Charter (Freedom of Transit) provided as follows:

> The Organization may undertake studies, make recommendations and promote international agreement relating to the simplification of customs regulations concerning traffic in transit, the equitable use of facilities required for such transit and other measures designed to promote the objectives of this Article. Members shall co-operate with each other directly and through the Organization to this end.[260]

The ITO Charter appeared to be more explicit on the particular functions of the ITO relating to the promotion of freedom of transit than the Agreement on Trade Facilitation, which entrusts the Committee on Trade Facilitation with a general task to encourage and facilitate discussions on the implementation of the Agreement. Nevertheless, the mandate of the Committee is broad enough to address problems specific to any particular area of transit, such as pipeline gas, through a multilateral dialogue, the involvement of specialised organisations or by even commissioning a study on this issue by a subsidiary body (such as a group of ad hoc or permanent experts). Depending on how active the Committee is in discharging its functions, some disputes of a political or technical nature arising in the area of transit could be resolved or prevented through the application of the above mechanisms.

[258] Article 23, ibid.

[259] Vinogradov, Sergei, 'Challenges of Nord Stream: Streamlining International Legal Frameworks and Regimes for Submarine Pipelines', 9 *Oil, Gas & Energy Law Intelligence* (2011), www.ogel.org, accessed 15 February 2014 at 5.

[260] International Trade Organization Charter, above n 62.

VI. Summary and Concluding Remarks

This chapter analysed the key disciplines in the GATT relevant to third-party access and capacity establishment rights, including the principle of freedom of transit in Article V:2 (first sentence); the principle of non-discrimination in Articles V:2 (second sentence) and V:5; and the non-violation complaint under Article XXIII:1(b). While these provisions do not expressly address these rights, they also do not exclude their existence, particularly in circumstances where these rights become integrally related to an established primary right (namely, freedom of transit or non-discriminatory treatment).[261]

The interpretation of these provisions as implying third-party access and capacity establishment rights is supported by different instruments of interpretation. For example, the interpretation of Article V:2 (first sentence) in this way is in line with the principle of freedom of transit in general international law, informing the ordinary meaning of the generic term 'freedom', the context of this provision, the principle of evolutionary treaty interpretation and the object and purpose of the WTO Agreement to expand the regime of trade in goods. Moreover, it will be demonstrated in the next chapter that this interpretation is also supported by other principles of general public international law.

Finally, this chapter looked at other WTO instruments indirectly relevant to third-party access and capacity establishment, such as the GATS and a number of rules regulating conditions of transit and exceptions to transit obligations. It is important to keep those instruments in mind for the purposes of the practical implementation of third-party access and capacity establishment rights. However, in this study, these instruments play only a secondary, contextual role.

[261] In this respect, Article XXIII:1(b) protects reasonable expectations arising from any right established by the GATT.

V

Third-Party Access and Capacity Establishment Rights in Light of General Principles of International Law

I. Introduction

In this chapter, a number of additional principles of general international law are discussed that, taken together, support the proposition that the obligations to guarantee freedom of transit and to treat traffic in transit in a non-discriminatory manner under GATT Article V contain within themselves third-party access and capacity establishment rights, as without such inherent rights, in the context of pipeline gas, these obligations would be meaningless. Furthermore, in the previous chapter, it was explained that the strength of a 'non-violation complaint' under GATT Article XXIII:1(b), based on the expectation that freedom of gas transit will not be impaired or nullified as the result of restrictions on third-party access and/or capacity establishment rights, will depend on the existence of international principles supporting these rights.

The principles of general international law discussed in this chapter do not establish third-party access or capacity establishment rights directly, nor are they derived from international instruments specifically regulating trade in energy. Given that there are scant rules in public international law addressing these rights,[1] it is unlikely that these rules have entered the corpus of general international law, applicable to all nations and all subject matters. Nevertheless, a number of fundamental principles of general international law are indirectly

[1] See, inter alia, Article 7(4) of the ECT, dealing with capacity establishment, and the reference to the doctrine of 'essential facility', relevant to third-party access, in the WTO Telecom Reference Paper. Energy Charter Secretariat, 'The Energy Charter Treaty and Related Documents: A Legal Framework for International Energy Co-operation' (Brussels: Energy Charter Secretariat, 2004); World Trade Organization, Reference Paper on Telecommunications Services (incorporated into some WTO Members' GATS Schedules of Specific Commitments) (Telecom Reference Paper), www.wto.org/english /tratop_e/serv_e/telecom_e/tel23_e.htm, accessed 30 March 2012.

relevant to third-party access and capacity establishment. These principles are: the principles of 'effective or integrated rights', 'economic cooperation', the principle that 'negotiations (on the terms and modalities of gas transit) must be conducted in good faith' (derived from the '*pactum de contrahendo*' nature of GATT Article V:2) and 'the prohibition of abuse of rights (*abus de droit*)'. These principles are the balancing tools that provide guidance on how the interests of a transit state and a transit-dependent state in the context of pipeline gas could be balanced and reconciled.[2] These principles are especially relevant in circumstances where the text of a given treaty provision, such as GATT Article V:2 (first and second sentence), is vague. This lack of clarity may result from an intentional silence left by the drafters, the evolution of international law or technological developments.

As elaborated on in this chapter, the principle of effective or integrated rights indicates that, in circumstances in which the practical implementation of a primary right established by a treaty (*in casu* freedom of transit or non-discrimination under GATT Article V) depends inherently on certain ancillary rights, such as third-party access and/or capacity establishment, the primary right must be operationalised by giving effect to those ancillary rights. The principles of economic cooperation and the obligation to negotiate in good faith provide guidance as to how this implementation must be achieved – that is, through good faith cooperation/negotiations. Finally, the principle of abuse of rights delimits the scope of the regulatory autonomy of a transit state vis-à-vis transit-dependent states in their efforts to operationalise obligations under GATT Article V.

The next section provides a detailed overview of these principles of general international law, their sources, content and relevance to third-party access and capacity establishment rights. Section III addresses a number of practical questions specific to pipeline gas, namely: (i) the scope of a capacity establishment obligation; (ii) whether compliance with the principle of freedom of transit in WTO law requires a transit state to provide a *compulsory* third-party access to existing pipelines; and (iii) the relationship between third-party access and capacity establishment rights in implementing freedom of gas transit. Finally, Section IV summarises the findings in this chapter.

[2] While these principles are not the only balancing instruments in international law (others would include 'proportionality', 'equity' and 'margin of appreciation'), they appear to be the most relevant and settled principles. Moreover, in any event, for the purposes of this study, there is no need to discuss all existing balancing principles as some of the principles would merely duplicate each other.

II. Relevant Principles of General International Law

A. *The Principle of Effective or Integrated Rights*

The proposition that the obligations to guarantee freedom of transit and to treat traffic in transit in a non-discriminatory manner under GATT Article V:2 contain within themselves third-party access and capacity establishment rights is supported by the principle of effective or integrated rights, which appears to have crystallised into an independent principle of general international law. Although this principle has rarely been mentioned directly in international jurisprudence, it will be demonstrated in this section that it has been applied implicitly on many occasions. The principle of effective rights denotes the basic idea that rights established by law must be effective, and is pertinent in circumstances where, on the basis of the treaty text, it is unclear whether a primary right established by that treaty encompasses additional, ancillary rights. As discussed further, the principle of effective rights compels the holistic interpretation of primary and ancillary rights in international law when these rights *are integrally related* and the recognition of an inherent ancillary right is *essential* to giving proper effect to the primary right. In the context of pipeline gas, a pertinent example of such an integral relationship is the inter-link between freedom of transit and third-party access or capacity establishment rights, as without recognising these ancillary rights, freedom of transit cannot be operationalised. The basic idea promoted by the principle of effective rights finds support in other, more general principles of public international law, such as the principles of good faith and *pacta sunt servanda*.[3] Nevertheless, the

[3] The principles of good faith and *pacta sunt servanda* are commonly recognised principles of general international law. The preparatory work of the PCIJ Statute, Article 38(1)(c), explicitly mentions the principle of good faith as the example of the legal instruments to be covered by this provision. See Cheng, Bin, *General Principles of Law as Applied by International Courts and Tribunals* (New York: Cambridge University Press, 2006) at 25. Mitchell calls the principle of good faith a general principle of law or a principle of customary international law. Mitchell, Andrew D., *Legal Principles in WTO Disputes* (New York: Cambridge University Press, 2011) at 113. The UN Charter, Article 2.2, establishes that '[a]ll Members ... shall fulfill *in good faith* the obligations assumed by them in accordance with the present Charter'. Charter of the United Nations of 1945, http://www.un.org/en/documents/charter/, accessed 10 January 2014. The preamble to the VCLT acknowledges that the principle of good faith is universally recognised; and it further establishes, in Article 26 (*Pacta sunt servanda*), that '[e]very treaty in force is binding upon the parties to it and must be performed by them in good faith'. Vienna Convention on the Law of Treaties of 1969, 1155 *United Nations Treaty Series* (1980) 331. The principle of good faith was expressly recognised in *Nuclear Tests* as a principle of

existence of these other principles does not undermine the relevance of the principle of effective rights to our subject. The latter has its independent genesis in municipal law (in particular the law of servitudes), from which it was later borrowed by international lawyers.

The principle of effective rights has its roots in the Roman law doctrine of servitude, and it was originally applied in the context of real property and transit or passage rights.[4] The authoritative commentary on Roman law reads: 'A man who has a right to draw water is also presumed to have an *iter* [passage], to enable him to exercise his right.'[5] Furthermore, the same source provides: 'If a man is left an *iter* . . . and cannot exercise his right unless he does some construction work, then . . . he is allowed to make a pathway by means of digging and laying foundations.'[6] In the same way, the principle of effective rights finds support in an analogous common law doctrine of 'easements', which in the case of a grant envisages the transfer of certain ancillary rights implied in that grant.[7]

general international law, where the ICJ stated as follows: 'One of the basic principles governing the creation and performance of legal obligations, whatever their source, is the principle of *good faith*. Trust and confidence are inherent in international co-operation, in particular in an age when this co-operation in many fields is becoming increasingly essential.' *Nuclear Tests (New Zealand v. France)*, Judgment 20 December 1974: I.C.J. Reports 1974, p. 457, para. 49, emphasis added. Similarly, in *Pulp Mills*, the ICJ has explicitly recognised the principle of *pacta sunt servanda* as a principle of customary international law. *Pulp Mills on the River Uruguay (Argentina v. Uruguay)*, Judgment of 20 April 2010: I.C.J. Reports 2010, p. 14, para. 145.

4 As noted earlier, the doctrine of servitude has been recognised in most civil and common law jurisdictions. See Rheinstein, Max, 'Observations et Conclusions du Gouvernement de la République Portugaise sur les Exceptions Préliminaire du Gouvernement de l'inde (Annexe 20), Étude Comparative sur le Droit D'Accès aux Domaines Enclaves par le Professeur Rheinstein', in *Case Concerning Right of Passage over Indian Territory*, Judgment of 12 April 1960: I.C.J. Reports 1960, pp. 6, 714; Buckland, W. W. and McNair, Arnold D., *Roman Law & Common Law: A Comparison in Outline*, 2nd ed., F. H. Lawson (ed.) (Cambridge: Cambridge University Press, 1952) at 127; Zimmermann, Reinhard, Visser, Daniel and Reid, Kenneth, *Mixed Legal Systems in Comparative Perspective: Property and Obligations in Scotland and South Africa* (New York: Oxford University Press, 2004) at 735; and Gordley, James and von Mehren, Arthur T., *An Introduction to the Comparative Study of Private Law: Readings, Cases, Materials* (New York: Cambridge University Press, 2006) at 196.

5 Mommsen, T. et al. (eds.), *The Digest of Justinian*, vol. 1 (Pennsylvania: University of Pennsylvania Press, 1985) at 259.

6 Ibid. at 251.

7 See, inter alia, the explanation by the United Kingdom's Law Commission of the rules established by the *Wheeldon v. Burrows* case, 'easements of necessity', and 'easements of intended use'. In *Wheeldon v. Burrows*, Lord Justice Thesiger explained as follows: 'on the grant by the owner of a tenement . . . there will pass to the grantee *all those continuous and apparent easements* . . ., or, in other words, *all those easements which are necessary to the*

The doctrine of easements is closely connected to the legal maxim 'whoever grants a thing is deemed to grant that without which the grant itself would be of no effect' ('*cuicunque aliquis quid concedit concedere videtur et id sine quo res ipsa esse non potuit*').[8] Broom illustrated this maxim as follows: 'as incident to the liberty to sink pits, the right[s] to affix to the land all machinery necessary to drain the mines, and draw the coals from the pits . . . were lawfully made'.[9]

At the beginning of the twentieth century, scholars and practitioners that saw an analogy between servitudes and international rights governing the passage through foreign territory (such as freedom of navigation over international rivers) started transposing this principle into international law. For example, in the *North Atlantic Coast Fisheries* arbitration, Counsel for the United States, prominent US lawyer Mr. Elihu Root referred to a similar principle when asserting certain rights of the United States relating to fishery in British territorial waters, which, in his view, were derived from the doctrine of servitude.[10] A renowned

reasonable enjoyment of the property granted, and which have been and are at the time of the grant used by the owners of the entirety for the benefit of the part granted.' Importantly, the condition that the implied right *must be used* at the time of a grant is less significant when the two other forms of easements are created. For example, for the purposes of the 'easement of necessity', it is sufficient that the *necessity* to use an inherent (implied) right existed or was foreseeable. Ancillary rights implied in the 'easement of intended use' will likely to be enforceable when they are *essential* for the *intended purpose* of the grant; the enforcement will, therefore, turn on whether that intent can be established. United Kingdom's Law Commission, 'Easements, Covenants and Profits à Prendre', Consultation Paper No 186 (2008) at 39–54, emphasis added.

[8] See Broom, Herbert, *A Selection of Legal Maxims: Classified and Illustrated*, 10th ed., R. H. Kersley (ed.) (Universal Law Publishing, 2006, reprinted) at 309; United Kingdom's Law Commission, above n 7 at 49. Some authors have referred to this maxim as the fundamental principle of justice applicable in international law. See Farran d'Olivier, C., 'International Enclaves and the Question of State Servitudes', 4 *International and Comparative Law Quarterly* (1955) 294 at 304. See also a closely related maxim: 'when the law commands a thing to be done, it authorizes the performance of whatever may be necessary for exercising its command' ('*quando aliquid mandatur, mandatur et omne per quod pervenitur ad illud*'). Broom, ibid. at 313.

[9] Broom, above n 8 at 310.

[10] In particular, Mr. Root referred to Artopaeus, who stated: 'The right created by the servitude shall not be extended beyond the compass explicitly granted; this does not impede the dominant party from taking the measures necessary for the exercise of its right. For *when a certain right is granted, the measures necessary for its exercise must also be given*.' See Root, Elihu, *North Atlantic Fisheries Arbitration at the Hague: Argument on Behalf of the United States*, Bacon, R. and Scott, J. B. (eds.) (Cambridge, MA: Harvard University Press, 1917) at 287, emphasis added. The *North Atlantic Coast Fisheries* case is discussed in this chapter, Section II.B.3(ii).

English scholar, John Westlake, after his exhaustive study on the international regimes of rivers, came to the following conclusion:

> [W]herever the right itself is admitted to exist . . . '[i]t seems that this right draws after it *the incidental right* of using *all the means which are necessary to the secure enjoyment of the principal right* itself. Thus the Roman law, which considered navigable rivers as public or common property, declared that the right to the use of the shores was incident to that of the water, and that the right to navigate a river involved the right to moor vessels to its banks, to lade and unlade cargoes, &c. The public jurists apply this principle of the Roman civil law to the same case between nations . . . '.[11]

Another prominent scholar, Elihu Lauterpacht, who analysed rights inherent in a treaty-based 'freedom of navigation' in the Shatt-Al-Arab region between Iran and Iraq, similarly stated that this 'freedom' could imply an ancillary 'right to *construct* jetties':

> The proposition canvassed here is limited to a situation in which one State, having been granted rights in an area subject to a neighbour's control, is obliged *for the purpose of taking due and proper advantage of the rights so granted*, to perform additional acts *necessary to implement the original right.*[12]

In the area of 'freedom of navigation', the principle of effective rights was implicitly recognised by the World Court when analysing rights inherent in the primary right at issue. In *Oscar Chinn*, the PCIJ examined the consistency of the Belgian government's financial assistance to a national transport company operating in Congo with the principles of 'commercial equality' and 'freedom of trade and navigation', embodied in the Convention of Saint-Germaine of 1919. For this purpose, the Court defined the term 'freedom of navigation' in a broad manner, including: 'freedom of movement for vessels, freedom to enter ports, and to make

[11] Westlake, John, *International Law*, Part I (Cambridge: Cambridge University Press, 1904) at 159, emphasis added.

[12] See Lauterpacht, Elihu, 'River Boundaries: Legal Aspects of the Shatt-Al-Arab Frontier', 9 *International and Comparative Law Quarterly* (1960) 208 at 231, emphasis added. The principle of effective rights has also been referred to in Brownlie, Ian, *Principles of Public International Law*, 7th ed. (New York: Oxford University Press, 2008) at 378 (although the author states that 'the grant of rights [of passage] should be reasonably effective', he denies the application of the doctrine of servitude as such in international law); Reid, Helen D., *International Servitudes in Law and Practice* (Chicago, IL: University of Chicago Press, 1932) at 6; and (centuries earlier) Von Pufendorf, Samuel, *De Jure Naturae Et Gentium Libri Octo*, vol. 2, Scott, J. B. (ed.), Oldfather, C. H. and Oldfather, W. A. (trans.) (Oxford: Clarendon Press, 1934) at 605.

use of plant and docks, to load and unload goods and to transport goods and passengers', as well as some aspects of 'freedom of commerce'.[13]

In another similar PCIJ case, the Court was demarcating the boundary between the competences of the European Commission of the Danube to protect freedom of navigation in relation to the Danube's specific sector, on one hand, and those of the Romanian authorities, on the other, established by a range of legal instruments, including the Treaty of Paris of 1856. The Court found that both the European Commission and the territorial authorities possessed 'some powers' in relation to this sector.[14] However, the question arose whether the broad principle of 'freedom of navigation' could also cover the right of access to ports, even though the applicable law was silent on this issue. The Court recognised this right, stating as follows:

> The conception of navigation includes, primarily and essentially, the conception of the movement of vessels with a view to the accomplishment of voyages. . . . [F]reedom of navigation *is incomplete* unless shipping can actually reach the ports. . . . The Commission's powers therefore extend to navigation into and out of the port, as well as through the port.[15]

In *Navigational and Related Rights*, the ICJ had to determine, inter alia, the scope of Costa Rica's right of free navigation on the Nicaraguan section of the San Juan River and in particular whether this right implies Nicaragua's obligation to notify Costa Rica of the regulations it makes with respect to the navigational regime at issue. Notably, the 1858 Treaty of Limits that established the primary right of Costa Rica does not contain any explicit notification requirements. Nevertheless, the ICJ stated that '[i]f the various purposes of navigation are to be achieved, it must be subject to some discipline, a discipline which depends on proper notification of the relevant regulations', and consequently recognised Nicaragua's obligation to notify its regulations.[16]

The principle of effective rights appears to be closely linked to the principle of 'integrated service', developed by the Panel in *China – Electronic Payment Services*. To recall, in this case, when interpreting

[13] See *Oscar Chinn*, Collection of Judgments, Orders and Advisory Opinions, Judgment, Series A/B. – No. 63, 12 December 1934 at 83.

[14] See *Jurisdiction of the European Commission of the Danube between Galatz and Braila*, Collection of Advisory Opinions, Series B. – No. 14, 8 December 1927 at 60–61.

[15] Ibid. at 64–65, emphasis added.

[16] See Dispute regarding Navigational and Related Rights (*Costa Rica v. Nicaragua*), Judgment of 13 July 2009: I.C.J. Reports 2009, p. 213, paras. 95, 97. This case was also discussed in Chapter III, Section IV.B.2.

China's specific commitments, the Panel merged several individual services into one 'integrated service', since, in its view, these services were integral elements of a system necessary for the payment card transaction to be completed. In other words, by merging different services into one 'integrated service', the Panel gave effect to China's obligations under the GATS, which otherwise would have been impractical.[17] This approach is, therefore, not different from the interpretative methods judges used in the previous cases.

The principle of effective rights can also be derived from a 'functional' interpretation of treaties, which places emphasis on the particular aspects of the surrounding area of international regulation, such as the particular features of the industry affected by treaty obligations that are at stake in a dispute.[18] In practice, such an interpretation may require the recognition of certain implicit rights inherent in the effective operation of the industry concerned.

For example, in *Aramco Arbitration*, the Arbitral Tribunal was examining Aramco's implied right to export oil overseas under the Concession Agreement between this company and the Kingdom of Saudi Arabia in the context of the 'integral activity of a company operating an oil concession'.[19] It has noted that Aramco's exclusive rights, according to that Agreement, 'must be understood in the plain, ordinary and usual sense, which is the sense accepted in the oil industry'.[20] The Tribunal emphasised that '[t]he interpretation of contracts is not governed by rigid rules; it is rather an art, governed by principles of logic and common sense, which purports to lead to an adaptation, as reasonable as possible, of the provisions of a contract to the facts of a dispute'.[21] Consequently, it

[17] See Panel Report, *China – Electronic Payment Services*, paras. 7.58–7.59. See this case discussed in Chapter IV, Section III.A.4.

[18] The term 'functional interpretation' is borrowed from Gardiner, Richard, *Treaty Interpretation* (New York: Oxford University Press, 2008) at 166. The functional approach to treaty interpretation appears to have been adopted by the Panel in the recent *EU – Seals* case. In this case, when analysing the consistency of the EU's prohibition of the marketing of seals and seal products in the EU with the non-discrimination standard under Article 2(1) of the TBT Agreement, the Panel stated as follows: 'For the purpose of our analysis, we must therefore assess the "design, structure, and expected operation" of the EU Seal Regime, as well as any *other relevant features of the market*, which may include the particular characteristics of *the industry at issue*.' See Panel Report, *EC – Seal Products*, para. 7.157, emphasis added; and the TBT Agreement, *WTO Legal Texts*, http://www.wto.org/english/docs_e/legal_e/legal_e.htm, accessed 5 March 2014.

[19] See *Saudi Arabia v. Arabian American Oil Company (Aramco Arbitration)*, in E. Lauterpacht (ed.), 27 *International Law Reports* (1963) 117 at 176.

[20] Ibid. at 179. [21] Ibid. at 172.

found that Aramco had the right to a complete and integral operation, including export.[22]

Similarly, in *Oil Platforms*, the ICJ noted: 'it would be a natural interpretation of the word "commerce" in Article X, paragraph 1, of the Treaty of 1955 that it includes . . . not merely the immediate act of purchase and sale, but also *the ancillary activities integrally related to commerce*', such as storage and transportation.[23] Judge Simma, in his Separate Opinion in the same case, also used the method of functional interpretation of the term 'commerce'. He stated: 'Trade in oil has to be viewed in light of the realities of that trade.'[24]

The ICJ ruling in *Oil Platforms* lends an interesting perspective on the principle of effective rights. In this dispute, the Court had to decide, inter alia, whether the destruction by the US military forces of Iranian oil platforms[25] violated 'freedom of commerce' between Iran and the United States, guaranteed under the Treaty of 1955. The important fact to keep in mind is that at that time the platforms were inoperative and no direct commerce in crude oil between those countries was a priori possible due to the existence of the US trade sanctions. The ICJ noted:

> [W]here a State destroys another State's means of production and transport of goods destined for export, or means ancillary or pertaining to such production or transport, there is in principle an interference with the freedom of international commerce. In destroying the platforms, whose function, taken as a whole, was precisely to produce and transport oil, the military actions made commerce in oil, at that time and from that source, impossible, and to that extent prejudiced freedom of commerce.[26]

However, at the same time, the ICJ appears to make a distinction between injury to 'potential commerce' (in an abstract sense) and injury to 'actual commerce'. It stated: 'the possibility must be entertained [that freedom of commerce] could actually be impeded as a result of acts entailing the destruction of goods destined to be exported, or capable of affecting their transport . . . with a view to export'.[27] Ultimately, the Court found that, under the given circumstances, the destruction of oil

[22] Ibid. at 202, 227–228.

[23] See *Case Concerning Oil Platforms (Islamic Republic of Iran v. United States of America)*, Judgment of 6 November 2003: I.C.J. Reports 2003, p. 161, para. 80.

[24] *Case Concerning Oil Platforms (Islamic Republic of Iran v. United States of America)*, Separate Opinion of Judge Simma, Judgment of 6 November 2003: I.C.J. Reports 2003, p. 161, p. 324, para. 29.

[25] Like pipelines, oil platforms are the means to supply energy to the market.

[26] *Oil Platforms*, above n 23, para. 89.

[27] Ibid. para. 92.

platforms could not hamper 'freedom of commerce' between the United States and Iran.[28]

In the light of the decision of the ICJ in *Oil Platforms*, an important question is whether the fact that a transit-dependent state is denied third-party access or capacity establishment rights could undermine the effectiveness of the *actual* freedom of transit, considering that there is no pipeline communication or the flow of gas between the states involved. This question directly relates to the problem of the correlation between primary and inherent ancillary rights in international law. It has to be recalled that, in the cases discussed earlier, the World Court interpreted freedom of navigation in an integral manner as including all other inherent rights without which this freedom would be incomplete. In *Oil Platforms*, the ICJ does not appear to depart from this basic understanding of 'effectiveness'. In this case, the Court arrived at its conclusion not because of the existing technical obstacle to producing oil by inoperative platforms, but rather because there could not be *any* US–Iran trade in oil under the given political circumstances.[29] Consequently, the ICJ judgment does not exclude the possibility that measures impeding access to transit facilities, or their establishment, could undermine the effectiveness of freedom of transit (whether this transit is regarded as actual or potential from a technical viewpoint).

Moreover, the very distinction between the effect of a measure on potential and actual benefits granted under treaties, suggested in *Oil Platforms*, appears to be somewhat blurred in WTO jurisprudence. In *Colombia – Ports of Entry*, apart from transit-related claims, the Panel also had to rule on whether the denial of access to certain 'ports of entry' amounted to a violation of the GATT's prohibition of import restrictions under Article XI:1. Citing a range of other WTO rulings, the Panel has emphasised that in WTO law the legality of a challenged measure depends not on whether the *actual* trade flow is impaired, but on whether that measure is designed in such a way that it has the *potential*

[28] Ibid. para. 98. This decision was not unanimous. Judge Simma, in his Separate Opinion, disagreed with the distinction between 'potential' and 'actual' commerce, stating as follows: 'the key issue is not damage to commerce in practice but the violation of *the freedom to engage in commerce*, whether or not there actually was any commerce going on at the time of the violation.' *Oil Platforms*, Separate Opinion of Judge Simma, above n 24, para. 25 and 34, emphasis original.

[29] In fact, one of the ICJ requirements was that Iran should prove that the platforms could have been repaired before the imposed US sanctions. In general, the Court admitted that a measure directed against the means of production or transportation could indeed violate 'freedom of commerce'. See *Oil Platforms*, above n 23, paras. 89 and 93.

to 'negatively affect the competitive opportunities' and trade between WTO Members.[30] Consequently, with respect to this claim, the Panel found a violation of the GATT.[31]

This aspect of the Panel's ruling in *Colombia – Ports of Entry* addresses the problem of the effectiveness of international obligations in a similar way to how it was addressed in all the disputes discussed earlier. In all these cases, when a successful implementation of a primary right/obligation at issue depended on obtaining access to certain inherent ancillary rights not mentioned in a treaty, the adjudicators, in one way or another, integrated the inherent ancillary rights in that primary right (see Table 2).

It is understood that these cases aimed primarily at the interpretation of particular treaty provisions. This, however, does not detract from the relevance of the basic principle of effective rights, originating in municipal law and elaborated on in a consistent line of judicial decisions and scholarly writings, for general legal relations. Principles of general international law are routinely used for interpreting provisions in different treaties, including through systemic integration. For example, just in the WTO legal framework, the Appellate Body and panels have referred to the principle of *abus de droit* when interpreting provisions of different covered agreements, namely GATT Article XX and Article 7 of the TRIPS Agreement.[32]

The aforementioned evidence of recognition of the principle of effective rights sufficiently attests to its crystallisation in international law. Moreover, the legal sources referenced earlier indicate that this principle applies in the context of access to fixed facilities. For example, by rejecting an inherent ancillary right of access to ports (including a port's infrastructure), a state impairs the effective implementation of the primary right to freedom of navigation or market access. Along the same lines, would a state measure depriving a gas exporter of access to pipelines (or in the absence of the latter, placing obstacles to their construction) not have a similar effect? From an economic or business perspective, both ports and pipelines are, in a way, the 'gates' to the market.

Furthermore, in the area of energy transit, a relevant expression of the principle of effective rights is ECT Article 7(4), which establishes that in

[30] See Panel Report, *Colombia – Ports of Entry*, paras. 7.236, 7.240, 7.253–256.

[31] Ibid. para. 7.275.

[32] See Appellate Body Report, *US – Shrimp*, para. 158; Panel Report, *US – Section 211 Appropriations Act*, para. 8.57.

Table 2: *Relationship between Primary and Inherent Ancillary Rights in Selected Cases*

Oscar Chinn (PCIJ)	Jurisdiction of the European Commission of the Danube (PCIJ)	Aramco Arbitration	Oil Platforms (ICJ)	Colombia – Ports of Entry (WTO Panel)
Primary rights: Freedom of Navigation	Freedom of Navigation	Oil Concession	Freedom of International Commerce	'Unrestricted Importation' (Market Access)
Recognised inherent rights: – Freedom of movement for vessels; – Freedom to enter ports; – Freedom to make use of plant and docks, to load and unload goods and to transport goods and passengers; – Some aspects of freedom of commerce.	– The movement of vessels with a view to the accomplishment of voyages; – Access to ports.	Export rights	Freedom to use the means of production and transport of goods destined for export, as well as the means ancillary or pertaining to such production or transport.	Access to ports of entry in such a way that does not limit the competitive opportunities of WTO Members.

the event that energy transit cannot be achieved on commercial terms by means of existing facilities, the Contracting Parties '*shall not place obstacles in the way of new capacity being established*'.[33] While the effect of the ECT on the formation of general international law should be approached cautiously, considering the treaty-law nature of the ECT, it undoubtedly indicates the existing trends in the area of energy governance. Moreover, some attempts to give effect to transit obligations in the context of network-bound energy trade by promoting the development of pipeline infrastructure were previously seen in a number of regional economic agreements.[34]

Finally, the principle of effective rights, to some extent, also relates to the principle of effective treaty interpretation ('*ut res magis valeat quam pereat*'). The latter is often derived from the VCLT's requirements in Article 31(1) that a treaty be interpreted 'in good faith' and 'in the light of its object and purpose'.[35] Describing the principle of effective treaty interpretation, Pauwelyn states that '[w]hen a treaty is open to two interpretations one of which does and the other does not enable the treaty to have appropriate effects . . . the former interpretation should be adopted'.[36] In *US – Offset Act*, the Appellate Body notes:

> As we have stated on many occasions, the internationally recognised interpretive principle of effectiveness should guide the interpretation of the WTO Agreement, and, under this principle, provisions of the WTO Agreement should not be interpreted in such a manner that whole clauses or paragraphs of a treaty would be reduced to redundancy or inutility.[37]

[33] Energy Charter Secretariat, 'The Energy Charter Treaty and Related Documents', above n 1. The details of the regulation of energy transit in the ECT are discussed in Chapter VI, Section III.A.2.

[34] See Chapter III, Section III.D.

[35] See Gardiner, above n 18 at 159 and 189; Pauwelyn, Joost, *Conflict of Norms in Public International Law: How WTO Law Relates to Other Rules of International Law* (New York: Cambridge University Press, 2003) at 201 at 248; Weeramantry, Romesh J., *Treaty Interpretation in Investment Arbitration* (Oxford: Oxford University Press, 2012) at 143–144.

[36] Pauwelyn, above n 35 at 248.

[37] See Appellate Body Report, *US – Offset Act (Byrd Amendment)*, para. 271; see also Appellate Body Report, *US – Shrimp*, para. 131. In international law, there appears to be a general presumption of effectiveness of legal rights and obligations. In its very first case, the ICJ noted: '[I]t would indeed be incompatible with the generally accepted rules of interpretation to admit that a provision of this sort occurring in a special agreement should be devoid of purport or effect.' *The Corfu Channel Case*, Judgment of 9 April 1949: I.C.J. Reports 1949, p. 4 at 24.

However, a word of caution must be given with regard to the full convergence of these principles. While in the absence of a specific pronouncement by international adjudicators it might be difficult to understand which of these principles is applied in a given case,[38] the principles of effective rights and effective treaty interpretation have different origins and have served different purposes in international law. They must, therefore, be distinguished.

The principle of effective treaty interpretation prevents an interpreter from adopting a reading that would result in reducing whole clauses or paragraphs of a treaty to redundancy or inutility[39] and requires the interpreter to read all applicable provisions of a treaty in a way that gives meaning and effect to all of them, harmoniously.[40] Thus, it provides guidance on how *various elements of a treaty* itself must be used to ensure its effectiveness. However, it does not allow 'the imputation into a treaty of words that are not there or the importation into a treaty of concepts that were not intended'.[41]

In contrast, as explained, the principle of effective rights was transposed into international law from Roman civil law, in particular the doctrine of servitudes. It appears to be an autonomous principle, which may fill gaps in treaties (including WTO law) through the principle of systemic integration. The only limiting condition on its importation into the interpreted treaty is that it does not clash with the rights and obligations established by the latter (*lex specialis*). While WTO law, in DSU Article 19(2), prohibits adding to or diminishing the rights and obligations provided in the covered agreements, this provision should not be misconstrued as placing WTO rules (especially, when they are vague or broad) in

[38] For example, in the cases analysed in this section, the adjudicators did not mention explicitly either the principle of effective rights or the principle of effective treaty interpretation, with the exception of *China – Electronic Payment Services*, in which the latter principle was mentioned in a different context. Nevertheless, one could argue confidently that, at least, in *Oscar Chinn* and *European Commission of the River Danube*, the adjudicators based their decisions on the principle of effective rights, as what was stated in those decisions regarding the integrity of rights pertaining to river navigation has strong similarities with what Westlake had stated earlier when he drew an analogy between freedom of navigation and Roman law.

[39] See Appellate Body Report, *US – Gasoline*, para. 61; *The Corfu Channel Case*, above n 37 at 24.

[40] Appellate Body Report, *Argentina – Footwear (EC)*, para. 81.

[41] Appellate Body Report, *India – Patents (US)*, para. 45.

clinical isolation from other sources of public international law. This would be contrary to DSU Article 3(2).[42]

This analysis of doctrinal sources, international jurisprudence and state practice, as expressed in the ECT and some regional economic agreements regulating trade in energy, provides sufficient support for the existence of the principle of effective rights in international law. It was demonstrated that the principle of effective rights requires the recognition of inherent ancillary rights when they are essential for implementing the primary right granted by a treaty. In this regard, the integral relationship between primary and inherent ancillary rights is not dependent on whether the former can or cannot be utilised due to temporary technical barriers (such as the absence of a pipeline or a pipeline capacity). In light of *Colombia – Ports of Entry*, if a lack of access to an inherent ancillary right reduces the competitive opportunities of a WTO Member, this would appear to be a sufficient cause of action under WTO law.

However, while the principle of effective rights promotes the effective implementation of freedom of transit in the context of pipeline gas, it does not explain how this freedom must be implemented. In this regard, the principle of economic cooperation and a general duty to negotiate in good faith discussed in the following section can provide further guidance.

B. *The Principle of Economic Cooperation and the* 'Pactum de Contrahendo' *Nature of GATT Article V:2*

1. The principle of economic cooperation and '*pactum de contrahendo*' in international law, their relevance to the interpretation of GATT Article V:2

The principle of economic cooperation can provide guidance to the interpreters of treaty provisions that contain an element of uncertainty as to how they must be operationalised in particular practical contexts. In the context of gas transit, the examples of such provisions are GATT Article V:2, first and second sentences. On one hand, these provisions contain a clear and precise requirement for a transit state to ensure a non-discriminatory freedom of transit via the most convenient routes.[43] On the other hand, they are silent on the terms and modalities that

[42] See Dispute Settlement Understanding, *WTO Legal Texts*, above n 18; Appellate Body Report, *US – Gasoline*, para. 158.

[43] GATT 1994, *WTO Legal Texts*, above n 18.

must be employed in order to satisfy this requirement. With a view to avoiding an interpretation of GATT Article V that either imposes unrealistic burdens on WTO Members or reduces the effectiveness of its obligations, a balanced approach should be taken.[44] This approach may entail falling back on the principle of economic cooperation whereby all the ambiguities in GATT Article V:2 as to the particular terms and modalities of transit must be resolved by good faith cooperative efforts of WTO Members. Alternatively, GATT Article V:2 can be regarded as a '*pactum de contrahendo*' by which Members have established a firm commitment to achieve certain specific results (namely a non-discriminatory freedom of transit), while leaving the precise method of achieving those results for a future agreement, the negotiations on which must be conducted in a manner consistent with international law. Both of these legal instruments are discussed in this section.

The principle of economic cooperation is a fundamental principle of the UN.[45] It is reflected in Article 55 of the UN Charter (International Economic and Social Cooperation), which provides:

> With a view to the creation of conditions of stability and well-being which are necessary for peaceful and friendly relations among nations ... the United Nations shall promote: ... b. solutions of international economic ... and related problems.[46]

[44] Recall that the interpretation of a treaty in a way that leads to results that are 'manifestly absurd or unreasonable' is inconsistent with the VCLT. See Panel Report, *US – Gambling*, para. 6.49.

[45] This principle has different corollaries which scholars have discussed in particular practical contexts, such as 'rights and duties of neighbourhood' and '*voisenage*'. These two principles also appear to be recognised as sources of general international law. See Brownlie, above n 12 at 377; Sauser-Hall G., 'Les Droits Des Etats Sur La Force Motrice Des Cours D'Eau Internationaux', 83 *Recueil des Cours* (1953) 539 at 554–555; and Rousseau, Charles, *Droit International Public*, Tome III (Paris: Sirey, 1977) at 273. Benvenisti refers to a duty of states to cooperate when discussing the general obligations of sovereigns as 'trustees of humanity'. Benvenisti, Eyal, 'Sovereigns as Trustees of Humanity: On the Accountability of States to Foreign Stakeholders', 107 *American Journal of International Law* (2013) 295 at 327–328. See a general overview of the principle of 'good neighbourliness' in Boute, Anatole, 'The Good Neighbourliness Principle in EU External Energy Relations: The Case of Energy Transit', in D. Kochenov and E. Basheska (eds.), *The Principle of Good Neighbourliness in the European Legal Context* (Leiden: Brill Nijhoff, 2015), 354.

[46] See also Article 1(3) of the Charter, declaring, inter alia, economic cooperation and development of friendly relations as the UN objective. Charter of the United Nations of 1945, above n 3.

The 'Declaration on Principles of International Law Concerning Friendly Relations and Cooperation among States', which the UN unanimously adopted, also refers specifically to the duty of states to cooperate and promote economic growth throughout the world and especially that of developing countries.[47] This Declaration, according to one of its drafters, 'represented an endeavour: "... to state the essential legal content of seven fundamental principles embodied in the Charter of the United Nations"', including the principles of cooperation and good faith in the fulfilment of the obligations of the UN Charter.[48] The duty of the state to cooperate was also reiterated in UN GA Resolution 53/101 on 'Principles and Guidelines for International Negotiations'.[49]

In the WTO, the principle of economic cooperation is reflected in DSU Article 3(10) (good faith efforts to resolve disputes).[50] As noted earlier, the preamble and Article 11(16) of the Agreement on Trade Facilitation state specifically that Members must 'endeavour to cooperate and coordinate with one another with a view to *enhancing freedom of transit*'.[51] Among the areas of cooperation and coordination, Article 11(6) lists understandings on charges, formalities and legal requirements, as well as the practical operation of transit regimes.[52]

The obligation to cooperate in operationalising GATT Article V, while it may seem broad on its face, in the context of gas transit, can be informed and reinforced by various commitments made by states at a political level in the form of, inter alia, declarations and promises to pursue cooperative and collaborative efforts to develop pipelines and access to gas markets. To recall, the UN GA Resolution 63/210 welcomes international cooperation in developing pipelines and recognises the need for extensive cooperation in determining the ways in which reliable

[47] United Nations, GA Resolution 2625 of 24 October 1970, 'Declaration on Principles of International Law Concerning Friendly Relations and Co-operation among States in Accordance with the Charter of the United Nations', in United Nations, *General Assembly Resolutions*, http://www.un.org/documents/resga.htm, accessed 15 November 2015.

[48] Koskenniemi, Martti, 'General Principles: Reflexions on Constructivist Thinking in International Law', in M. Koskenniemi (ed.), *Sources of International Law* (Aldershot: Ashgate Darmouth, 2000) 360 at 363 (citing I. Sinclair).

[49] United Nations, GA Resolution 53/101 'Principles and Guidelines for International Negotiations', above n 47.

[50] Dispute Settlement Understanding, *WTO Legal Texts*, above n 18.

[51] World Trade Organization, Ministerial Conference, Agreement on Trade Facilitation, Ministerial Decision of 7 December 2013, WT/MIN(13)/36, WT/L/911 (11 December 2013)/ Preparatory Committee on Trade Facilitation, Agreement on Trade Facilitation, WT/L/931 (15 July 2014), emphasis added.

[52] Ibid.

energy transportation to international markets could be achieved.[53] The Almaty Ministerial Conference recognised that '[p]ipelines provide a cost-effective means of transport for both oil and natural gas' and that land-locked and transit developing countries should cooperate in constructing pipelines 'along the most cost-effective and most suitable or shortest routes, taking into account the interests of parties concerned'.[54] While establishing the commitments to 'strive to establish efficient transit transport systems in both land-locked and transit developing countries', the Declaration, which resulted from this Conference, expected these commitments to be fulfilled through 'cooperative and collaborative efforts'.[55] The Vienna Declaration, which updated the Almaty's Programme of Action, also recognises that '[t]here is inadequate physical infrastructure in ... pipelines,' and that '[m]issing links need to be addressed urgently'.[56]

Similar declarations and promises have been made by the signatories of the 1991 European Energy Charter and the 2015 International Energy Charter. They declared, in these documents, the importance of, among other things, facilitating access to transport infrastructure, promoting access to international markets, as well as the development of commercial international energy transmission networks and their interconnection, including gas pipelines.[57]

It was explained earlier that soft law instruments (especially UN resolutions and declarations) often reflect common intentions of the international community (including WTO Members) regarding the preferred way to resolve particular problems of common concern.[58] Therefore, it should not come as a surprise that such instruments may influence, in one way or another, the understanding of legal rights and obligations evolving in positive law (such as WTO law), and how those

[53] United Nations, GA Resolution 63/210 of 19 December 2008, 'Reliable and Stable Transit of Energy and Its Role in Ensuring Sustainable Development and International Cooperation', above n 47.

[54] See United Nations, 'The UN Report of the International Ministerial Conference on Land-Locked and Transit Developing Countries and Transit Transport Co-operation', and the attached Almaty Declaration, A/CONF.202/3 (28 and 29 August 2003) at 17.

[55] Ibid. at 24–25. See also Chapter III, Section III.C.3.

[56] United Nations, GA Resolution 69/137 of 12 December 2014, 'Programme of Action for Landlocked Developing Countries for the Decade 2014–2024', above n 47, Annex II at 9–10.

[57] See Energy Charter Secretariat, 'The Energy Charter Treaty and Related Documents', above n 1 at 217; International Energy Charter, http://www.energycharter.org/process/international-energy-charter-2015/, accessed 15 September 2015.

[58] See Chapter I, Section III.B.2(iv).

rights and obligations must be applied in particular factual circumstances. Finally, the cooperative approach to facilitating freedom of transit adopted by WTO Members in the new Agreement on Trade Facilitation should also be taken into account in interpreting GATT Article V.[59]

As noted, GATT Article V:2 (first and second sentences) can also be regarded as a '*pactum de contrahendo*'. Hishashi Owada (the former ICJ president) defines '*pactum de contrahendo*' as 'an agreement between parties creating a binding obligation to conclude a future agreement on a particular subject'.[60]

Importantly, an obligation '*pactum de contrahendo*' must be distinguished from an obligation '*pactum de negotiando*' 'by which two or more parties assume a binding obligation to enter into future negotiations with an intention to conclude a future treaty . . . [which] does not go as far as to commit the parties to come to a final agreement'.[61] In other words, the obligation *pactum de contrahendo* can be characterised as an 'obligation of result', whereas the obligation *pactum de negotiando* is an 'obligation of conduct'.[62]

However, in practice, there may be a fine line between these two categories of obligations.[63] As with any treaty obligation, a *pactum de contrahendo* can be explicit or implicit; it must be derived from the wording of a treaty provision interpreted in its context and in the light of the treaty's object and purpose.[64] In this regard, Owada states: 'it is the content of an instrument, ie its creation of any legal obligation, and not its form or appellation that determines whether it constitutes

[59] See Article 11(16) of the Agreement on Trade Facilitation, above n 51.

[60] See Owada, Hisashi, 'Pactum de Contrahendo, Pactum de Negotiando', in *Max Planck Encyclopaedia of Public International Law* (2010), www.mpepil.com, accessed 30 March 2012 at 1 (paras. 1 and 3).

[61] Ibid. at 2 (para. 5). See also McNair, Arnold D., *The Law of Treaties* (Oxford: Clarendon Press, 1961) at 27–29.

[62] On the legal implications of *pactum de contrahendo*, see Degan, Vladimir D., *Sources of International Law* (The Netherlands: Kluwer Law International, 1997) at 514.

[63] Kunoy states: 'A mere bifurcation between obligations of conduct and obligations of result to determine the content of international commitments may be inappropriate and even misleading. Some obligations of conduct may be more goal-oriented than others.' Kunoy, Bjørn, 'The Ambit of *Pactum de Negotiatum* in the Management of Shared Fish Stock: A Rumble in the Jungle', 11 *Chinese Journal of International Law* (2012) 689 at 714. See also Owada, above n 60 at 2 (para. 9); and Crawford, J., Special Rapporteur, 'Second Report on State Responsibility', A/CN.4/498 and Add.1–4 (17 March, 1 and 30 April, 19 July 1999) paras. 89–90.

[64] Article 31(1) of the Vienna Convention on the Law of Treaties, above n 3.

a *pactum de contrahendo* or *pactum de negotiando*'.[65] The more precise the obligation is and its intended outcome (such as *in casu* 'there *shall be* freedom of transit . . .' or 'no distinction *shall be* made [with respect to 'traffic in transit']'), the more enforceable it is likely to be under international law.[66] It is therefore hardly controversial that the obligations under GATT Article V:2 (first and second sentences) are obligations of result (i.e. '*pactum de contrahendo*'), as they require a precise outcome.[67]

Regardless of the extent and intensity of the obligation to negotiate, states must comply with a general duty to conduct negotiations in good faith. This duty has been recognised as a matter of general international law in the international jurisprudence and academic literature.[68]

[65] Owada, above n 60 at 3 (paras. 9 and 15). [66] Ibid. at 1 (para. 3), emphasis added.

[67] See a similar view in Azaria, Danae, *Treaties on Transit of Energy via Pipelines and Countermeasures* (Oxford: Oxford University Press, 2015) at 65.

[68] See Panizzon, Marion, *Good Faith in the Jurisprudence of the WTO: The Protection of Legitimate Expectations, Good Faith Interpretation and Fair Dispute Settlement* (Oxford: Hart Publishing, 2006) at 73 (Panizzon states that 'the duty to negotiate in good faith forms part of customary law and is considered binding upon all nations'); and Barnidge, Robert P., 'The International Law of Negotiations as a Means of Dispute Settlement', 36 *Fordham International Law Journal* (2013) 545. The obligation to negotiate in good faith appears to be introduced in international law by the Arbitral Tribunal in *Tacna-Arica Arbitration* between Chile and Peru dated 1925. See McNair, *The Law of Treaties*, above n 61 at 29. The ICJ recognised a general obligation to negotiate in good faith in *North Sea Continental Shelf* in the context of the delimitation of adjacent continental shelves. See *North Sea Continental Shelf*, Judgment of 20 February 1969: I.C.J. Reports 1969, p. 3 paras. 85–87. In *Pulp Mills*, the ICJ clarified this obligation as follows: '[T]he mechanism for co-operation between States is governed by the principle of good faith. ... [The latter] applies to all obligations established by a treaty, including procedural obligations which are essential to co-operation between States.' *Pulp Mills*, above n 3, para. 145. See also *Legality of the Threat or Use of Nuclear Weapons*, Advisory Opinion of 8 July 1996, I.C.J. Reports 1996, p. 226, paras. 98–105; and *Application of the International Convention on the Elimination of All Forms of Racial Discrimination (Georgia v. Russian Federation)*, Judgment of 1 April 2011: I.C.J. Reports 2011, p. 70, paras. 156–161. Finally, in private law, parties also often make use of various analogies of the obligation to negotiate in good faith in: (i) arbitration clauses (where consultations are often a precondition for a binding dispute settlement); (ii) situations where specific terms and modalities of a primary contract are left for a subsequent contract (see normally production-sharing agreements); and (iii) when a 'fundamental change in circumstance' arises. There is a strong view in the literature that in these contexts (and provided that certain technical conditions are met), the obligation to hold genuine and good faith negotiations must be enforced. See, in this regard, a comparative analysis of common and civil law systems in Chapman, Simon, 'Multi-tiered Dispute Resolution Clauses: Enforcing Obligations to Negotiate in Good Faith', 27 *Journal of International*

In the WTO, a general duty to negotiate in good faith has been recognised in the context of compliance with the Article XX (General Exceptions) requirements and a non-violation complaint under Article XXII:2 of the Agreement on Government Procurement (equivalent to GATT Article XXIII:1(b)).[69] It is noteworthy that neither of these provisions expressly mentions the principle of 'good faith' or an obligation to negotiate. This suggests that the adjudicators in those disputes derived the obligation to negotiate in good faith from general international law.

In *US – Shrimp (Article 21.5 – Malaysia)*, the Appellate Body supported the existence of this obligation by referring to '*the decided preference for multilateral approaches voiced by WTO Members and others in the international community* in various international agreements for the protection and conservation of endangered sea turtles' (i.e. the legitimate objective pursued by the United States (Respondent) pursuant to GATT Article XX).[70] A similar preference for cooperative 'multilateral approaches' has been expressed with respect to the development of gas pipeline networks, and can similarly guide WTO adjudicators in their interpretation of GATT Article V.

The principle of economic cooperation and the '*pactum de contrahendo*' nature of the obligations under GATT Article V:2 direct WTO Members to engage in good faith cooperation on those particular technical details of transit that are not explicitly addressed in this provision but are crucial for the implementation of WTO transit obligations. In other words, in light of these instruments, certain technical gaps in the WTO's transit regulation must be interpreted as an implicit agreement to fill these gaps through on-going cooperation and, if necessary, by concluding implementation agreements.[71]

Arbitration (2010) 89; and Botchway, Francis N. 'Can the Law Compel Business Parties to Negotiate?', 3 *Journal of World Energy Law & Business* (2010) 286 at 299–302.

69 In relation to GATT Article XX, in *US – Shrimp (Article 21.5 – Malaysia)*, the Appellate Body held as follows: '[T]o avoid "arbitrary or unjustifiable discrimination", the United States had to provide all exporting countries "similar opportunities to negotiate" an international agreement. ... [To this end] the United States, in our view, would be expected *to make good faith efforts* to reach international agreements that are comparable from one forum of negotiation to the other.' Appellate Body Report, *US – Shrimp (Article 21.5 – Malaysia)*, paras. 122–124, emphasis added. In *Korea – Procurement*, the Panel noted that the failure of a WTO Member to negotiate a treaty in good faith could give rise to a non-violation complaint. See Panel Report, *Korea – Procurement*, paras. 7.93–102.

70 Appellate Body Report, *US – Shrimp (Article 21.5 – Malaysia)*, para. 122, emphasis added.

71 See the term 'implementation agreement' defined in Kojima, Chie and Vereshchetin, Vladlen S., 'Implementation Agreements', in *Max Planck Encyclopaedia of Public International Law* (2010), www.mpepil.com, accessed 30 March 2012.

This reading of transit obligations under GATT Article V in light of the applicable principles of general international law appears to be in line with a well-settled practice in the area of the international regulation of transit to set out basic principles in a framework agreement, while leaving the specifics to other more detailed agreements. Given that transit, especially network-bound transit, is an extremely complex matter, it cannot be regulated completely by a comprehensive multilateral treaty, such as the WTO Agreement. As discussed in Chapter III, among the examples of multilateral treaties explicitly supporting this practice is the 1919 Versailles Treaty Article 23(e) and LOSC Article 125.[72] Furthermore, recall that the 1921 Barcelona Convention on Freedom of Transit, upon which GATT Article V was modelled, did not intend to provide all-inclusive rules on international transit. As its drafters explained during the Barcelona Conference, they were 'inspired by just those principles of freedom in the loftiest sense, and of equal respect for the rights and interests of every nation, which the Commission ... has never failed to maintain and assert'.[73]

However, despite this practice and in line with what was explained earlier with respect to the interpretation of Article V:2 (first sentence), resorting to implementation agreements should not be misconstrued as providing room for *de novo* bargaining or renegotiating the obligations undertaken.[74] It is rather a progressive dialogue on the practical implementation of the *already* existing rights, governed by the existing law, such as WTO transit rules and the principle of freedom of transit in general international law, as complemented by other principles discussed in this chapter.

Having said this, it is equally important to mention that the commitment to engage in economic cooperation/negotiations with a view to establishing freedom of transit does not necessarily translate into an obligation to reach an agreement on all technical terms and modalities of transit (such as the route and transit charges) on which the parties involved may well disagree. This disagreement can result from legitimate differences in views, as well as an intentional frustration of transit,

[72] See Chapter III, Section III.B.

[73] See United Nations Secretariat, 'Question of Free Access to the Sea of Land-Locked Countries', A/CONF.13/29 (14 January 1958) at 322; Chapter III, Section III.B.1.

[74] Without doubt, under most circumstances, trade partners may contract out from their obligations by exchanging other concessions. This, however, depends on a mutual agreement and cannot be done unilaterally.

disguised as unreasonable requirements that a transit-dependent state would never be able to fulfil.

If the former is the case, a transit state can hardly be compelled to conclude an implementation agreement against its will, even though GATT Article V:2 requires the achievement of a specific result (i.e. a non-discriminatory freedom of transit). As the Appellate Body explained in *US – Shrimp (Article 21.5 – Malaysia)*:

> Requiring that a multilateral agreement be *concluded* by the United States in order to avoid 'arbitrary or unjustifiable discrimination' in applying its measure would mean that any country party to the negotiations with the United States, whether a WTO Member or not, would have, in effect, a veto over whether the United States could fulfill its WTO obligations. Such a requirement would not be reasonable.[75]

Nevertheless, it is important to identify some basic safeguard mechanisms that could be applied when a transit state abuses its right or does not have a genuine intention to operationalise the freedom of transit obligations under GATT Article V.[76] For this purpose, the procedural requirements will first be examined that the parties must satisfy in genuine, good faith negotiations on the terms and modalities of gas transit. This issue is addressed in the next section. Second, it was noted earlier that GATT Article V:2 requires the negotiators to achieve a specific result (namely, a non-discriminatory freedom of transit), which must impose additional constraints on what a transit state could demand in the process of the negotiations. To this end, the boundaries must be set defining the legitimate regulatory autonomy of a transit state in the process of negotiations on the terms and modalities of gas transit contemplated under GATT Article V, read in the light of the principles of general international law. Setting these boundaries is especially important, given that the negotiations on an energy transit project would likely deal with complex facts, including regulatory (in the sense

[75] Appellate Body Report, *US – Shrimp (Article 21.5 – Malaysia)*, para. 123. See also *Lake Lanoux Arbitration* in E. Lauterpacht (ed.), 27 *International Law Reports* (1957) 100 at 127–128.

[76] This study acknowledges that, in principle, a transit-dependent state can also abuse its freedom of transit, granted under WTO law and complemented by general international law. However, this abuse is less relevant to our discussion, since a transit state retains its sovereign right to impose 'necessary' or 'reasonable' regulations on traffic in transit or a pipeline transit project and, in extreme cases, it can even ban any transit through its territory. See the discussion of instruments that transit states can use to pursue legitimate interests under WTO law in Chapter IV, Section V.

of access to networks, property rights, competition policies), social and environmental considerations, as well as bargaining on commercial matters. The latter issue is addressed in Section II.B.3.

2. The meaning and content of the obligation to negotiate in good faith

The understanding of the obligation to negotiate in good faith has been evolving.[77] The early international case law developed a presumption that the obligation to negotiate does not necessarily amount to an obligation to reach an agreement, which can be perceived as a significant limitation on its enforceability.[78] While in contemporary international law this basic axiom still holds,[79] the particular aspects of this obligation have been elaborated significantly.

It was explained in the previous section that it might not be an easy task to draw a fine line between an 'obligation of conduct' (traditionally associated with the obligation to negotiate or *pactum de negotiando*) and an 'obligation of result' (i.e. *pactum de contrahendo*). The more precise the obligation is and its intended outcome (such as a non-discriminatory freedom of transit under GATT Article V:2), the more enforceable it is likely to be under international law.

In *Legality of the Threat or Use of Nuclear Weapons*, the ICJ went as far as to expressly acknowledge that when a treaty establishes an obligation to reach *a precise* result of negotiations, there may exist an obligation not only to pursue those negotiations in good faith, but also to bring them to a conclusion.[80] By contrast, in most cases where

[77] For an overview, see Degan, above n 62 at 503–514.

[78] See *Railway Traffic between Lithuania and Poland*, Collection of Judgments, Orders and Advisory Opinions, Advisory Opinion, Series A/B. – No. 42, 15 October 1931 at 12.

[79] See *Pulp Mills*, above n 3, para. 150; Appellate Body Report, *US – Shrimp (Article 21.5 – Malaysia)*, paras. 123–124; *Wintershall A.G. et al vs. the Government of Qatar* (Partial Award on Liability), 28 *International Legal Materials* (1989) 798 at 814.

[80] In its Advisory Opinion, the ICJ interpreted Article VI of the 1968 Treaty on the Non-proliferation of Nuclear Weapons, which reads as follows: 'Each of the Parties to the Treaty undertakes to pursue negotiations in good faith on effective measures relating to cessation of the nuclear arms race at an early date and to nuclear disarmament, and on a treaty on general and complete disarmament under strict and effective international control.' The Court stated: 'The legal import of that obligation goes beyond that of *a mere obligation of conduct*; the obligation involved here is *an obligation to achieve a precise result* – nuclear disarmament in all its aspects – by adopting a particular course of conduct, namely, the pursuit of negotiations on the matter in good faith.' See *Legality of the Threat or Use of Nuclear Weapons*, above n 68, para. 99, emphasis added.

international tribunals did not find the obligation to negotiate enforceable, the reasons for such a decision lay either in the soft law language of the relevant treaty provision (or another legal instrument) creating this obligation or in the ambiguity as to what should have been the precise aim or result of the negotiations.[81]

As to the content of the obligation to negotiate in good faith, in *Elimination of All Forms of Racial Discrimination*, the ICJ clarified that 'the concept of "negotiations" ... requires – at the very least – *a genuine attempt* by one of the disputing parties *to engage in discussions* with the other disputing party, *with a view to resolving the dispute*'.[82] According to the ICJ, the obligation to negotiate in this manner is considered exhausted only when the process either brought about fruitful results or failed due to objective reasons.[83] The latter is essentially a question of fact that an international tribunal will have to consider on a case-by-case basis.[84]

In other cases, the World Court held that, should states engage in negotiations in good faith, they must 'pursue them as far as possible, with a view to concluding [an] agreement'.[85] In *North Sea Continental Shelf*, the ICJ stated:

[81] For example, in *Railway Traffic between Lithuania and Poland*, the PCIJ had to rule, inter alia, on whether, on the basis of the Resolution of the Council of the League of Nations of 10 December 1927, Lithuania and Poland (the parties to a dispute) had an obligation to enter into negotiations with a view to opening a specific railway connection between these countries. The relevant part of the resolution provided as follows: 'Recommends the two Governments to enter into direct negotiations as soon as possible in order to establish such relations between the two neighbouring States as will ensure "the good understanding between nations upon which peace depends".' Given this broad language containing no specific requirements, the Court did not find a violation of the obligation to negotiate. *Railway Traffic between Lithuania and Poland*, above n 78 at 11–12. See a similar finding in *Wintershall A.G. et al vs. the Government of Qatar*, above n 79 at 32–33. With respect to the obligation to negotiate in good faith under national contract law, Chapman cites the judgment of Allsop P. in *the United Group* case, stating that '[w]hat the phrase "good faith" signifies in a particular context and contract will depend on that context and that contract'. In this case, the Court of Appeal in New South Wales (Australia) upheld the obligation to undertake genuine and good faith negotiations preconditioning the right to commence arbitration proceedings, based on a multi-tiered dispute resolution clause. Chapman, above n 68 at 96.

[82] *Elimination of All Forms of Racial Discrimination*, above n 68, para. 157, emphasis added.

[83] Specifically, the Court stated: 'Manifestly, in the absence of evidence of a genuine attempt to negotiate, the precondition of negotiation is not met. However, where negotiations are attempted or have commenced, the jurisprudence of this Court and of the Permanent Court of International Justice clearly reveals that *the precondition of negotiation is met only when there has been a failure of negotiations, or when negotiations have become futile or deadlocked*.' See ibid., para. 159, emphasis added.

[84] Ibid., para. 160. [85] *Railway Traffic between Lithuania and Poland*, above n 78 at 12.

> [T]he parties are under an obligation to enter into negotiations with a view to arriving at an agreement, and not merely to go through a formal process of negotiation ... they are under an obligation so to conduct themselves that the negotiations are meaningful, which will not be the case when either of them insists upon its own position without contemplating any modification of it.[86]

Along the same lines, in *US – Shrimp (Article 21.5 – Malaysia)*, the obligation to negotiate in good faith was defined as a *serious* and *continuous* process – not a 'one-off' exercise.[87] Consequently, in the absence of this 'genuine', 'serious and continuous' attempt, or if there is indeed a blunt refusal to negotiate, inconsistent with the explicit requirements or the spirit of an applicable treaty provision, the obligation to negotiate in good faith will not be satisfied.[88]

In *Lake Lanoux Arbitration*, the Arbitral Tribunal interpreted the treaty-based obligation to negotiate an agreement between an upstream riparian state (France) and a downstream state (Spain) prior to altering the natural flows of a river running through both states for the purposes of constructing electric power facilities in the territory of France. It stated as follows:

> In reality, the engagements thus undertaken by States [to negotiate an agreement] take very diverse forms and have a scope which varies according to the manner in which they are defined and according to the procedures intended for their execution; but the reality of the obligations thus undertaken is incontestable and sanctions can be applied in the event, for example, of an unjustified breaking off of the discussions, abnormal delays, disregard of the agreed procedures, systematic refusals to take into consideration adverse proposals or interests, and, more generally, in cases of violation of the rules of good faith.[89]

The Tribunal ultimately concluded that France (Respondent) complied with the obligation, since it showed a flexible position that had undergone essential transformations throughout the process of

[86] *North Sea Continental Shelf*, above n 68 para. 85.

[87] See Appellate Body Report, *US – Shrimp (Article 21.5 – Malaysia)*, para. 115.

[88] In this regard, in *Elimination of All Forms of Racial Discrimination*, the ICJ found that Georgia did not genuinely or specifically attempt to engage in negotiations with Russia and, therefore, did not satisfy the precondition to commencing a dispute before the Court, established by a dispute settlement provision in the 1965 International Convention on the Elimination of All Forms of Racial Discrimination upon which Georgia relied in its substantive claims. On this basis, the Court upheld the objection of Russia to its jurisdiction. *Elimination of All Forms of Racial Discrimination*, above n 68, paras. 180, 184.

[89] *Lake Lanoux Arbitration*, above n 75 at 128.

negotiations. Importantly, the Tribunal noted that while France was obliged to examine the schemes of possible water flows proposed by Spain, it had the right to give preference to its own scheme, provided that it took into consideration, in a reasonable manner, the interests of the other party.[90] Although this interpretation of the obligation to negotiate certainly gave wide policy space to a state in whose territory the construction works were to be carried out, it also placed the burden on that state to ensure that: (i) the basic right to negotiate a mutually acceptable solution was not denied in the first place; and (ii) the views of the other party were not disregarded.

Other commonly recognised principles guiding good faith negotiations include obligations to:

- take due account of the importance of engaging, in an appropriate manner, in international negotiations with the states whose vital interests are directly affected by the matters in question;
- set a purpose and object of negotiations that is fully compatible with the principles and norms of international law;
- adhere to the mutually agreed framework for conducting negotiations;
- endeavour to maintain a constructive atmosphere during negotiations and to refrain from any conduct which might undermine the negotiations and their progress;
- facilitate the pursuit or conclusion of negotiations by remaining focused throughout on the main objectives of the negotiations; and
- use best endeavours to continue to work towards a mutually acceptable and just solution in the event of an impasse in negotiations.[91]

The failure to engage in good faith negotiations or the violation of these principles may result in an internationally wrongful act, enabling an aggrieved party to seek redress by resorting to legal remedies available under the law of state responsibility.[92] For example, in *Pulp Mills*, the ICJ established the violation of the obligation to engage in negotiations in respect of Uruguay's (Respondent) authorisation to construct pulp mills in its territory before the expiration of the period of negotiations with Argentina (Complainant) required by applicable law.[93] Depending on the gravity of the infringement, in general international law, such

[90] Ibid. at 140–142.

[91] United Nations, GA Resolution 53/101 'Principles and Guidelines for International Negotiations', above n 47.

[92] Owada, above n 60, para. 4. See also Barnidge, above n 68 at 548.

[93] *Pulp Mills*, above n 3 paras. 158, 282.

remedies can take the form of 'non-forcible countermeasures', or a termination or suspension of the operation of the treaty at issue in whole or in part.[94] WTO law prescribes its own remedies in the form of the suspension of concessions.[95]

However, assuming the parties engaged in the negotiations on the implementation of a pipeline transit project observe the principles guiding the conduct of international negotiations, the next question that must be answered is where should the boundaries lie between permissible and inappropriate demands of a transit state? As was explained, this question is important, as GATT Article V:2 establishes the obligations of result, as opposed to a mere obligation of conduct. This monograph suggests that the principle of the prohibition of abuse of rights (*abus de droit*) can provide further guidance. It may assist in determining the scope of the regulatory autonomy of a transit state vis-à-vis transit-dependent states in their efforts to operationalise obligations under GATT Article V.[96]

3. The regulatory autonomy of transit states in light of the principle of *abus de droit*

(i) The principle of *abus de droit* in international law In his landmark work on the international law of transit, Elihu Lauterpacht referred to the principle of abuse of rights (*abus de droit*) as a legal instrument supporting a general right to transit via a foreign territory. While Lauterpacht acknowledged the relatively indefinite nature of this principle, potentially affecting its use, he argued that this was not a 'sufficient

[94] See Degan, above n 62 at 511–512 (citing the Opinion of the Arbitration Commission of the International Conference on the Former Yugoslavia); Barnidge, above n 68 at 562; and Article 49 of the ILC Articles on State Responsibility in International Law Commission, 'Responsibility of States for Internationally Wrongful Acts with Commentaries', *Yearbook of the International Law Commission*, vol. 2 (Part 2) (2001). Owada also notes in this connection that 'a party's failure to meet its obligations under a *pactum de negotiando* or *pactum de contrahendo* should be evaluated as any other breach of treaty obligations' (citing VCLT Article 60, which, in the case of a material breach of a treaty by one of the parties, entitles the other party affected by such breach to terminate that treaty or suspend its operation in whole or in part). Owada, above n 60 para. 44; Vienna Convention on the Law of Treaties, above n 3.

[95] See Chapter VI, Section II.C.1.

[96] One could argue that this principle finds its expression in GATT Article V:4, which prohibits charges and regulations that are not 'reasonable'. As explained further (in Section II.B.3(ii)), the determination of whether the regulation in question is consistent with the principle of *abus de droit* has often been made through the assessment of its 'reasonableness'.

reason for the view that any inequitable conduct such as is implied in the strangulation of international commerce or the very life of a State as the result of denial of transit is a matter of indifference to the law'.[97]

The principle of abuse of rights is a corollary of the principle of good faith and is a commonly recognised principle of general international law.[98] In *US – Shrimp*, the Appellate Body clarified as follows:

> [The principle of good faith], at once a general principle of law and a general principle of international law, controls the exercise of rights by states. One application of this general principle, the application widely known as the doctrine of *abus de droit*, prohibits the abusive exercise of a state's rights and enjoins that whenever the assertion of a right 'impinges on the field covered by [a] treaty obligation, it must be exercised bona fide, that is to say, reasonably'.[99]

The principle of abuse of rights thus 'delimits the exercise of legal power by states' by balancing their rights and obligations vis-à-vis other states.[100] Similarly, Byers links the principle of abuse of rights to the balancing of interests when the law does not clearly define the interaction of rights and obligations.[101] Cheng provides the following definition of the principle:

> [W]henever . . . the owner of a right enjoys a certain discretionary power, this must be exercised in good faith, which means that it must be exercised reasonably, honestly, in conformity with the spirit of the law and with due regard to the interests of others.[102]

[97] Lauterpacht, Elihu, 'Freedom of Transit in International Law', 44 *Transactions of the Grotius Society* (1958–1959) 313 at 337. In a similar context, in *Right of Passage over Indian Territory*, while supporting Portugal's right of passage to its enclaves in the Indian territory, Judge Koo stated in his Separate Opinion: 'it is inconceivable in international law that one sovereignty exists only by the will or caprice of another sovereignty.' *Case Concerning Right of Passage over Indian Territory*, Separate Opinion of Judge V. K. Wellington Koo, Judgment of 12 April 1960: I.C.J. Reports 1960, p. 6, p. 54 at 64.

[98] In this regard, Cheng stated: 'The principle of good faith which governs international relations controls also the exercise of rights by States. The theory of abuse of rights . . . is merely an application of this principle to the exercise of rights.' Cheng, above n 3 at 121. Similarly, for Mitchell, the principle of abuse of rights is a specific form of good faith. Mitchell, above n 3 at 108.

[99] Appellate Body Report, *US – Shrimp*, para. 158. See also Appellate Body Report, *US – Cotton Yarn*, footnote 53.

[100] Panizzon, above n 68 at 30–31.

[101] Interestingly, Byers discusses this principle, inter alia, in the context of the obligation to negotiate in good faith. Byers, Michael, 'Abuse of Rights: An Old Principle, a New Age', 47 *McGill Law Journal* (2002) 390 at 411, 421.

[102] Cheng, above n 3 at 134.

Cheng sees the raison d'être of the principle of abuse of rights in a well-established presumption that 'law cannot precisely delimit every right in advance'.[103] In agreement with Cheng, Byers, Hersch Lauterpacht and Panizzon regard the principle of abuse of rights as a 'promoter of changes in the law when, perhaps, negotiations [on a treaty] fail to address a new development'.[104] Lauterpacht argued that 'the doctrine thus conceived can be regarded as one of great potentialities in the process of judicial legislation adjusting the law to new conditions and preventing unfair or anti-social use of rights'.[105]

The principle of abuse of rights originated in municipal law, in particular the law of torts.[106] Since the beginning of the twentieth century, this principle has been gradually incorporated into treaties and other international instruments, as well as recognised by scholars and judicial authorities.[107] For example, the principle of abuse of rights, together with the principle of good faith, is incorporated into LOSC Article 300 (Good faith and abuse of rights).[108]

In light of this, it is hardly controversial that the principle of abuse of rights is a principle of general international law that can be used in the interpretation of WTO law through systemic integration. However, what does this principle require in the context of gas transit? The next sections examine the relevant substantive content of this principle and how it can be applied as a balancing tool to distinguish a legitimate transit regulation or requirement from an abuse of the right to regulate.

(ii) The relevant substantive content of the principle of *abus de droit* The principle of abuse of rights does not appear to have been used in international case law to fill in gaps in treaties regulating gas transit, let alone third-party access or capacity establishment rights. Nevertheless, it was mentioned earlier that the very purpose of this principle is to balance rights and obligations in a situation in which the existing law becomes

[103] Ibid. at 132.

[104] Byers, above n 101 at 413–414, 429; Panizzon, above n 68 at 33; Lauterpacht, Hersch, *The Function of Law in the International Community* (USA: Archon Books, 1966) at 286.

[105] Lauterpacht, *The Function of Law*, above n 104 at 286. [106] Ibid. at 292–297.

[107] For a detailed overview of the process of crystallisation of the principle, see ibid. at 286–297; Byers, above n 101 at 392–410; and Kiss, Alexandre, 'Abuse of Rights', in *Max Planck Encyclopaedia of Public International Law* (2010), www.mpepil.com, accessed 30 March 2012.

[108] United Nations Convention on the Law of the Sea of 1982, http://www.un.org/depts/los/convention_agreements/texts/unclos/unclos_e.pdf, accessed 30 December 2012. See also Chapter VI, Section III.A.3.

vague, due to, inter alia, its evolution or new technological developments and one of the parties to a treaty possesses discretionary powers. Therefore, there is no prima facie reason it could not be used to inform the nature of transit obligations in GATT Article V.

In doctrine and international jurisprudence, the test of abuse of rights has often been carried out by examining whether the right at issue was exercised in a 'reasonable' and 'bona fide' manner.[109] In Cheng's view, a 'reasonable' and 'bona fide' use of a right is:

> one which is appropriate and necessary for the purpose of the right (*i.e.*, in furtherance of the interests which the right is intended to protect). It should at the same time be fair and equitable as between the parties and not one which is calculated to procure for one of them an unfair advantage in the light of the obligation assumed. ... [It] is regarded as compatible with the obligation.[110]

It was also noted earlier that, in Cheng's view, a 'reasonable' regulation, consistent with the principle of abuse of rights, would be in conformity with the spirit of the law, which in essence is the object and purpose of a given treaty.[111] Thus, according to Cheng, a 'reasonable' regulation, consistent with the principle of abuse of rights, can be described as:

- appropriate and necessary for the purpose of the right;
- not conferring an unfair advantage on the relevant stakeholders in the regulating state; and
- compatible with obligations undertaken under the treaty at issue (*in casu* GATT Article V), including its object and purpose.[112]

Some of these criteria have been applied in various international disputes in which adjudicators assessed the 'reasonableness' of challenged regulations. While, in some of these cases, international tribunals interpreted particular treaty terms based on VCLT principles, the final outcome of the interpretation reflected closely the basic substantive content of the principle of abuse of rights Cheng and other scholars have described.

For example, one of the earliest applications of the 'reasonableness' test, intertwined with the principle of abuse of rights, can be traced back to the *North Atlantic Coast Fisheries* arbitration. In this case, the Arbitral

[109] Byers, above n 101 at 411. [110] Cheng, above n 3 at 125. [111] Ibid. at 134.

[112] See also other literature sources on this subject suggesting that a situation of abuse of rights can arise when 'a State exercises its rights in such a way that another State is hindered in the exercise of its own rights and, as a consequence, suffers injury'; or 'a right is exercised intentionally for an end which is different from that for which the right has been created'. Kiss, above n 107 at 1; Panizzon, above n 68 at 30–31.

Tribunal had to rule on: (i) whether certain fishery rights in British territorial waters granted by Great Britain to the United States under a number of Conventions (such as the 1818 Treaty), recognising the independence of the United States and accomplishing the territorial settlement between both parties, reserved for Great Britain a right to reasonable domestic regulation; and (ii) if so, whether that reasonable regulation was permitted without the accord and concurrence of the United States.[113] It is important to mention that the applicable Conventions did not make either an explicit or implicit reference to such a right, reasonable or otherwise.[114]

As a general matter, the Tribunal held that if a treaty was silent on the right to regulate, the presumption was in favour of such a right as a basic attribute of sovereignty, unless the claiming party proved otherwise.[115] The Tribunal, however, noted that the right to regulate fishery rights was limited by the obligation to execute the treaty in good faith, which imposed on the right a condition of 'reasonableness'.[116] In the Tribunal's view, this condition meant that the regulation 'must be made bona fide and must not be *in violation* of the said Treaty'; the interference with the fishery rights in a manner that was *inappropriate* or *unnecessary* for the protection and preservation of the fisheries would not satisfy this condition.[117] The Tribunal thus recognised the two important elements of a 'reasonable' regulation: appropriateness/necessity, and compatibility with obligations undertaken.

In *Navigational and Related Rights*, the ICJ had to determine the scope of Nicaragua's power to regulate the use of the part of the San Juan River under its sovereignty in which Costa Rica had the right of free navigation, based on the 1858 Treaty of Limits. Although the treaty at issue did not address specifically the scope of Nicaragua's regulatory autonomy, the Court has developed the following basic characteristics of a 'reasonable' regulatory measure that Nicaragua could apply to navigation:

(1) it must only subject the activity to certain rules without rendering impossible or substantially impeding the exercise of the right of free navigation;
(2) it must be consistent with the terms of the Treaty . . .;

[113] *North Atlantic Coast Fisheries (United States v. Great Britain)*, Permanent Court of Arbitration, 7 September 1910, http://www.pca-cpa.org, accessed 30 September 2011 at 1–3, 7–8.
[114] Ibid. at 8. [115] Ibid. at 8. [116] Ibid. at 14. [117] Ibid. at 16.

(3) it must have a legitimate purpose, such as safety of navigation, crime prevention and public safety and border control;
(4) it must not be discriminatory . . .;
(5) it must not be unreasonable, which means that its negative impact on the exercise of the right in question must not be manifestly excessive when measured against the protection afforded to the purpose invoked.[118]

Based on these characteristics, a 'reasonable' regulation must be compatible with the right established (such as freedom of navigation) as well as other treaty obligations, must have a legitimate objective, must strike an appropriate balance between the established right and the right to regulate for the protection of legitimate objectives, and must not be discriminatory.

In *Iron Rhine Arbitration*, the principle of abuse of rights also appears to have been invoked through the concepts of a 'reasonable' and 'bona fide' regulation. In this case, under the 1839 Treaty of Separation between Belgium and the Netherlands, the Netherlands granted certain transit rights to Belgium, such as rights relating to the construction of a railway connection between Belgium and Germany for commercial purposes through the territory of the Netherlands, at Belgium's expense and on the terms agreed between the parties.[119] The railway was constructed but, since World War II, it was not fully utilised until modern times. When, in the 1990s, Belgium and the Netherlands reopened the dialogue on the utilisation of the railway, they failed to agree on the allocation of the costs necessary for adapting it to modern (long-term) use and particular conditions to be attached to such use. One of the principal points of the disagreement was over which party would bear the costs of compliance with various environmental standards and conditions required by the Netherlands.[120] Belgium argued that:

> [T]he exercise of jurisdiction by the Netherlands over the Iron Rhine railway [was] limited by the Netherlands' obligations under international law and in particular the obligations of *good faith* and *reasonableness*. . . . Exercise of its rights . . . [of Belgium to have the Iron Rhine railway

[118] *Navigational and Related Rights*, above n 16, para. 87.

[119] For details, see Article XII of the Treaty and its interpretation by the Tribunal in *Belgium v Netherlands (Iron Rhine Arbitration)*, Permanent Court of Arbitration, 24 May 2005, www.pca-cpa.org, accessed 30 March 2012, paras. 32 and 62–96.

[120] Ibid., paras. 16–25.

> reactivated in the Dutch territory] must not be rendered 'unreasonably difficult' by, among other things, the 'highly expensive' environmental protection measures.[121]

The position of the Netherlands was that its requirements were 'reasonable' and did not constitute an abuse of rights.[122]

While the case was mainly decided on the basis of applicable treaties, including the Treaty of Separation, the Tribunal established two main conditions for a measure to be considered 'reasonable'. On one hand, the Tribunal agreed with Belgium that according to the generally accepted principles of good faith and reasonableness, any measures the Netherlands was to prescribe for the reactivation of the railway *would not render unreasonably difficult* the exercise of Belgium's transit right.[123] On the other hand, the Tribunal held that the Netherlands did not forfeit more sovereignty than that necessary for the railway to be built and operated to allow a commercial connection between Belgium and Germany.[124] In this regard, based on relevant principles of international environmental law incorporated into the applicable treaties through systemic integration and evolutionary interpretation,[125] the Tribunal found that environmental protection measures fell within the scope of the sovereign jurisdiction of the Netherlands and must be treated as an integral component of the project to modernise the railway.[126] The question of which party was to bear specific portions of the project expenses was ultimately resolved on the basis of criteria derived from the applicable legal instruments.[127]

In *Iron Rhine Arbitration*, the Arbitral Tribunal thus appears to recognise that a 'reasonable' measure must not defeat the object and purpose of a right established by international law by rendering its exercise unreasonably difficult. Interestingly, it incorporated legitimate sovereign concerns (such as environmental protection) into a general right to regulate, which was not affected by the applicable treaty law obligations.

The term 'reasonable' regulation or measure is also often used in treaties. Nevertheless, international adjudicators have interpreted it consistently with some of the criteria of a 'reasonable' regulation in general international law (in particular appropriateness/necessity for the purpose of the right). For example, in *AES Summit Generation Ltd et al vs. the*

[121] See ibid., para. 23 (and also 162), emphasis added. [122] Ibid., paras. 25 and 203.
[123] Ibid. paras. 163 and 220. [124] Ibid. para. 87.
[125] See the principle of 'evolutionary interpretation' discussed in Chapter IV, Section III.A.4.
[126] *Iron Rhine Arbitration*, above n 119, paras. 58–59, 220. [127] Ibid. paras. 220–236.

Republic of Hungary, the Arbitral Tribunal examined the legality of the reintroduction by Hungary of administrative electricity pricing, resulting in price cuts for electricity generated by investors, under certain investment rules, including ECT Article 10(1). This provision provides that 'no Contracting Party shall in any way *impair by unreasonable* or discriminatory *measure* their [investment's] management, maintenance, use, enjoyment or disposal'.[128] The Arbitral Tribunal interpreted the standard of a 'reasonable' measure as requiring *an appropriate correlation* between *a rational public policy objective* and *the measure* adopted to achieve it. It held as follows:

> There are two elements that require to be analyzed to determine whether a state's act was *unreasonable*: the *existence of a rational policy*; and the *reasonableness of the act of the state in relation to the policy*. . . . A rational policy is taken by a state following a logical (good sense) explanation and with the aim of addressing a public interest matter. . . . Nevertheless, a rational policy is not enough to justify all the measures taken by a state in its name. A challenged measure must also be *reasonable*. That is, there needs to be *an appropriate correlation between the state's public policy objective and the measure adopted to achieve it*. This has to do with the nature of the measure and the way it is implemented.[129]

Similarly, in a number of WTO cases, the panels determined whether the measure was 'reasonable' within the meaning of particular WTO provisions, such as GATT Article X:3(a) (Publication and Administration of Trade Regulations). This provision requires that each Member administer in a uniform, impartial and *reasonable* manner all its laws, regulations, decisions and rulings.[130] In *Dominican Republic – Cigarettes*, the Panel defined the term 'reasonable', inter alia, as meaning: '"not irrational . . . ", "proportionate", . . . "within the limits of reason, not greatly less or more

[128] Energy Charter Secretariat, 'The Energy Charter Treaty and Related Documents', above n 1.

[129] *AES Summit Generation Ltd et al vs. the Republic of Hungary*, ICSID, 23 September 2010, available at: http://www.italaw.com/sites/default/files/case-documents/ita0014_0.pdf, accessed 3 July 2013, paras. 10.3.7–10.3.9, emphasis added (see also paras. 4.1–4.26). A similar interpretation was adopted by a different Tribunal in the context of an analogous provision in the BIT between the Netherlands and the Czech Republic, in *Saluka Investments BV (the Netherlands) v. the Czech Republic*. See *Saluka Investments BV (the Netherlands) v. the Czech Republic*, Partial Award, UNCITRAL, 17 March 2006, http://www.italaw.com/sites/default/files/case-documents/ita0740.pdf, accessed 3 July 2013, paras. 457–460; and the BIT between the Netherlands and the Czech Republic, Article 3(1), http://www.unctadxi.org/templates/DocSearch.aspx?id=779, accessed 3 July 2013.

[130] GATT 1994, *WTO Legal Texts*, above n 18.

than might be thought likely or appropriate"', which implies a logical relationship between the measure and the objective it pursues.[131]

In *US – Shrimp*, the Appellate Body used the chapeau of GATT Article XX (General Exceptions) as a litmus test for determining whether the United States (Respondent) *abused* exceptions provided in the GATT and, hence, acted 'unreasonably' or inconsistently with the principle of *abus de droit* as expressed in the chapeau.[132] To recall, Article XX allows WTO Members to adopt measures that fall under one of the sub-paragraphs of this provision, such as measures relating to the conservation of exhaustible natural resources. According to the chapeau, these measures must not be applied in a manner that would constitute *a means of arbitrary* or *unjustifiable discrimination* between countries where the same conditions prevail or a disguised restriction on international trade.[133] In *US – Shrimp*, the Appellate Body held that the chapeau is, in fact, but one expression of the principle of good faith, expressed through the doctrine of abuse of rights.[134] Unfortunately, it is not clear from the report what purpose this doctrine served in the Appellate Body's interpretation of the chapeau, in particular whether, in the Appellate Body's view, a measure that constitutes a means of arbitrary or unjustifiable discrimination or a disguised restriction on international trade is inconsistent with this specific provision, or the principle of *abus de droit* in general.[135] As noted earlier, some scholars and other international tribunals have considered that a measure that confers an unfair advantage, such as arguably an 'arbitrary or unjustifiable discrimination' within the meaning of Article XX, is inconsistent with the principle of *abus de droit*.

In any event, in the context of WTO transit rules, the question of whether a transit state can impose discriminatory conditions on the

[131] Panel Report, *Dominican Republic – Import and Sale of Cigarettes*, para. 7.385; see also Panel Report, *US – COOL*, para. 8.850. See this term defined in the same way in the context of Section 2.2(b) of the Telecom Reference Paper in Panel Report, *Mexico – Telecoms*, para. 7.182.

[132] Appellate Body Report, *US – Shrimp*, para. 160.

[133] See GATT 1994, *WTO Legal Texts*, above n 18. See also Chapter IV, Section V.A.2.

[134] Appellate Body Report, *US – Shrimp*, para. 158.

[135] Some passages in *US – Shrimp* suggest that the former is the case. For example, the Appellate Body notes: '[O]ur task here is to interpret the language of the chapeau, seeking additional interpretative guidance, as appropriate, from the general principles of international law. . . . We address, in other words, whether the application of this [US] measure constitutes an *abuse* or *misuse* of the provisional justification made available by Article XX(g).' Ibid. paras. 158, 160, emphasis added.

traffic in transit consistently with the principle of abuse of rights is moot, as the requirement of non-discrimination is incorporated into the text of the GATT (inter alia Article V:2, second sentence). Consequently, most types of discrimination against foreign traffic in transit would fall afoul of the GATT itself. In addition, GATT Article V(3) and Articles 11(1) and 11(6) of the Agreement on Trade Facilitation prohibit transit regulations that are 'unnecessary', which is another criterion for determining whether the measure constitutes an abuse of rights in general international law.

This overview of the doctrinal sources and international jurisprudence demonstrates that a 'reasonable' regulation, consistent with the principle of *abus de droit*: (i) must be appropriate and necessary for achieving a legitimate public policy objective; (ii) must not discriminate in an unfair manner; and (iii) must be compatible with the obligations undertaken, such as the obligation to provide a non-discriminatory freedom of transit under GATT Article V:2 (first and second sentences). While the first two criteria are addressed by WTO law itself, the third criterion can provide useful guidance to WTO adjudicators assessing the 'reasonableness' of transit regulations and requirements (namely, their consistency with the principle of *abus de droit*). The next section discusses the application of this criterion in the context of gas transit.

(iii) Application of the principle of *abus de droit* to gas transit If the negotiations between a transit state and a transit-dependent state truly aim at implementing obligations established under GATT Article V and other relevant principles of general international law, they must only address the terms and modalities of transit, and not the right to transit per se. In this way, the parties will ensure that the transit-related requirements are compatible with the obligation to provide a non-discriminatory freedom of transit and the principle of *abus de droit*. The terms and modalities of transit may include a particular route through which the flow of gas will be allowed into the transit state's territory, the allocation of pipeline capacity, transit charges for gas transportation and compliance with technical and environmental standards (including gas quality specifications). However, the transit state cannot impose demands and conditions irrelevant to the object of negotiations (i.e. the conditions of transit), as GATT Article V requires a non-reciprocal freedom of transit, which can only be subjected to reasonable regulations and limited exceptions.[136]

The proposition that a legitimate regulation of transit can only concern its terms and conditions and must not interfere with the existence of the right per se was upheld explicitly in a number of international cases. It was mentioned earlier, that, in *Corfu Channel*, the ICJ found that Albania was in breach of its international obligation to allow the innocent passage of the British fleet through an international strait located in its territorial waters.[137] The Court noted that 'Albania ... would have been justified in issuing regulations in respect of the passage of warships through the Strait, but not in *prohibiting such passage* or in *subjecting it to the requirement of special authorization*.'[138]

Along the same lines, in *Right of Passage over Indian Territory*, Judge Fernandes proposed a similar demarcation between freedom of passage over land (transit) and its legitimate regulation. In this case, the ICJ had to establish the relationship between Portugal's alleged right of passage to its land-locked enclaves in India and India's right to regulate this passage.[139] The Court upheld Portugal's right on the basis of a crystallised local custom, in relation to private persons, civil officials and goods, subject to the regulation and control of India.[140] While the Court's judgment as such does not elaborate on the exact boundary between the respective rights of Portugal and India, Judge Fernandes, in his Dissenting Opinion, clarified this boundary by looking at the distinction between the right and its authorisation in municipal laws. He noted that the latter distinction is common to all branches of law, but it is particularly illustrative in public law, such as administrative law.[141] For Judge

[136] See Articles V:2, V:3, V:4, XX and XXI of the GATT. GATT 1994, *WTO Legal Texts*, above n 18. See these provisions discussed in Chapter IV. On the object of negotiations, see also United Nations, GA Resolution 53/101 'Principles and Guidelines for International Negotiations', above n 47.

[137] See *Corfu Channel Case*, above n 37 at 23 and 32; and Chapter III, Section IV.B.1.

[138] Ibid. at 29, emphasis added.

[139] *Case Concerning Right of Passage over Indian Territory*, Judgment of 12 April 1960: I.C.J. Reports 1960, p. 6.

[140] Ibid. at 38.

[141] In his Opinion, the term 'authorisation' is defined as: '[T]he administrative act by which the authority removes in each case the limitations imposed by rules of law upon the exercise, by a given subject, of a right or power *already belonging* to the subject himself to exercise a certain activity ... [S]uch exercise ... [is] assessed in the light of the public interest which it is the duty of the authority to safeguard'. *Case Concerning Right of Passage over Indian Territory*, Dissenting Opinion of Judge Fernandes (translation), Judgment of 12 April 1960: I.C.J. Reports 1960, p. 6, p. 123, at 130, emphasis original.

Fernandes, the authorisation does not create a new right, such as the right of passage; it is a conditioning act – an act of control.[142] Consequently, in light of the ICJ judgment in *Corfu Channel* and the Opinion of Judge Fernandes in *Right of Passage over Indian Territory*, the important limitation on any transit state's powers to regulate transit is that normally those powers would extend only to the *conditions* of transit – not the matter of transit per se.

In LOSC Article 125, the right of transit is delimited from its sovereign regulation in a similar way. The authoritative commentary on this Convention explains in this regard that:

> [F]reedom of transit as such does not depend on the conclusion of an agreement between the land-locked and the transit State. Only the terms and modalities for exercising freedom of transit [that is, conditions of transit] may be the subject of agreements between the States concerned at a bilateral or multilateral level. . . . [T]he transit State may not refuse the passage across its territory of . . . 'traffic in transit' . . . if the terms and modalities agreed upon for the exercise of freedom of transit have been complied with.[143]

Likewise, in the context of general international law, Elihu Lauterpacht argues:

> [T]here is room for the view that States are not entitled arbitrarily to determine that the enjoyment of a right of transit is excluded by considerations of security. What they may do is . . . to indicate one route of transit in preference to another, or possibly, to allow the use of the route subject only to certain conditions. But it must be doubted whether the discretion of the States stretches beyond this.[144]

In circumstances in which a violation of the obligation to negotiate is claimed to result from unreasonable demands or conditions that are inconsistent with international law, a tribunal would be expected to objectively assess this claim by looking at the conditions put on the negotiating table and determining whether they are 'reasonable' or whether they constitute an abuse of rights by a transit state. In WTO law, this obligation of panels would stem from their judicial function as defined in DSU Article 11: 'a panel should make an objective assessment of the matter

[142] Ibid.

[143] See Dupuy, René-Jean and Vignes, Daniel (eds.), *A Handbook on the New Law of the Sea*. vol. 1 (Boston: Hague Academy of International Law, Martinus Nijhoff Publishers, 1991) at 517–518. See a similar view in Melgar, Beatriz H., *The Transit of Goods in Public International Law* (Leiden: Brill Nijhoff, 2015) at 197.

[144] Lauterpacht, 'Freedom of Transit', above n 97 at 340.

before it, including an objective assessment of the facts of the case and the applicability of and conformity with the relevant covered agreements'.[145]

There is extensive experience of different international tribunals exercising similar functions. For example, in *Iron Rhine Arbitration*, not only did the Arbitral Tribunal assess the 'reasonableness' of the environmental requirements of the Netherlands (transit state) for the construction of a railway line between Belgium and Germany, but it also assisted the parities in determining a fair allocation of costs for that purpose.[146]

In *US – Shrimp (Article 21.5 – Malaysia)*, the Appellate Body engaged in the assessment of the effort of the United States to reach a multilateral solution to a 'trade-and-environment' dispute with its counterparts by comparing such efforts with those already accomplished with other WTO Members. The Appellate Body stated: 'we used the Inter-American Convention to show that "consensual and multilateral procedures are available and feasible for the establishment of programmes for ... [meeting the legitimate regulatory targets established by the United States]".'[147] Consequently, in *US – Shrimp (Article 21.5 – Malaysia)*, the Appellate Body selected a factual reference for what could be a desirable or reasonable result of the negotiations at issue.

In *Mexico – Telecoms*, the Panel analysed whether Mexico met its obligations under Section 2.2(b) of the Telecom Reference Paper. This provision required Mexico (Respondent) to ensure that its major supplier of telecommunications services Telmex provided interconnection, which in essence is a right analogous to 'third-party access', to US suppliers 'at cost-oriented rates that are ... reasonable, having regard to economic feasibility'.[148] The Panel found Mexico in violation of this provision. It noted that 'more than one costing methodology could be used to calculate "cost-oriented" rates'. However, after reviewing evidence provided by the United States (Complainant) and recommendations developed by the International Telecommunications Union with respect to standard costs of providing interconnection services, the Panel came to the conclusion that, under any methodology that it examined, the interconnection rates Telmex charged to US suppliers were substantially higher than the actual costs.[149] In light of these examples, it is not an

[145] Dispute Settlement Understanding, *WTO Legal Texts*, above n 18.

[146] *Iron Rhine Arbitration*, above n 119, paras. 220–236.

[147] See Appellate Body Report, *US – Shrimp (Article 21.5 – Malaysia)*, para. 128.

[148] Panel Report, *Mexico – Telecoms*, para. 7.18. Some aspects of the Telecom Reference Paper are addressed in Chapter VI, Section III.B.3.

[149] Panel Report, *Mexico – Telecoms*, paras. 7.168–7.216.

impossible task for a WTO panel to assess both parties' positions with a view to determining whether a disagreement that arose during negotiations is the result of legitimate differences in opinions on the appropriate conditions of transit or an abuse of rights.[150]

This assessment can be significantly facilitated by the fact that there exist best practices, standards and various model agreements regulating the terms and modalities of pipeline transit, which have been developed or compiled by specialised international bodies and agencies, as well as the industry itself.[151] The fact that gas pipeline networks exist in most countries can also contribute to this assessment. For example, a prohibition on the construction of additional pipeline capacity along the route where other pipelines observing similar technical and environmental standards already exist may raise a question of whether this measure is legitimate. Likewise, restrictions on access to existing pipelines should not be applied with a view to affording protection to the domestic industry. If, under national law, only domestic gas producers or shippers can obtain access to the pipeline network, this distinction may give rise to a violation of GATT Article V:2 (second sentence), which prohibits discrimination based on the place of origin or the ownership of goods.[152] All of these factors can be taken into account in considering whether the negotiations on establishing gas transit between WTO Members comply with WTO law and the principles of general international law discussed in this chapter.

The analysis of the principles of effective or integrated rights and economic cooperation, the principle that negotiations must be conducted

[150] This does not necessarily mean that a panel would have to pronounce on whether one of the parties to the negotiations acted in bad faith. Indeed, it is a well-known principle of international law that bad faith is not presumed. See *Lake Lanoux Arbitration*, above n 75 at 126; and Appellate Body Report, *EC – Sardines*, para. 278. It may be counterproductive to require an aggrieved party to prove malice on the part of the infringer. Such an unrealistic threshold for the burden of proof would make the principle of abuse of rights inoperative and has not been set in international case law dealing with this principle or the obligation to negotiate in good faith. Along the same lines, Mitchell argues that the application of the doctrine of abuse of rights will be more successful if the claim concerns objective factors (such as the consequences of the abuse – injury or the violation of a treaty obligation), rather than subjective factors (i.e. intentions). Mitchell, above n 3 at 121. See also Byers, above n 101 at 412.

[151] See, inter alia, Energy Charter Secretariat, 'Bringing Gas to the Market. Gas Transit and Transmission Tariffs in Energy Charter Treaty Countries: Regulatory Aspects and Tariff Methodologies' (Brussels: Energy Charter Secretariat, 2012); and Energy Charter Secretariat, 'Model Intergovernmental and Host Government Agreements for Cross-Border Pipelines', 2nd ed. (Brussels: Energy Charter Secretariat, 2007).

[152] GATT 1994, *WTO Legal Texts*, above n 18.

in good faith and *abus de droit* has demonstrated that there are ways to reconcile legitimate sovereign concerns of a transit state and the interests of states depending on transit. While these principles do not directly address third-party access and capacity establishment, they require that states recognise the effectiveness of the obligation to provide freedom of transit and engage in a process of good faith negotiations with a view to implementing this freedom, subject to procedural and substantive obligations under international law. Depending on the facts of each case, this process is likely to lead the states involved to operationalise freedom of transit by giving effect to either a third-party access or capacity establishment right or both. These conclusions support the interpretation of WTO transit obligations in the previous chapter. There are, however, a few other practical questions relating to pipeline gas transit that must be addressed before concluding this chapter. These questions are discussed in the following sections.

III. Practical Questions Arising in the Context of Effective Freedom of Gas Transit

A. The Scope of the Capacity Establishment Obligation: Does Freedom of Transit Require the Opening of any Particular Route or the Making of Investment in a Transit State's Territory?

To avoid any misinterpretation of obligations derived from a capacity establishment claim, some observations must be made regarding its scope. First, the proposition that, under international law, states must give effect to the principle of freedom of transit, even in the absence of the infrastructure necessary for its implementation, does not result in an obligation for those states to open a particular route, such as a specific transit corridor for the construction of a pipeline. This can be inferred from the PCIJ judgment in *Railway Traffic between Lithuania and Poland.* One of the matters the Court addressed in this case was whether a general obligation 'to secure and maintain freedom of communications and of transit' in Article 23(e) of the 1919 Versailles Treaty obliged Lithuania to repair and open a specific railway line (i.e. the Landwarow–Kaisiadorys railway sector).[153] This question was answered in the negative.[154]

[153] *Railway Traffic between Lithuania and Poland*, above n 78 at 14–15.

[154] In particular, the Court held: 'it is impossible to deduce from the general rule contained in Article 23(e) of the Covenant an obligation for Lithuania to open the Landwarow–Kaisiadorys railway sector for international traffic, or for part of such

The Advisory Opinion of the Court, however, should not be understood as depriving the principle of 'freedom of communications and transit' in the Versailles Treaty of any practical effect. Although far from requiring a transit state to open a specific transit route, this principle could well contain a specific obligation – namely the obligation to operationalise 'freedom of communications and transit'.[155]

One of the reasons Lithuania successfully escaped the obligation to give effect to the principle of freedom of transit under the Versailles Treaty was that the question formulated for the PCIJ was too narrow and missed the real point. The actual concern of Poland, as well as of the whole League of Nations at that time, seems to have been whether Lithuania could refuse to have *any* railway communication with Poland at all (as opposed to the connection via the disputed specific line) and how both parties were supposed to operationalise the principle established by Article 23(e) of the Versailles Treaty. Instead, the question was put in this way: 'Do the international engagements in force oblige Lithuania, in the present circumstances, to open for traffic the Landwarow–Kaisiadorys railway sector?'[156] Judge Anzilotti raised this point in his Separate Declaration. For this judge, Lithuania's refusal to make 'freedom of communications and transit' effective would be inconsistent with the Treaty, even though compliance with this principle would logically imply the construction of certain transit facilities.[157]

In light of *Railway Traffic between Lithuania and Poland*, the correct formulation of the claim is the key to success in a dispute related to the enforcement of freedom of transit. Freedom of transit is a basic principle, which as such does not establish an obligation to provide access to any particular 'route'. In GATT Article V:2, this is further reinforced by the obligation to guarantee transit on the '*most convenient* routes', as opposed to '*all* routes'.[158]

Second, there has been a suggestion that the obligation to allow capacity establishment, inter alia, under GATT Article V:2 (first sentence), would necessarily imply an obligation for a transit state to allow the 'making of

traffic.' Ibid. at 14–15. On Article 23(e) of the Versailles Treaty, see Chapter III, Section III.B.1.

155 Lauterpacht, 'Freedom of Transit', above n 97 at 349–350.

156 *Railway Traffic between Lithuania and Poland*, above n 78 at 7.

157 Ibid. at 19. The PCIJ declined to rule on this issue. Ibid. at 15.

158 GATT 1994, *WTO Legal Texts*, above n 18; Panel Report, *Colombia – Ports of Entry*, para. 7.401.

investment' in its territory.[159] It is noteworthy that in investment law the right to 'make an investment' usually depends either on particular treaty obligations granting market access to foreign investors or a specific authorisation.[160] Along the same lines, the argument has also been advanced that, unlike other areas of transit where the means of transit just pass through the foreign territory, the pipeline gas transit, effected through capacity establishment, would imply the permanent occupation of that territory.

Both of these arguments refer to capacity establishment in too narrow a sense. This right implies neither the obligation to allow the 'making of investment', nor the permanent occupation of territory.[161] The principle of freedom of transit, embodied, inter alia, in GATT Article V:2, leaves a transit state broad discretion not only as to how exactly it should operationalise this principle but, should it agree to expand the existing pipeline infrastructure, also as to who will build, own and operate this infrastructure – it simply does not address these matters.[162] The sources of funding for a pipeline project can also be decided between the parties involved. For example, the funding could be provided by an international donor, such as the World Bank, in the form of a project-finance loan to the transit state or the transit-dependent state or to a pipeline operator selected by the parties. In other words, there could be different ways in which a pipeline-construction project could be implemented leaving the transit state's sovereignty intact. In the end, what appears to be of key importance for the effective implementation of the principle of freedom of transit is that gas transit *is* established on reasonable terms, as opposed to *how* exactly this should be done. To this end, the parties involved have different options as to which terms and modalities or which particular 'route' should be used.

Finally, on the other side of the spectrum, the question is often raised (relevant to both capacity establishment and third-party access) whether a transit state has an obligation to operationalise freedom of transit only if there is no alternative transit or export route for a transit-dependent

[159] See the term 'Make Investments' defined in Article 1(8) of the ECT, discussed in Chapter III, Section III.C.2(i).

[160] See Wälde, Thomas W., 'International Investment under the 1994 Energy Charter Treaty', in T. W. Wälde (ed.), *The Energy Charter Treaty: An East–West Gateway for Investment and Trade* (London: Kluwer Law International, 1996) 251 at 277.

[161] For example, the negotiators on ECT Article 7(4) (i.e. capacity establishment) also made a distinction between the 'making of investment' and merely the creation of new capacity. See European Energy Charter Conference Secretariat, 'Room Document 10', Plenary Session, 24–26 March 1993 at 3.

[162] This choice can, however, be limited by other international law commitments, such as investment law and obligations under the GATS.

state. Such a position does not appear to be in line with the principle of freedom of transit. Here, an analogy could be drawn with the ICJ judgment in *Corfu Channel* discussed earlier. In this case, when ruling on whether Albania (Respondent) violated its obligation to allow the innocent passage of the British fleet through the North Corfu Channel located in its territorial sea, the ICJ rejected Albania's argument that the channel was not an *international* strait (highway), necessary for maritime passage, because of its secondary importance and the existence of other routes. In the Court's view, even an alternative route could be considered a useful route for international maritime traffic.[163] As noted, Elihu Lauterpacht did not find any compelling reasons for refusing to apply the aforementioned reasoning of the Court concerning navigation through territorial straits to the issue of passage over the territory of a state.[164]

B. The Problem of Compulsory Third-Party Access to Pipelines

In Chapter IV, the argument was made that a broad principle of freedom of transit under GATT Article V:2 does not require a transit state to take positive measures to enforce a third-party access right against a private company operating pipeline facilities in its territory. Given, however, that in many jurisdictions pipeline networks are privatised or managed by companies *de jure* independent from a state, this matter requires a more thorough examination. This section explains why imposing a compulsory third-party access right on a domestic pipeline operator as a means to implement transit obligations would not always be acceptable to the transit state and indeed cannot be required under WTO law. In such situations, the establishment of new pipeline capacity becomes the only practical way to operationalise freedom of gas transit.

1. Compulsory third-party access and legal bases for WTO Members' responsibility for internationally wrongful acts

While a *negotiated* third-party access right could be one of the options to operationalise freedom of gas transit, *compulsory* third-party access

[163] *Corfu Channel Case*, above n 37 at 28–29.

[164] Lauterpacht, 'Freedom of Transit', above n 97 at 333. Hyde expressed a somewhat different opinion, arguing that, depending on the level of 'necessity', the claim to transit could be stronger or weaker. Hyde, Charles C., *International Law Chiefly as Interpreted and Applied by the United States*, vol. 1, 2nd ed. (Boston: Little Brown and Company, 1951) at 618.

will not always be a practical solution acceptable to both a transit state and a transit-dependent state. Although pipelines, as ports or motorways, are important gateways to the market,[165] they have important differences from other kinds of infrastructure.[166] For example, if a motorway is given on concession to a private operator, in most cases it will be open to public use as a condition of this concession. The purpose of such a concession agreement is not simply to privatise a road, but to transfer commercial risks relating to its operation and maintenance from the government to concessionaires in exchange for the right of the concessionaire to seek reasonable profit from this undertaking. By contrast, pipelines have a different legal status and purpose. In liberalised energy markets with compulsory third-party access rights based on national legislation, such as in the EU, pipelines are intended to be turned into common 'highways' (similar to general/public infrastructure)[167] for the transportation of natural gas (comparable to motorways).[168] At the same time, in a monopolised market with predominantly one user of the pipeline, such as in Russia,[169] pipelines may be regarded as a strategic (and/or privately-owned) facility, which may serve the national purposes and can even be dedicated to a specific gas field. In such a market, the requirement of compulsory third-party access could trigger concerns relating to the transit state's energy security or infringe the property rights of the investors who built the pipeline for private exploitation. In the latter scenario, the pipeline resembles more a private railway line to

[165] This point was argued earlier in this chapter, Section II.A.

[166] However, see Ehring and Selivanova using the analogy with a motorway to show that by preventing access to motorways by their private operators, freedom of transit could be illegitimately denied. Ehring, Lothar and Selivanova, Yulia, 'Energy Transit', in Y. Selivanova (ed.), *Regulation of Energy in International Trade Law: WTO, NAFTA and Energy Charter* (The Netherlands: Kluwer Law International, 2011) 49 at 69.

[167] In *EC and certain member States – Large Civil Aircraft*, the Panel clarified the term 'general infrastructure' under Article 1.1(a)(1)(iii) of the SCM Agreement as 'infrastructure that is not provided to or for the advantage of only a single entity or limited group of entities, but rather is available to all or nearly all entities'. Panel Report, *EC and certain member States – Large Civil Aircraft*, para. 7.1036.

[168] In practice, this is done through a comprehensive set of rules governing the functioning of the energy market, such as third-party access to transmission, distribution and storage facilities, unbundling and network regulation, as well as competition policies preventing companies controlling gas infrastructure from abuse of their dominant position. See Chapter III, Section III.D.

[169] It was explained earlier that, in Russia, Gazprom owns the Unified Gas Supply System (the main gas transmission system in Russia). Gazprom, http://www.gazprom.com/about/, accessed 10 January 2013.

transport coal from the mining site to a port rather than a road used for public purposes.[170]

This study takes the view that *compulsory* third-party access rights cannot be implied lightly in GATT Article V, including the freedom of transit and prohibition of non-discrimination obligations. Under the rules of state responsibility, a state's responsibility can be invoked when that state commits an internationally wrongful act by violating its international commitments.[171] In certain limited circumstances, an act that does not violate a treaty per se, but nonetheless nullifies the benefits granted by that treaty, can also trigger responsibility.[172] It was explained, however, that a purely private measure falls outside WTO law.[173]

Based on these considerations, to determine whether a restriction imposed by a pipeline operator on third-party access to a pipeline gives rise to a WTO Member's responsibility under the DSU, two questions must be answered: (i) can this conduct be attributed to a WTO Member; and (ii) if so, does it violate any WTO rules?

2. When can private conduct be attributed to a WTO Member?

In general international law, the link between private conduct and an internationally wrongful act of a state, violating a treaty or general international law, is established through the rules of 'attribution', which determine whether a private party is the sole author of the act or whether it is merely the means by which the act is performed by the

170 See World Trade Organization, 'Communication from Egypt and Turkey – Discussion Paper on the Inclusion of the Goods Moved via Fixed Infrastructure into the Definition of Traffic in Transit', TN/TF/W/179 (4 June 2012) at 4.

171 See Article 1 of the ILC Articles on State Responsibility. International Law Commission, 'Responsibility of States for Internationally Wrongful Acts', above n 94. See also DSU Article 3(2) and GATT Article XXIII(1)(a) in *WTO Legal Texts*, above n 18. The responsibility can result from a positive act or omission (an obligation to act which has not been met) on the part of the state – a distinction that does not change the legal consequences of the act, at least in the WTO context. Appellate Body Report, *US – Corrosion-Resistant Steel Sunset Review*, para. 81; Latty, Franck, 'Actions and Omissions', in J. Crawford et al. (eds.), *The Law of International Responsibility* (New York: Oxford University Press, 2010) 355 at 359–361.

172 See GATT 1994 Article XXIII(1)(b), *WTO Legal Texts*, above n 18. In a somewhat analogous manner, VCLT Article 18 obliges states – parties to a treaty – to refrain from acts which would defeat the object and purpose of that treaty prior to its entry into force. Vienna Convention on the Law of Treaties, above n 3.

173 See Chapter II, Section III.C.1, and Chapter IV, Sections III.A.1 and III.A.2(i).

state.[174] According to these rules, a state can be held liable for the internationally wrongful conduct of a private person if the latter is in fact acting on the instructions of, or under the direction or control of, the state in carrying out the conduct.[175] Furthermore, if an entity enjoying a legal personality separate from a state is empowered with specific functions akin to those normally exercised by organs of the state, the state can be held liable for that entity's act which is inconsistent with international law (regardless of whether the state *de jure* or *de facto* controls the execution of these functions or provides any instructions to the entity).[176]

In WTO law, the attribution link between private conduct and a state is established in a similar manner. Pursuant to the first rule of attribution, the private action will be attributed to a WTO Member if that Member directs a private entity to act inconsistently with WTO law. In this respect, WTO law contains a broad notion of a 'government measure', encompassing *any* action that has a substantial connection, affiliation or collaboration with the government.[177] Even non-binding (quasi-binding) recommendations of a state, which provide sufficient incentives/disincentives to private actors largely dependent on governmental

[174] Condorelli, Luigi, and Kress, Claus, 'The Rules of Attribution: General Considerations', in J. Crawford et al. (eds.), *The Law of International Responsibility* (New York: Oxford University Press, 2010) 221 at 221.

[175] See Article 8 of the ILC Articles on State Responsibility. International Law Commission, 'Responsibility of States for Internationally Wrongful Acts', above n 94.

[176] Momtaz, Djamchid, 'Attribution of Conduct to the State: State Organs and Entities Empowered to Exercise Elements of Governmental Authority', in J. Crawford et al. (eds.), *The Law of International Responsibility* (New York: Oxford University Press, 2010), 237 at 244–245. In this regard, Article 5 of the ILC Articles on State Responsibility provides: 'The conduct of . . . *entity* which is not an organ of the State under article 4 but *which is empowered by the law of that State to exercise elements of the governmental authority* shall be considered an act of the State under international law, provided the person or entity is acting in that capacity in the particular instance.' The ILC commentary to this article reads that '[t]he generic term "entity" reflects the wide variety of bodies . . . They may include public corporations, semi-public entities, public agencies of various kinds and even, in special cases, private companies, provided [that the requirements of Article 5 are met].' International Law Commission, 'Responsibility of States for Internationally Wrongful Acts', above n 94 at 42–43, emphasis added.

[177] See Article 3.3 of the Dispute Settlement Understanding, *WTO Legal Texts*, above n 18 (setting the objective of the DSU to settle promptly 'situations in which a Member considers that any benefits accruing to it directly or indirectly under the covered agreements are being impaired by measures taken by another Member'); Panel Report, *Japan – Film*, paras. 10.43–56; Appellate Body Report, *US – Corrosion-Resistant Steel Sunset Review*, para. 85; and Zedalis, Rex J., 'When Do the Activities of Private Parties Trigger WTO Rules?', 10 *Journal of International Economic Law* (2007) 335 at 361.

action, can be attributed to a WTO Member.[178] Thus, in light of this rule, if a transit state formally directs or encourages a private pipeline operator to reject third-party access of a foreign gas supplier, this restriction could, in principle, be attributed to the transit state.

In the absence of direct instructions from a transit state, pursuant to the second rule of attribution, the transit state can still be held liable for private conduct if a pipeline operator acts on behalf of the government of the transit state. With respect to certain obligations, the GATT presumes that an entity acts as a 'government' if it is an STE.[179] These obligations, however, have limited relevance to transit. The negotiating history of GATT Article V suggests that certain transit obligations, such as those relating to transportation charges, were intended to cover charges by government-owned modes of transportation (such as government-owned railways).[180] While this clarification is useful to determine the responsibility of a government-owned pipeline operator under the GATT, it does not explain in which circumstances private or quasi-private entities (inter alia, a pipeline operator or owner) could be considered as acting on behalf of a transit state for the purposes of Members' responsibility under WTO law.

WTO jurisprudence clarifies (albeit in the context of subsidy regulation) that, in order to act as a 'government', an entity must perform a 'governmental function' – that is to be vested with *governmental authority*.[181] This performance can be deduced from, among others, an

[178] Appellate Body Report, *US – Corrosion-Resistant Steel Sunset Review*, para. 88; Panel Report, *Japan – Film*, paras. 10.43–51. Zedalis notes that in most GATT/WTO cases interpreting the attribution link between private conduct and non-binding governmental actions the focus was on 'the effectiveness of the measure employed, not the extent or the nature of the governmental action'. Zedalis, above n 177 at 343.

[179] See GATT Article XVII (State-Trading Enterprises) and the explanatory Ad Note to GATT Articles XI, XII, XIII, XIV and XVIII, which clarifies that throughout these provisions the terms 'import restrictions' or 'export restrictions' include restrictions made effective through state-trading operations. GATT 1994, *WTO Legal Texts*, above n 18. For a detailed assessment of Article XVII, albeit in another energy-related context, see Pogoretskyy, Vitaliy, 'Energy Dual Pricing in International Trade: Subsidies and Anti-dumping Perspectives', in Selivanova, Y. (ed.), *Regulation of Energy in International Trade Law: WTO, NAFTA and Energy Charter* (The Netherlands: Kluwer Law International, 2011) 181 at 193.

[180] See Economic and Social Council, 'Preparatory Committee of the International Conference on Trade and Employment', E/PC/T/C.II/54/Rev.1 (28 November 1946) at 10; and World Trade Organization, Secretariat, 'Article V of GATT 1994 – Scope and Application', Negotiating Group on Trade Facilitation, TN/TF/W/2 (12 January 2005) at 6.

[181] The meaning of the term 'government' (and its corollary 'public body') was clarified in Appellate Body Report, *United States – Definitive Anti-dumping and Countervailing*

express statutory delegation of authority by the law of the regulating Member, normal governmental practice or the actual performance of such a function.[182]

These two rules of attribution indicate that, unless a third-party access right is denied upon the instructions/quasi-binding recommendation of a transit state or the regulation of this right is within a 'governmental function' in the transit state's jurisdiction (inter alia, when a certain pipeline operator, through legislation or otherwise, is entrusted or directed by that state to regulate the third-party access to a national or generally-available pipeline network), *compulsory* third-party access cannot be imposed on a private entity acting independently from the government. In other words, if the third-party access right is not affected by the state and has nothing to do with a governmental function, it should be considered a purely private, commercial matter, outside WTO transit rules. The latter situation can arguably occur when this right is not allowed in relation to a private pipeline constructed by investors for commercial exploitation, which is not intended to fulfil the function of general infrastructure.[183]

3. Does WTO law require a transit state to compel a private pipeline operator to provide third-party access?

Some authors have argued that, when capacity constraints (bottlenecks) arise, the third-party access right follows from the freedom of transit and the obligation of non-discrimination under GATT Article V:2.[184] Notably, however, such a categorical proposition lacks support either in the text of

Duties (China), paras. 290, 310 and 318. See also Appellate Body Report, *US – Carbon Steel (India)*, paras. 4.9–4.10.

182 See Appellate Body Report, *United States – Definitive Anti-dumping and Countervailing Duties (China)*, paras. 297 and 317–318.

183 For example, as explained earlier, under the Agreement between Georgia and Azerbaijan on transit through the SCP System, parties explicitly agreed that the pipeline must not be regulated as a public utility, will not involve the provision of services to the public at large and is not intended or required to operate in the service of the public benefit or interest. See Article II(8) of the 2001 Agreement between Georgia and Azerbaijan relating to Transit, Transportation and Sale of Natural Gas in and beyond the Territories of Georgia and Azerbaijan through the SCP System (Mutual Representations, Warranties and Covenants), in British Petroleum, *Legal Agreements*, http://www.bp.com/sectiongenericarticle.do?categoryId=9029334&contentId=7053632, accessed 10 December 2012. This agreement was discussed in Chapter III, Section III.E.

184 Ehring and Selivanova, above n 166 at 70–71; Azaria, *Treaties on Transit of Energy via Pipelines and Countermeasures*, above n 67 at 64, 67; Grewlich, Klaus W., 'International Regulatory Governance of the Caspian Pipeline Policy Game', 29 *Journal of Energy & Natural Resources Law* (2011) 87, footnote 91.

the GATT or general international law, neither of which directly addresses third-party access rights. This argument can only be applicable, if at all, in liberalised energy markets. In such markets, if, despite a national law obligation to provide compulsory third-party access to pipelines, a pipeline operator discriminates between the gas of domestic origin and foreign traffic in transit, it can be argued that a transit state (in the territory of which such discrimination occurs) should be responsible for not enforcing its laws, under, inter alia, GATT Article V:2 (second sentence).[185]

However, if, according to a transit state's national legislation, the pipeline network is not destined for general use but is owned or operated for a private purpose by a private entity or an entity in law or in fact separate from the government of the transit state, there appears to be no basis in the GATT for holding the transit state liable for that entity's rejection of third-party access rights. To argue otherwise would mean that under GATT rules (whether Article V or a non-violation complaint), a transit state must provide instructions to its pipeline operators to release a certain portion of their pipeline capacity for general use or enact laws compelling third-party access. However, this would be a far-fetched argument, essentially suggesting the existence of competition-type of obligations under the GATT. It must be recalled that international law regulates anti-competitive conduct only to a very limited extent.[186] Not surprising, commentators analysing this issue argued that, in most circumstances, WTO rules do not discipline infringements of competition law and that the regulation of restrictive business practices should be sought outside the WTO regime.[187]

[185] Arguably, depending on whether the pipeline operator in question is a state or private company, the transit state's responsibility can result from an action or omission on its part.

[186] See Chapter II, Section III.C.1.

[187] Roessler, for example, after examining the GATT case law on non-violation complaints, comes to the conclusion that it would be 'an illusion to think that panels or the Appellate Body would eagerly embrace the idea of handing out licenses to retaliate against restrictive business practices without any prior normative guidance by the membership of the WTO'. Roessler, Frieder, 'Should Principles of Competition Policy Be Incorporated into WTO Law through Non-violation Complaints?', 2 *Journal of International Economic Law* (1999) 413 at 420. Along the same lines, García-Castrillón states that both violation and non-violation complaints relating to restrictive business conduct will present evidentiary challenges and that bilateralism appears to be the only solution to most competition problems in the WTO regime. García-Castrillón, Carmen, O., 'Private Parties under the Present WTO (Bilateralist) Competition Regime', 35 *Journal of World Trade* (2001) 99 at 121.

It is hardly possible to identify an international norm (including norms of general international law) imposing an explicit *compulsory* third-party access obligation, apart from those in EU law.[188] Yet, even in the EU's liberalised gas market, the third-party access right contains exceptions, among others, for the purpose of attracting investment in major new infrastructure developments relating to the construction of gas pipelines.[189] As noted earlier, the ECT, in Understanding IV:1(b)(i), explicitly rules it out.[190]

In light of the foregoing, the argument that WTO transit rules provide for a compulsory third-party access right in the context of pipeline gas transit is problematic from the perspectives of both rules on state responsibility (i.e. attribution) and WTO law itself. As has been argued in this study, in this technically complex area, the transit state must be left some discretion as to how it prefers to implement its transit obligations, provided that the choice of implementation is in line with transit principles set out in WTO law and general international law. In this connection, the transit state could decide to grant a non-discriminatory third-party access right to domestic and foreign gas shippers under its national legislation[191] or agree on a third-party access right based on the negotiated terms and conditions.

IV. Summary and Concluding Remarks

This chapter examined principles of general international law that support the interpretation of GATT Article V:2, first and second sentences, as encompassing third-party access and capacity establishment rights. Alternatively, these principles could strengthen a 'non-violation

[188] See Wälde, Thomas W. and Gunst, Andreas J., 'International Energy Trade and Access to Energy Networks', 36 *Journal of World Trade* (2002) 191 at 209; and Cossy, Mireille, 'Energy Trade and WTO Rules: Reflexions on Sovereignty over Natural Resources, Export Restrictions and Freedom of Transit', in Herrmann, C. and Terhechte, J. P. (eds.), 3 *European Yearbook of International Economic Law* (2012) 281 at 298–299.

[189] See Article 30 in Regulation (EC) No. 715/2009 of the European Parliament and the Council of 13 July 2009 on Conditions for Access to the Natural Gas Transmission Networks and Repealing Regulation, OJ 2009 L 211, 36.

[190] Energy Charter Secretariat, 'The Energy Charter Treaty and Related Documents', above n 1. This provision is discussed in Chapter VI, Section III.A.2.

[191] Similarly, during the negotiations on the ECT, it was clarified that, in practice, a third party in the sending or receiving country could have third-party access to the transit state's pipeline either on the basis of the transit state's legislation or if it owned that pipeline. See European Energy Charter Secretariat, 'Note for the Attention of S. Fremantle by C. Jones' 583 (CLJ/JI) (Brussels, 30 August 1993).

complaint' challenging the measure restricting these rights. These principles are the principles of effective or integrated rights and economic cooperation, the obligation to negotiate in good faith (derived from the '*pactum de contrahendo*' nature of GATT Article V:2) and the prohibition of abuse of rights (*abus de droit*). They require that a transit state recognise inherent ancillary rights essential for taking due and proper advantage of the primary rights established by a treaty, such as freedom of transit and non-discrimination under GATT Article V. This can be achieved by engaging in good faith negotiations with a transit-dependent state on particular terms and modalities of transit, which, in the context of pipeline gas, would inevitably concern third-party access and/or capacity establishment – the only practical ways to operationalise freedom of gas transit. Furthermore, in light of these principles, a transit state must regulate transit in such a way that this regulation is compatible with its international obligations (including WTO commitments), does not defeat the purpose of these obligations, is appropriate and necessary for achieving legitimate regulatory objectives and does not confer an unfair advantage on its domestic industry. The regulation of transit will likely be found compatible with the international obligations of a transit state if it addresses the conditions of transit and not the right to transit per se, unless a general or security exception applies.

Despite the fact that these principles of general international law impose certain limitations on the regulatory autonomy of a transit state, they also leave the latter a wide margin of discretion as to how its transit obligations should be operationalised to establish gas transit. In particular, it was argued in this study that the transit state cannot be required to provide a compulsory third-party access right to a foreign gas shipper, especially in the absence of the pipeline capacity or infrastructure that it controls. A transit state is not obliged to open any particular route or all routes for transit, but must rather make available the route that is 'most convenient'. In this respect, the determination of 'convenience' will have to balance the interests of a transit state and a transit-dependent state, and be based on the ability of the route to satisfy the needs of 'international transit'.[192] Finally, subject to its particular obligations under investment law and the GATS, a transit state does not have to allow foreign investment in its territory to build a pipeline.

[192] See this issue discussed in detail in Chapter IV, Section III.A.2(ii); and this chapter, Section III.A.

What appears to be required from a transit state under international law (including WTO law) is that it allows gas transit via its territory on reasonable terms and conditions. This obligation can be satisfied through various means, including the allocation of a portion of a pipeline to foreign bidders or engaging in good faith negotiations on a pipeline construction project, funded by foreign investors, international donors or a transit state itself, if it so wishes. In this regard, the parties involved can negotiate on the right to operate a pipeline or its elements by rendering pipeline transportation services in a particular territory. They can decide whether the pipeline at issue will have an 'open access' system or will have a private pipeline regime. The GATT and the principles of general international law discussed in this chapter do not appear to establish a hierarchy between third-party access and capacity establishment rights. Both rights can be effected to operationalise freedom of transit, provided that a transit state affords fair and equal opportunity to use its 'routes' to both domestic and foreign gas shippers that are similarly situated.

The interpretation of GATT Article V:2 in light of these principles of general international law will likely strike an appropriate balance between, on one hand, the effective freedom of gas transit and, on the other hand, the right to its legitimate regulation. When interpreted in this manner, this provision, albeit vague on its face, turns into an enforceable legal instrument capable of creating precise legal rights and obligations in the context of pipeline gas transit.

In the last substantive chapter of this study, two roles the WTO could play in establishing freedom of gas transit are analysed: (i) a dispute settlement forum for enforcing rights essential to effective freedom of gas transit, such as potentially third-party access and capacity establishment rights; and (ii) a negotiating platform for the improvement of current WTO transit rules.

VI

Freedom of Gas Transit in the WTO: Dispute Settlement or Legislative Reform?

I. Introduction

At the outset of this study, it was explained that its key objective is to examine whether and how WTO law regulates third-party access and capacity establishment rights. These rights are essential to effective pipeline gas transit, which, in turn, is crucial for the development of an international gas market, energy security and the sustainable development of WTO Members. While WTO transit rules do not mention third-party access and capacity establishment rights, it was explained in the previous chapters why these rights are implied in the obligation to provide freedom of transit and accord non-discriminatory treatment under GATT Article V:2, first and second sentences, and what this means in practice.

This chapter is divided into two parts (Sections II and III) discussing separate, albeit related issues. Section II concludes the analysis of the WTO's regulation of third-party access and capacity establishment rights by examining how these rights can be enforced in a WTO dispute settlement proceeding. Section III assesses the need for a legislative reform of WTO transit rules in order to regulate the particular features of pipeline gas transit more effectively, including third-party access and capacity establishment rights. Section IV is the conclusion.

II. Enforcing Third-Party Access and Capacity Establishment Rights in the WTO Dispute Settlement: A Viable Option?

A. What Claims Can Be Brought before a WTO Panel Relating to Restrictions on Third-Party Access and Capacity Establishment Rights?

In Chapter I, a hypothetical scenario was discussed involving WTO Members A, B, C and D (see Figure 1). To recall, in this scenario, A (a developing, land-locked country) requested B's (a transit state) permission to transit gas to the market of C via the pipeline network 'Red

Stream' operated in B's territory by company Y. To the extent that there is no sufficient pipeline capacity in the Red Stream for the transit of A's gas to C, A offered to B to establish additional capacity, along the same route, at A's expense. B, however, refused to operationalise A's freedom of gas transit by giving effect to either a third-party access right to the Red Stream or a capacity establishment right. In the light of what has been discussed in this study, the question can be asked whether B's blunt refusal to allow gas transit via its territory, let alone the refusal to negotiate with A or C on the terms and modalities of this transit, is consistent with the GATT and whether, in this respect, A and/or C could challenge B's denial of third-party access and/or capacity establishment rights in the WTO.

It was explained in Chapter IV that, in the context of gas transit, the provisions of the GATT most relevant to these rights are: (i) GATT Article V:2 (first sentence), requiring that a WTO Member allow freedom of transit via its territory; (ii) GATT Article V:2 (second sentence), establishing a non-discrimination obligation with respect to 'traffic in transit'; and (iii) GATT Article XXIII:1(b), which, in certain circumstances, permits WTO Members to claim a nullification or impairment of GATT obligations by a measure that may be WTO-consistent but, nevertheless, frustrates the reasonable expectations of the complaining party derived from the GATT. Thus, A and/or C could challenge B's measure in a WTO dispute settlement proceeding based on the following claims:

a. B acted inconsistently with GATT Article V:2 (first sentence) because it refused to negotiate on the terms and conditions of gas transit from A to C, thereby rejecting the access to its territory of goods in transit and failing to provide 'freedom of transit ... via the routes most convenient for international transit, for traffic in transit', as required by this provision.
b. B acted inconsistently with GATT Article V:2 (second sentence) by treating the traffic of gas of domestic origin in a 'distinct' (i.e. more favourable) manner compared to the traffic of gas in transit. In particular, B granted to its domestic company Y the right to operate and use all gas transportation facilities under its jurisdiction, while refusing to negotiate with A on the terms and conditions under which the traffic in transit could cross its territory from that Member to C via the most convenient route.
c. Alternatively, the application of these measures by B nullifies or impairs, within the meaning of GATT Article XXIII:1(b), the benefits

of A (or C) resulting from B's obligations under GATT Articles V:2, first sentence and second sentence.

Furthermore, depending on additional facts, a transit dispute described in the hypothetical scenario could give rise to other claims under more specific provisions of GATT Article V, such as Article V:3, prohibiting 'unnecessary delays or restrictions' and Article V:4, allowing the imposition of 'reasonable' charges only.[1] These claims could be filed, for example, if, instead of refusing to negotiate on the terms and conditions of gas transit, B, as a precondition of transit: (i) demanded that A or its gas shipper X comply with technical standards that would not contribute to the attainment of any legitimate objective of B; or (ii) extorted unreasonable transit fees.

The following section provides a brief overview of procedural requirements that must be satisfied in order to bring a justiciable claim before a WTO panel. It is demonstrated in that section that, from a procedural perspective, there are no legal obstacles in the DSU that would prevent A and/or C from challenging B's measure restricting gas transit by virtue of denying third-party access and capacity establishment rights. Assuming that this challenge is successful, the last questions addressed in Section II.C of this chapter are:

- How could A and/or C enforce third-party access and capacity establishment rights against B in a WTO dispute settlement proceeding;[2] and
- What practical benefits could these Members gain from their challenge of B's measure, apart from a declaratory judgement that B violated its transit obligations under the GATT?

This study does not intend to delve into a general debate on the practical effectiveness of the WTO dispute settlement system, which has been a subject of separate studies.[3] Nevertheless, a few remarks on

[1] As was explained in Chapter IV, Sections III.A.3 and V.C, similar obligations are set out in the Agreement on Trade Facilitation, which, however, has not yet entered into force.

[2] Recall that, pursuant to Article 3(2) of the DSU, one of the key functions of the WTO dispute settlement system is 'to *preserve the rights and obligations of Members under the covered agreements*'. Dispute Settlement Understanding, *WTO Legal Texts*, http://www.wto.org/english/docs_e/legal_e/legal_e.htm, accessed 15 February 2014, emphasis added.

[3] See Bown, Chad P. and Pauwelyn, Joost (eds.), *The Law, Economics and Politics of Retaliation in WTO Dispute Settlement* (New York: Cambridge University Press, 2010); and Bronckers, Marco and Baetens, Freya, 'Reconsidering Financial Remedies in WTO Dispute Settlement', 16(2) *Journal of International Economic Law* (2013) 281.

how the implementation and enforcement phases of the WTO dispute settlement proceedings would unfold if A, C or both of these Members were to succeed in challenging the measure in question before a panel must be made.

B. *Overview of Procedural Requirements that Must Be Satisfied to Bring a Claim before a WTO Panel*

From a procedural perspective, the starting point in examining whether there are any legal obstacles to making a claim based on a violation of third-party access and/or capacity establishment rights is DSU Article 3.3, which determines the jurisdiction of WTO panels. This provision broadly states that a Member can bring a 'claim' before a panel if it 'considers that any benefits accruing to it directly or indirectly under the covered agreements are being impaired by measures taken by another Member'.[4] Thus, if a Member considers that a measure taken by another Member either directly or indirectly impairs *any* benefits it obtained under WTO agreements, such as GATT Article V, the Member can challenge this measure in the WTO.

It was explained earlier that the notion of a 'measure taken by another Member' is broad and may encompass: (i) 'violation' and 'non-violation complaints';[5] (ii) governmental acts and omissions;[6] (iii) measures that are clearly binding and those that, although on their face are non-binding, create sufficient incentives or disincentives inducing market participants to act in a particular manner inconsistent with WTO law.[7] Furthermore, WTO Members have challenged 'measures' that constituted both *de jure* violations of a WTO provision (i.e. a WTO-inconsistent measure written into the national law of a Member) and *de facto* violations (i.e. a measure that is in fact inconsistent with WTO law).[8]

[4] Dispute Settlement Understanding, *WTO Legal Texts*, above n 2.

[5] See GATT, Article XXIII(1)(a) and (b), *WTO Legal Texts*, above n 2; Chapter IV, Section III.C. In theory, WTO law also allows the so-called situation complaint. The meaning of this type of complaint has not been clarified, and it has never been used. See GATT, Art. XXIII(1)(c), ibid.; and Roessler, Frieder, 'Should Principles of Competition Policy Be Incorporated into WTO Law through Non-violation Complaints?', 2 *Journal of International Economic Law* (1999) 413 at 417.

[6] Appellate Body Report, *US – Corrosion-Resistant Steel Sunset Review*, para. 81.

[7] Panel Report, *Japan – Film*, paras. 10.43–51.

[8] See Panel Report, *US – COOL*, paras. 7.298–7.299.

The important conditions imposed on a 'measure' falling within the WTO panels' jurisdiction is that: (i) it must be attributable to a WTO Member;[9] and (ii) it must nullify or impair benefits under a 'covered agreement' (i.e. agreements listed in Appendix 1 of the DSU).[10]

If a particular claim, such as that challenging a restriction on third-party access and/or capacity establishment rights, satisfies these conditions, it must also comply with other procedural requirements of the DSU, among others, those set out in Article 6(2). This provision requires that the request for the establishment of a panel 'identify the specific measures at issue' and 'provide a brief summary of the legal basis of the complaint sufficient to present the problem clearly'.[11]

In the context of restrictions on gas transit, the fulfilment of the requirements of Article 6(2) can become an issue, since, in the course of the negotiations on the terms and modalities of gas transit, various acts or omissions of a transit state can independently or cumulatively violate different aspects of WTO law and may not always be transparent. For example, a transit state may *de facto* restrict freedom of transit by extorting 'unreasonable' transit fees from a transit-dependent state or by imposing 'unnecessary' regulatory requirements as a pre-condition for the right to transit. Furthermore, by applying these measures, the transit state may treat domestic and foreign gas shippers in a 'distinct' (i.e. discriminatory) manner. Thus, a clear formulation of each claim, explaining its legal basis in WTO law, as well as the accurate description of a specific measure at issue, would be instrumental to a successful complaint.[12] Needless to say,

[9] See Appellate Body Report, *US – Corrosion – Resistant Steel Sunset Review*, para. 81; Chapter V, Section III.B.

[10] A number of DSU provisions clearly indicate that the WTO dispute settlement system 'serves to preserve the rights and obligations of Members under the *covered agreements*'. See DSU Articles 1(1), 3(2), 7(1) and 11. Dispute Settlement Understanding, *WTO Legal Texts*, above n 2 (emphasis added). These provisions have been interpreted as a limitation on WTO panels' jurisdiction. Appellate Body Report, *EC – Poultry*, paras. 79, 81; Panel Report, *EC and certain member States – Large Civil Aircraft*, para. 7.89; and Pauwelyn, Joost, *Conflict of Norms in Public International Law: How WTO Law Relates to Other Rules of International Law* (New York: Cambridge University Press, 2003) at 443–444.

[11] If a claim in the request for the establishment of a panel does not satisfy the requirements under Article 6(2), it will be found to fall outside the Panel's terms of reference (i.e. the Panel's jurisdiction) within the meaning of DSU Article 7. See Dispute Settlement Understanding, *WTO Legal Texts*, above n 2; and Appellate Body Report, *US – Carbon Steel*, para. 124.

[12] Among the examples of cases in which the panel rejected claims as formulated in too broad a manner, see the preliminary rulings in *China – Raw Materials*. In this case, the United States (Complainant) provided an 'open-ended' list of measures in its panel

these claims must be substantiated by solid evidence, since the complaining party will bear the burden of proof.[13]

The evidentiary burden is likely to be higher if a WTO-inconsistent transit restriction takes an unwritten form, for example, an informal refusal to negotiate. Challenging such measures may require a careful planning of the litigation strategy and the preparation of a considerable amount of evidence. Nevertheless, even such non-transparent measures can be challenged successfully. For example, in *Argentina – Import Measures*, the Complainants (the EU, Japan and the United States) challenged Argentina's so-called Restrictive Trade-Related Requirements imposed on importers and other economic operators as a condition to import into Argentina. This measure was not contained in any law, regulation, administrative act or official publication and was enforced by various bodies of the Argentine government through informal talks with the representatives of business. Given the non-transparent nature of the measure and the lack of direct evidence on the existence of each individual requirement, the Complainants decided to challenge them as a single global measure, and had to rely on various indirect evidence, including communications addressed to Argentine officials by private companies, public statements by Argentine officials and notes posted on websites of the Argentine government, articles in newspapers and magazines, data from industry surveys and reports prepared by market intelligence entities. The Complainants ultimately succeeded, and the measure was found to be inconsistent with, inter alia, GATT Article XI:1.[14]

As this discussion demonstrates, the DSU imposes certain basic procedural requirements on WTO Members that must be satisfied in challenging a restriction on third-party access or capacity establishment rights as inconsistent with WTO law. This Agreement, however, leaves

request by using the term 'among others'. The Panel stated as follows: 'the Complainants cannot be allowed to include additional measures other than those listed and identified by bullet points in the panel requests. Such an "open ended" list would not contribute to the "security and predictability" of the WTO dispute settlement system as required by Article 3.2 of the DSU'. Panel Report, *China – Raw Materials*, Addendum, Annex F-1, paras. 10–13.

[13] On the principles of the burden of proof applied in the WTO dispute settlement, see Appellate Body Report, *US – Wool Shirts and Blouses* at 12–17.

[14] See Appellate Body Report, *Argentina – Import Measures*, paras. 5.147–150. See this case discussed in Romero, Hugo and Piérola, Fernando, 'Unwritten Measures: Reflections on the Panel Reports in *Argentina – Measures Affecting the Importation of Goods*', 10(1) *Global Trade and Customs Journal* (2015) 54.

a wide discretion to Members as to which benefits or rights they consider worth litigation. If a WTO Member can prove that, in light of particular facts, the restriction by a WTO Member (a transit state) of third-party access or capacity establishment rights prevents it from exercising its rights under GATT Article V:2 (first and second sentences), there are no formal procedural obstacles in the DSU to such claims.

C. *Enforceability of Third-Party Access and Capacity Establishment Rights in the WTO Dispute Settlement Proceedings*

This section concludes the analysis of the regulation of third-party access and capacity establishment rights under WTO transit rules carried out in Chapters III–VI of this study. The important practical question it addresses is the enforceability of third-party access and capacity establishment rights in the WTO dispute settlement proceedings.

1. Key features of the implementation and enforcement phases of the WTO dispute settlement system

At the outset, note that, unlike other dispute settlement fora, such as investment arbitration, the WTO dispute settlement mechanism does not provide for compulsory compensation for damages.[15] Moreover, the responsibility for a violation of WTO obligations (in the form of a mutually agreed compensation or retaliation) is prospective in its nature and can be triggered only when a WTO Member fails to comply with the DSB recommendation, following a number of procedural steps outlined in this section.[16]

In cases where a panel or the Appellate Body finds that a certain measure of a WTO Member (such as a transit restriction) is WTO-inconsistent, they must recommend that the DSB request that Member to bring its measure into conformity with WTO law.[17] In addition, the DSU authorises WTO adjudicators to suggest ways in which the Member concerned could implement their recommendation.[18] In the context of a transit dispute, this suggestion could assist the parties involved in determining the route most convenient for transit, the terms and modalities of transit or, at least, the principles upon which the implementation of the transit obligations at issue must be achieved,

[15] Bronckers and Baetens, above n 3 at 289. [16] Ibid.

[17] See DSU Article 19(1) (Panel and Appellate Body Recommendations), Dispute Settlement Understanding, *WTO Legal Texts*, above n 2.

[18] Ibid.

and this could play a useful role in the prompt and satisfactory settlement of the dispute.[19] For this purpose, the panel could seek the advice of technical experts under DSU Article 13. The latter provision establishes a rather broad right of panels to seek information and technical advice from any relevant source, individual or body which it deems appropriate.[20]

However, the right to make these suggestions is discretionary, and WTO adjudicators have often left the choice of the means of the implementation of their decisions to respondents, even when complaining parties had explicitly asked them to utilise this right.[21] But, even if a panel or the Appellate Body did suggest the means, the respondent would still have wide policy space to choose the means it considers most appropriate.[22] In addition, in circumstances when it is impracticable for the respondent to comply with the DSB recommendation immediately, 'a reasonable period of time' for such compliance will be afforded. This period is determined through the proceedings established by DSU Article 21(3), and, in any event, it cannot exceed a period of fifteen months.[23]

Having said this, the discretion of a respondent as to how it wishes to bring its measure into conformity with WTO law is also not unlimited. The implementation results are subject to the scrutiny of the so-called 21(5) panel, whose function it is to review the implementation of the DSB recommendations.[24] It is noteworthy that, on average, in more than 80 per cent of cases, WTO Members have implemented the DSB recommendations. As Pauwelyn states, such a high level of compliance is the result of the 'equivalent retaliation' remedy (discussed later) and reputation or 'community costs' linked to non-compliance, since WTO

[19] The prompt and satisfactory settlement of WTO disputes is among the key objectives of the DSU. See DSU Articles 3(3) and 3(4), ibid. In *US – Zeroing (EC) (Article 21.5 – EC)*, the Appellate Body acknowledged that suggestions made by panels or the Appellate Body under DSU Article 19(1) 'may provide useful guidance and assistance to Members and facilitate implementation of DSB recommendations and rulings, particularly in complex cases'. Appellate Body Report, *US – Zeroing (EC) (Article 21.5 – EC)*, para. 466.

[20] Dispute Settlement Understanding, *WTO Legal Texts*, above n 2.

[21] On this 'discretionary' right, see Appellate Body Report, *US – Zeroing (EC) (Article 21.5 – EC)*; Panel Report, *US – Countervailing Measures on Certain EC Products*, para. 6.43. Among limited exceptions, see suggestions made by the Panel in Panel Report, *EC – Bananas III (Article 21.5 – Ecuador)*, paras. 6.154–6.159.

[22] In the end, DSU Article 19(1) authorises WTO adjudicators to make a mere *suggestion*. See Dispute Settlement Understanding, *WTO Legal Texts*, above n 2.

[23] Ibid.

[24] See Article 21(5) (Surveillance of Implementation of Recommendations and Rulings) of the DSU, ibid.

Members share a common goal of compliance with their WTO obligations.[25]

If the 21(5) panel concludes that the DSB recommendation was not implemented or the measure (current or amended) is still inconsistent with the relevant rules, the aggrieved Member may either negotiate a mutually acceptable compensation with the Member infringing its WTO obligations or (if these negotiations fail within twenty days after the date of the expiry of a reasonable period) request the DSB to authorise retaliation (i.e. the suspension of the application to that Member of concessions or other obligations under WTO agreements).[26]

The permissible quantity of retaliation must be determined by arbitrators pursuant to an arbitration proceeding under DSU Article 22(6).[27] For this purpose, the arbitrators must assess whether the method(s) of calculating retaliation proposed by the complainant (or an alternative method(s) suggested by the respondent) result in the 'equivalent' amount of the nullification or impairment in the sense of DSU Article 22(4)[28] and is sufficient to induce compliance with the DSB recommendation.[29]

[25] Pauwelyn, Joost, 'The Calculation and Design of Trade Retaliation in Context: What Is the Goal of Suspending WTO Obligations?', in C. P. Bown and J. Pauwelyn (eds.), *The Law, Economics and Politics of Retaliation in WTO Dispute Settlement* (New York: Cambridge University Press, 2010), 34 at 47–48. See also Davey, William J. 'Dispute Settlement in the WTO and RTAs: A Comment', in L. Bartels and F. Ortino (eds.), *Regional Trade Agreements and the WTO Legal System* (New York: Oxford University Press, 2006) 343 at 348.

[26] See DSU Articles 22(2) and 22(6) (Compensation and the Suspension of Concessions). Dispute Settlement Understanding, *WTO Legal Texts*, above n 2.

[27] Ibid.

[28] Note that arbitrators acting under DSU Article 22(6) have no jurisdiction to examine the qualitative aspect of the proposed retaliation (i.e. which particular concessions should be suspended), unless the method of retaliation contradicts the principles of the DSU, such as those set out in Article 22(3). The latter provision governs situations in which the complainant is authorised to cross-retaliate, inter alia, under a WTO agreement that is different from the agreement under which a violation was established. See DSU Articles 22(3), 22(6) and 22(7), ibid. See also Renouf, Yves, 'From Bananas to Byrd: Damage Calculation Coming Age?', in C. P. Bown and J. Pauwelyn (eds.), *The Law, Economics and Politics of Retaliation in WTO Dispute Settlement* (New York: Cambridge University Press, 2010) 135 at 136.

[29] While the DSU is not entirely clear as to which particular goal the retaliation mechanism was intended to fulfil (e.g. rebalancing of concessions, compensation or compliance), in a number of cases, arbitrators found that among these goals was 'to induce compliance', tempered by the requirement of 'equivalence' and the prohibition of 'punitive retaliation'. See, inter alia, Decision by the Arbitrators, *EC – Bananas III (US) (Article 22.6 – EC)*, para. 6.3; and a detailed discussion of this issue in Pauwelyn 'The Calculation and Design of Trade Retaliation in Context', above n 25. Jackson notes that among the goals of the WTO dispute settlement system, including its retaliation mechanism, are 'keeping the peace'

If none of the methods the parties propose is consistent with the DSU, the arbitrators will determine the quantity of retaliation pursuant to their own method.[30]

To make this determination, the arbitrators would, first, have to derive an understanding of the situation that would have prevailed if the illegality was removed. This implies the analysis of a sort of counterfactual as a comparator.[31] Such an analysis may have to be based on a benchmark period, must account for various factors affecting the counterfactual (e.g. competition in the importing Member) and, in complex cases, would have to rely on economic modelling.[32]

Second, the arbitrators would need to measure the level of nullification or impairment caused by the illegality, which, in most cases, has been done through the assessment of the loss in 'trade flows'.[33] Needless to say that such a determination would, to some extent, be based on assumptions and, like any question about future events, it cannot result in a precise calculation of the equivalent level of nullification or impairment.[34]

2. Practical implications for the parties involved in a gas transit dispute

Turning to the questions of how A and/or C could enforce their third-party access/capacity establishment rights against B and what practical benefits these Members could gain from their challenge of B's measure, it is conceivable that, under the pressure of a potential reputation damage or retaliation, B would have to look for ways to accommodate the interests of A and C. In addition, a well-elaborated suggestion made by the Panel under DSU Article 19(1) could crystallise the scope of the

and 'addressing the asymmetry of power in . . . [the] system'. Jackson, John H., 'Comment on Chapter 2', in C. P. Bown and J. Pauwelyn (eds.), *The Law, Economics and Politics of Retaliation in WTO Dispute Settlement* (New York: Cambridge University Press, 2010) 66 at 67.

[30] See Decision by the Arbitrators, *EC – Hormones (Canada) (Article 22.6 – EC)*, para. 12. See also Renouf, above n 28 at 136–137.

[31] See Sebastian, Thomas, 'The Law of Permissible WTO Retaliation', in C. P. Bown and J. Pauwelyn (eds.), *The Law, Economics and Politics of Retaliation in WTO Dispute Settlement* (New York: Cambridge University Press, 2010) 89 at 101–106.

[32] Ibid. at 108–109; and Renouf, above n 28 at 140–142. See also Decision by the Arbitrator, *US – Gambling (Article 22.6 – US)*, paras. 3.134–3.135.

[33] See Decision by the Arbitrators, *EC – Hormones (Canada) (Article 22.6 – EC)*, para. 41. See the review of other relevant cases in Sebastian, above n 31 at 108; and Renouf, above n 28 at 142–143.

[34] Decision by the Arbitrators, *EC – Hormones (Canada) (Article 22.6 – EC)*, para. 40.

disagreement and eventually eradicate it. Alternatively, if B failed to implement the DSB recommendation, A and/or C would be granted the right to retaliate, which they could utilise until B complies with its WTO transit obligations.

With respect to the last point, the pertinent questions are:

- Which counterfactuals would the arbitrators use to determine the level of nullification or impairment as well as the quantity of retaliation; and
- In circumstances in which a transit dispute (including the Article 22(6) arbitration) is initiated by both A and C, how would the arbitrators allocate the quantity of retaliation between these parties?

For example, would the counterfactual be the full amount of trade in gas between A and C, through the territory of B, lost due to B's transit restriction? Such a loss could be estimated in billions of US dollars per annum, given that the average capacity of a cross-border pipeline is between 20 and 30 Bcm per year and the average price for 1 mcm of gas, for example, in Europe significantly exceeds 350 US dollars.[35]

With respect to import restrictions, the assessment of a permissible retaliation normally aims at calculating the loss of export by the aggrieved party as a result of the illegality.[36] A similar assessment has not yet been conducted with respect to transit restrictions, in particular gas transit.[37] A significant difference between import and transit restrictions lies in the fact that a transit restriction is applied by a third (transit) state rather than an importing Member and may, therefore, equally affect both importing and exporting states.

Another complication in determining the counterfactual in the context of restrictions on third-party access/capacity establishment rights is that this counterfactual would have to assume the existence of available pipeline capacity or a pipeline – something that in reality does not exist

[35] Based on British Petroleum, 'Statistical Review of World Energy' (June 2013), www.bp.com/statisticalreview, accessed 21 August 2013 at 27, and own calculations. See also the examples of pipelines constructed or planned within the framework of the EU's Southern Gas Corridor in Pogoretskyy, Vitaliy, 'The Transit Role of Turkey in the European Union's Southern Gas Corridor Initiative: An Assessment from the Perspective of the World Trade Organization and the Energy Charter Treaty', 2(2) *Journal of International Trade and Arbitration Law* (Istanbul, 2013), 121 (also available at www.ogel.org, accessed 15 February 2014) at 4–8.

[36] See Decision by the Arbitrators, *EC – Hormones (Canada) (Article 22.6 – EC)*, paras. 41–42.

[37] As noted earlier, the only panel decision in the WTO dealing with transit is *Colombia – Ports of Entry*. At the time of writing, there was neither an Article 21(5) nor 22(6) proceeding initiated in this dispute.

and could only result from the negotiations on the conditions on gas transit between the Members concerned. The outcome of these negotiations may or may not necessarily be the most favourable for the complainant(s). The question thus arises as to which outcome (the most or the least) favourable the arbitrators must assume in determining the amount of nullification or impairment.

In *US – Gambling (Article 22.6 – US)*, the Arbitrators articulated the following basic principle addressing this question:

> [T]he counterfactual should, in our view, *reflect at least a plausible or 'reasonable' compliance scenario.* A counterfactual that would assume a compliance scenario that leads to an implausibly high level of nullification or impairment of benefits would lead to a suspension in excess of the level of nullification or impairment actually suffered. Conversely, a counterfactual that would underestimate the level of benefits accruing to the complaining party would risk leading to an unwarranted reduction of the level of suspension below the level that that complaining party is entitled to seek, namely 'equivalence'.[38]

Although this principle may seem too general, it, at least, suggests that the amount of nullification or impairment cannot be lower than the *minimum* level of trade that would have been 'reasonably' expected if the illegality were to be removed. This approach of reliance on the *minimum* benefit available in the counterfactual scenario finds support in some national legal systems. For example, Sebastian, when analysing the same question, cited English contract law principles, according to which:

> [W]here a contract contains a promise that can be performed in alternative ways, damages for breach are assessed on the assumption that the party in breach would have performed in the manner least advantageous to the party seeking compensation.[39]

The lesson he draws from this principle is that 'the retaliating state should not get more than it bargained for [in exchanging WTO commitments].'[40]

[38] Decision by the Arbitrator, *US – Gambling (Article 22.6 – US)*, para. 3.27, emphasis added.

[39] Sebastian, above n 31 at 105.

[40] Ibid. However, see a Separate Opinion of an Arbitrator in *US – Gambling (Article 22.6 – US)* stating that 'in a situation where different means of compliance might form the basis of a counterfactual in order to determine the level of nullification or impairment of benefits, the complaining party would not be prevented from selecting a counterfactual that may lead to a *higher* level of nullification or impairment than others, provided that such counterfactual is reasonable'. This Opinion was not shared by other Arbitrators in

Due to the legal nature of this study, it does not attempt to develop an economic model for determining counterfactuals for the purposes of estimating the level of nullification or impairment caused by restrictions on third-party access/capacity establishment rights in the context of pipeline gas transit. The loss of trade, adjusted to at least the minimum of what the complainants could have reasonably obtained, absent the refusal of the transit state to negotiate the terms and modalities of gas transit, would appear to be a sound basis for determining the counterfactual. In this regard, the loss of trade could be derived from statistics on gas production capacity in the exporting Member; the domestic consumption of the exporting Member; and the demand trend in the potential importing Member(s).[41]

The last question raised is how the level of nullification or impairment (the quantity of retaliation) should be allocated between A and C. It is likely that both Members would be involved in an Article 22(6) arbitration as third parties in each other's proceeding and will have the right to comment on this matter, thereby assisting the arbitrators in answering this question.[42] That said, A and C would have to be treated as separate parties, requesting the suspension of their own concessions granted to B. The arbitrators would thus have to quantify the effects of the measure on A and C separately, and determine the value of retaliation for each of these Members.[43]

However, due to the particular nature of a transit measure, it may be challenging for arbitrators to isolate its effects on imports from those on exports. As noted earlier, this measure could harm both exporters and importers equally. For example, if, as a result of B's transit restriction, the loss of A's gas exports equals 2 billion US dollars per year, the loss of C's gas imports can equal the same figure. This is considering that the actual

this dispute. Decision by the Arbitrator, *US – Gambling (Article 22.6 – US)*, para. 3.70, emphasis added.

[41] This data can be available in the complaining Members. It is also published by various international organisations and enterprises specialised in energy. See International Energy Agency, 'Natural Gas Market Review' (Paris: IEA Publications, 2009); and British Petroleum, 'Statistical Review of World Energy' (June 2013), above n 35 at 22–23.

[42] While the DSU does not specifically address the issue of third-party participation in Article 22 arbitration proceedings, the arbitrators have derived their right to authorise this participation from Article 12(1) of the DSU, which grants discretion to panels to decide on procedural matters not regulated by the Agreement. See Decision by the Arbitrators, *EC – Hormones (Canada) (Article 22.6 – EC)*, para. 7.

[43] See the same approach taken in Decision by the Arbitrator in *US – Upland Cotton (Article 22.6 – US I)*, paras. 4.199–4.202. See also Sebastian, above n 31 at 107–108 (citing Decisions by the Arbitrator in *US – Gambling (Article 22.6 – US)* and *US – Byrd Amendment (Article 22.6)*).

value of trade flow restricted by the measure in a given year is only $2 billion. In this situation, it is somewhat unclear whether the annual level of retaliation consistent with DSU principles would have to be $2 or 4 billion, divided between A and C.

This section has demonstrated that third-party access and capacity establishment are rights that are, in principle, enforceable through the WTO dispute settlement proceedings. The failure to bring a measure restricting these rights into conformity with WTO law can result in concrete practical consequences for a transit state. Apart from reputation damage, these consequences will include heavy retaliatory measures, imposed on the exports of the transit state to the complaining parties, estimated potentially in billions of US dollars. These measures will apply until the illegality is removed – namely until the transit state removes discrimination or engages in good faith negotiations on the terms and modalities of implementing its freedom of transit obligation.

III. How Can the Regulation of Third-Party Access and Capacity Establishment Rights in the WTO Be Improved?

The second main question discussed in this study is how WTO transit rules could be improved through a legislative reform to regulate third-party access and capacity establishment rights better. The possibility of reforming WTO transit rules is discussed here as one of the ways to enhance the role the WTO plays or could play in the international regulation of pipeline gas transit. It was explained in Chapter I that this question is analysed from two perspectives:

- the codification of the already existing principles of general international law relevant to transit – the principles of effective and integrated rights and economic cooperation, the obligation to negotiate in good faith with a view to implementing freedom of transit and *abus de droit* – in the WTO; and
- the expansion of GATS obligations regulating trade in energy services, based on the model of the additional commitments of Ukraine on pipeline transport, or the Telecom Reference Paper.

The latter approach would call for the progressive development of WTO law, encompassing some aspects of competition regulation.[44]

[44] The difference between the notions of 'codification' and 'progressive development' was explained in Chapter I, Section II.A.1.

A. *Codification Approach*

1. Why codification and how it can be implemented

(i) The advantages and disadvantages of 'codification' Chapter I discussed the importance of codification of international law (either within particular treaty regimes, such as the WTO, or a broader framework of the UN) in a manner consistent with the objectives and principles recognised by the international community. Regardless of whether this approach is driven by a corollary of the principle of systemic integration or a general goal of harmony between international law sources as expressed, inter alia, in the preamble to the UN Charter, it would likely ensure the harmonious co-existence of different branches of international law (e.g. trade law, the law of the sea, energy cooperation), as well as different levels of international governance (e.g. multilateral, regional or bilateral). To recall, the method of codification entails the articulation and documentation of the obligations already existing under international law, in particular, general international law.[45]

The question can be asked whether a reform of WTO transit rules through codification is warranted given the conclusion made earlier that transit states already have a number of enforceable obligations in this context. Despite this fact, codification of international law has generally fulfilled a number of important functions within the UN framework.

First of all, any treaty obligation, by its very nature expresses the *actual consent* of states undertaking it.[46] A concrete treaty obligation would thus add legitimacy to a principle of general international law integrated into this treaty (including through the principle of systemic integration), as well as may resolve possible differences between states regarding the precise meaning of this principle.[47] Second, codification ensures the equal participation of all members of the international community (such as all WTO Members) in the articulation of legal obligations that are binding on them. By contrast, not all current WTO Members existed or could engage in relevant 'state practice' (inter alia, because they were not active in the field of energy cooperation or network-bound trade) when certain relevant principles of general international law were being

[45] See Chapter I, Sections II.A.1 and III.B.3.

[46] Villiger, Mark E., *Customary International Law and Treaties: A Manual on the Theory and Practice of the Interrelation of Sources*, 2nd ed. (The Hague: Kluwer Law International, 1997) at 130.

[47] Ibid. at 131, 134.

crystallised.[48] Consequently, the codification of such principles would give an opportunity to all WTO Members to express their views on how these principles should be applied in the WTO legal system. Finally, as it is further explained, due to their broad nature, when principles are applied they may lead to different outcomes in light of specific facts, which may undermine the 'security and predictability' of international obligations – one of the institutional principles of the WTO.[49] Codification may eliminate this uncertainty by placing general principles into a proper and more specific legal framework. One important risk of codification that must be taken into account, however, is that this process may lead to the opposite result of weakening the principle of general international law that is being codified if some WTO Members were to denounce it or question its existence.[50]

If WTO Members were to decide to reform WTO transit obligations to regulate third-party access and capacity establishment rights in a more direct and effective way, based on the codification of principles of general international law discussed in Chapter V, the next questions to address are: (i) how should new (more elaborate) obligations be derived from such principles; and (ii) through which legal mechanisms can the results of this codification be converted into WTO law? These questions are addressed in the following sections.

(ii) The nature of principles and how it could affect the process of codification To understand how a new and more elaborate treaty obligation can be derived from existing principles of general international law, one must recall the nature of 'principles' in a legal system and how 'principles' differ from 'rules'. It was explained earlier, that, unlike rules, which are conclusive norms that apply in an 'all or nothing' manner, principles are general and non-conclusive norms, the application of which individually does not lead to any final or concrete result.[51] Whereas contradicting

[48] Ibid. at 137–138. See the discussion of the contribution of the so-called specially affected states to the development of customary international law in ibid. at 30–33.

[49] See Article 3(2) of the Dispute Settlement Understanding, *WTO Legal Texts*, above n 2.

[50] Villiger, *Customary International Law and Treaties*, above n 46 at 131–132.

[51] See Dworkin, Ronald, *Taking Rights Seriously* (London: Duckworth, 1977) at 35–36; Alexy, Robert, *A Theory of Constitutional Rights*, J. Rivers (trans.) (New York: Oxford University Press, 2010) at 45; Koskenniemi, Martti, 'General Principles: Reflexions on Constructivist Thinking in International Law', in M. Koskenniemi (ed.), *Sources of International Law* (Aldershot: Ashgate Darmouth, 2000) 360 at 370, 373. See Chapter I, Section III.A.

rules exclude or invalidate each other, competing principles can apply simultaneously to the extent that this is factually and legally possible.[52]

This fundamental distinction between principles and rules is important for the analysis of ways to codify the principles of general international law relevant to third-party access and capacity establishment rights in WTO law. Throughout this study, some rules for the application of these principles (i.e. 'conditional relations of precedence') were derived by balancing them against the principle of sovereignty. These rules include the following:

- To operationalise freedom of gas transit, a transit state must engage in cooperation/good faith negotiations with transit-dependent states with a view to reaching an agreement on the terms and modalities of this transit (in this regard, a number of procedural requirements that the parties must fulfil were discussed);
- A legitimate regulation of gas transit must concern the conditions of transit as opposed to the right of transit per se; and
- Transit conditions must comply with the obligations of a transit state under WTO law and general international law (for example, those on 'reasonable' charges and regulations).[53]

It was further explained that, in implementing its freedom of transit obligation, a transit state reserves wide policy space. For example, it may allow third-party access to an existing pipeline or agree to the construction of another pipeline based on specific terms and conditions. The extent of this policy space in turn implies some uncertainty as to how relevant competing principles would be balanced in different factual scenarios and how the outcome of this balancing would affect third-party access and/or capacity establishment rights of stakeholders. This uncertainty threatens to provide an insufficient market signal for investors in gas production and supply chains, hindering the development of an international gas market and keeping the energy security of WTO Members at risk.[54]

[52] See Dworkin, above n 51 at 24–27; Alexy, above n 51 at 47–48; Koskenniemi, 'General Principles', above n 51 at 374.

[53] In her monograph, Melgar also lists a number of similar 'rules', although she refers to them as general 'prohibitions' and 'limitations' imposed on transit and transit-dependent states. See Melgar, Beatriz H., *The Transit of Goods in Public International Law* (Leiden: Brill Nijhoff, 2015) at 325–327.

[54] Jenkins, David, 'An Oil and Gas Industry Perspective', in T. W. Wälde (ed.), *The Energy Charter Treaty: An East–West Gateway for Investment & Trade* (London: Kluwer Law International, 1996) 187 at 191.

Through codification, WTO Members could correct any uncertainty regarding the terms and conditions of gas transit by striking a balance between the interests of transit states and transit-dependent states with respect to third-party access and capacity establishment rights once and for all, and by recording the outcome of this balance in a WTO agreement. The advantage of having rules addressing these issues clearly articulated in a written document is that they would add relative certainty and predictability to the international trade system and provide some guidance to stakeholders on how the obligation to provide freedom of transit must be implemented in the context of pipeline gas.

This is not to suggest that WTO law could possibly regulate all aspects of pipeline gas transit, including some technical modalities of the implementation of third-party access and capacity establishment rights. The full implementation of gas transit obligations would require decisions on a number of complex practical issues, which cannot, and perhaps should not, be resolved at a multilateral level, such as: (i) precise technical conditions of third-party access i (e.g. the allocation of pipeline capacity, pipeline transportation fees, gas quality specification); and (ii) conditions relating to capacity establishment (e.g. project financing, procurement of various permits, licences and land easements, compliance with safety and environmental standards and pipeline operation). As explained in Chapter III, all these issues have been historically dealt with at bilateral or regional levels.

Moreover, there are separate legal regimes regulating the aspects of pipeline gas transit that fall beyond international trade. For example, the law of the sea addresses the rights of states to lay pipelines on the continental shelf and EEZ.[55] Various bilateral and regional investment treaties as well as the ECT regulate investment protection.[56] Some regional regimes, such as the EU-led Energy Community Treaty, regulate international competition policies.[57]

When the time is ripe, some of the aforementioned rules, such as investment and competition rules, could be gradually incorporated into WTO law.[58] However, in the short term, in the specific area of

[55] On the LOSC, see this chapter, Section III.A.3.

[56] On the ECT, see this chapter, Section III.A.2

[57] See Articles 18 and 19 of the 2005 Treaty Establishing the Energy Community, Energy Community Secretariat, Energy Community Treaty, http://www.energy-community.org, accessed 12 September 2013.

[58] On the prospects and benefits of incorporating investment rules in the WTO framework, see Sauvé, Pierre, 'Multilateral Rules on Investment: Is Forward Movement Possible?', 9

international gas trade, WTO Members could elaborate further the rules on the application of the *existing principles of international law* discussed in this study to the particular features of this trade. This would contribute to improving and maintaining Members' competitive opportunities in the international gas market.

Due to the broad nature of these principles, it would be impractical to propose a specific wording of their codification in the WTO legal system. This wording would naturally reflect the author's subjective view on what could be a fair balance between the interests of transit states and transit-dependent states. In the end, to the extent that WTO Members do not deviate from these principles, there could be different ways to codify them. By way of a mere illustration, the following sections analyse how these principles have been addressed in other legal regimes, namely the ECT and the LOSC. Some of the legislative approaches adopted under these treaties can be transposed to WTO law. These legal regimes were chosen because they deal with the rights of access to and use of transit infrastructure (including some aspects of third-party access and capacity establishment rights) in a manner more specific than WTO law. They also reflect the consensus among a large group of international community members on how these rights must be regulated at a multilateral level.[59]

(iii) Institutional mechanisms through which principles of general international law could be converted into WTO law Before concluding this general section, it is worth mentioning that there are different institutional mechanisms available in the WTO through which the results of the aforementioned codification could be converted into WTO law. They include:

- Amendment of the text of WTO agreements (including main text, footnotes, or an explanatory Ad Note);
- Modifications of schedules annexed to the GATT and the GATS;
- Decisions of the Ministerial Conference supplementing or modifying the WTO agreements, in the form of waivers, authoritative interpretations and other decisions; and

Journal of International Economic Law (2006) 325; and González, Anabel, 'The Rationale for Bringing Investment into the WTO', in S. J. Evenett and A. Jara, *Building on Bali: A Work Programme for the WTO* (A VoxEU.org eBook, Centre for Economic Policy Research, 2013) 67.

[59] At the moment of writing, the LOSC and the ECT had 166 and 54 Contracting Parties, respectively. See Appendix 2.

- Incorporation of new agreements, including plurilateral agreements, into WTO law.

Without delving into particular aspects of these institutional mechanisms, which have been discussed elsewhere,[60] it would suffice to note that they have different levels of procedural complexity and implications for WTO Members' rights and obligations. Among options to reform WTO rules relevant to energy transit that have been discussed in the academic literature are the adoption of a sectoral Framework Agreement on Energy, a Reference Paper on Energy Services and an interpretative instrument clarifying rights and obligations under GATT Article V.[61]

2. The model of third-party access and capacity establishment rights under the ECT

The origin of the Energy Charter and the ECT as well as some aspects of the ECT's regulation of third-party access and capacity establishment rights were explained in Chapter III.[62] The current chapter analyses these rights in a more comprehensive manner, including various conditions and requirements attached to the implementation of these rights.

As explained in Chapter I, this study does not deal with different safeguard mechanisms protecting the already established gas flows.

[60] See Nottage, Hunter and Sebastian, Thomas, 'Giving Legal Effect to the Results of WTO Trade Negotiations: An Analysis of the Methods of Changing WTO Law', 9 *Journal of International Economic Law* (2006), 989.

[61] See Lamy, Pascal, *The Geneva Consensus: Making Trade Work for All* (New York: Cambridge University Press) at 117–118; Marceau, Gabrielle, 'The WTO in the Emerging Energy Governance Debate', in J. Pauwelyn (ed.), *Global Challenges at the Intersection of Trade, Energy and the Environment* (Geneva: The Graduate Institute, Centre for Trade and Economic Integration, 2010) 25 at 38–40; Cottier, Thomas, et al., 'Energy in WTO Law and Policy', NCCR, Working Paper No 2009/25 (May 2009), http://phase1.nccr-trade.org/images/stories/projects/ip6/IP6%20Working%20paper.pdf, accessed 30 November 2013; Poretti, Pietro and Rios-Herran, Roberto, 'A Reference Paper on Energy Services: The Best Way Forward?', 4 *Oil, Gas & Energy Law Intelligence* (2006), www.ogel.org, accessed 15 February 2014; Evans, Peter C., 'Strengthening WTO Member Commitments in Energy Services: Problems and Prospects', in P. Sauvé and A. Mattoo (eds.), *Domestic Regulation and Services Trade Liberalization* (Washington, DC: World Bank Publications 2003) 167 at 177–185; Nartova, Olga, *Energy Services and Competition Policies under WTO Law* (Moscow: Infra-M, 2010) at 234–257; and Wälde, Thomas W. and Gunst, Andreas J., 'International Energy Trade and Access to Energy Networks', 36 *Journal of World Trade* (2002) 191 at 217. These proposals, however, do not make specific suggestions on how to resolve the problem of access to gas transit infrastructure in the context of trade in goods.

[62] See Chapter III, Section III.C.2.

These mechanisms include the prohibition on interrupting or reducing the existing energy flows, established by Articles 7(5) and 7(6) of the ECT and the conciliation procedures set up to resolve contractual disputes arising from gas transit under Article 7(7) of the ECT.[63] By contrast, if a dispute were to concern the application or interpretation of rights or obligations under the ECT relating to third-party access and capacity establishment, this dispute would likely fall within the scope of a state-to-state arbitration under Article 27 (Settlement of Disputes between Contracting Parties).[64]

Finally, this section does not discuss negotiations on the ECT Transit Protocol. These negotiations are still on-going and do not appear to have produced results that WTO Members could readily accept as a model for an international regulation of third-party access and capacity establishment rights.[65]

(i) General background The major distinction between the ECT and WTO law is that the former governs energy transit in a more specific and direct way. Having said this, the ECT does not derogate from WTO rules, nor does it introduce rights with respect to third-party access or capacity establishment that do not exist under the GATT. As explained throughout this study, similar rights could be derived from GATT Article V:2 when read in the light of general principles of international law. What the ECT does achieve with respect to third-party access and capacity establishment rights is its own particular balance between these rights, by setting forth detailed rules on their application.[66] This approach to regulate third-party access and capacity establishment in a direct way is probably the reason some commentators commended the ECT on being

[63] Energy Charter Secretariat, 'The Energy Charter Treaty and Related Documents: A Legal Framework for International Energy Co-operation' (Brussels: Energy Charter Secretariat, 2004). See the scope of this study explained in Chapter I, Section II.B.

[64] Ehring, Lothar and Selivanova, Yulia, 'Energy Transit', in Y. Selivanova (ed.), *Regulation of Energy in International Trade Law: WTO, NAFTA and Energy Charter* (The Netherlands: Kluwer Law International, 2011) 49 at 92; Wälde and Gunst, above n 61 at 214.

[65] See Chapter I, Section II.B. On the detailed overview of the negotiations on the Transit Protocol and its key provisions, see Ehring and Selivanova, above n 64 at 95–100; and Konoplyanik, Andrei A., 'A Common Russia–EU Energy Space (The New EU–Russia Partnership Agreement, *Acquis Communautaire*, the Energy Charter and the New Russian Initiative)', 7 *Oil, Gas & Energy Law Intelligence* (2009), www.ogel.org, accessed 15 February 2014.

[66] See Pogoretskyy, 'The Transit Role of Turkey', above n 35 at 18–19.

an innovative treaty that fills a textual gap in the GATT's regulation of issues essential to effective energy transit.[67]

A number of provisions in the ECT link this treaty with WTO law in the area of transit regulation.[68] Article 4 of the ECT explicitly establishes that '[n]othing in this Treaty [ECT] shall derogate, as between particular Contracting Parties which are members of the WTO, from the provisions of the WTO Agreement as they are applied between those Contracting Parties'.[69] Article 29 of the ECT (Interim Provisions on Trade-Related Matters) and the ECT Trade Amendment incorporate a number of WTO trade rules, including GATT Article V, into the ECT's legal framework, extending their application to ECT Members that have not yet joined the WTO, such as Azerbaijan, Belarus, Bosnia and Herzegovina, Turkmenistan and Uzbekistan.[70] Paragraph 3(a) of Annex D to the ECT (Interim Provisions for Trade Dispute Settlement) provides that, in ruling on any WTO provision incorporated in the ECT, an ECT panel must be guided by the interpretations given to the WTO Agreement within the WTO framework.[71] Thus, pursuant to Annex D, when interpreting GATT Article V, ECT panels would have to take into account clarifications given to this provision by WTO panels. Finally, ECT Article 7(1) (Transit) establishes an overarching obligation of each Member to

[67] See Clark, Bryan, 'Transit and the Energy Charter Treaty: Rhetoric and Reality', 5 *Web Journal of Current Legal Issues* (1998) at 3; Ehring and Selivanova, above n 64 at 82; Rakhmanin, Vladimir, 'Transportation and Transit of Energy and Multilateral Trade Rules: WTO and Energy Charter', in J. Pauwelyn (ed.), *Global Challenges at the Intersection of Trade, Energy and the Environment* (Geneva: The Graduate Institute, Centre for Trade and Economic Integration, 2010) 123 at 124; and Wälde and Gunst, above n 61 at 213.

[68] However, see contra in Azaria, Danae, 'Energy Transit under the Energy Charter Treaty and the General Agreement on Tariffs and Trade', 27 *Journal of Energy & Natural Resources Law* (2009) 559 at 560. This author argues that GATT Article V would play a limited role in the interpretation of ECT transit rules, since these rules do not explicitly refer to the GATT and the object and purpose of both treaties differ, as do their ethos and practices.

[69] See Energy Charter Treaty, Article 4, as modified by Article 2 of the Trade Amendment, which entered into force on 21 January 2010, above n 63. See also Energy Charter Secretariat, '1998 Trade Amendment', http://www.encharter.org/index.php?id=608&L=0, accessed 1 December 2013.

[70] It is noteworthy that, pursuant to paragraph B(10)(b) of Annex W to the Trade Amendment (Exceptions and Rules Governing the Application of the Provisions of the WTO Agreement), transit provisions of the Agreement on Trade Facilitation (when it enters into force) would apply among ECT Members, unless one of them requests that the Charter Conference (a political body of the Energy Charter) disapply or modify it. See ECT Article 29 and the Trade Amendment. Energy Charter Treaty, above n 63

[71] Annex D to the ECT and Article 3 of the Trade Amendment, ibid.

'take the necessary measures to facilitate the Transit of Energy Materials and Products *consistent with the principle of freedom of transit*'.[72] The negotiating history of the ECT clarifies that this language should be understood as an indication that ECT Article 7 was based on the basic principles of GATT Article V.[73] In light of this, it is likely that ECT transit rules would be interpreted consistently with analogous WTO provisions and that GATT Article V would be used as context for ECT Article 7 within the meaning of Article 31(1) of the VCLT.[74]

Before turning to the provisions of the ECT regulating third-party access and capacity establishment rights, a few remarks regarding specific terminology used in the Treaty must be made. The definition of 'transit' under the ECT is not substantially different from that under GATT Article V, although it contains certain exceptions and particular features. Article 7(10)(a) defines this term as:

> (i) the carriage through the Area of a Contracting Party, or to or from port facilities in its Area for loading or unloading, of Energy Materials and Products originating in the Area of another state and destined for the Area of a third state, so long as either the other state or the third state is a Contracting Party [and] (ii) the carriage through the Area of a Contracting Party of Energy Materials and Products originating in the Area of another Contracting Party and destined for the Area of that other Contracting Party, *unless the two Contracting Parties concerned decide otherwise and record their decision by a joint entry in Annex N.*[75]

It can be observed that the term 'transit' under the ECT covers the same transit scenarios as transit under GATT Article V.[76]

[72] Ibid., emphasis added.

[73] See Ehring and Selivanova, above n 64 at 83; as well as European Energy Charter Conference Secretariat, 'Note from the Chairman of Working Group II', Annex II, 8/91 (BP 2), Brussels, 12 September 1991; European Energy Charter Conference Secretariat, Working Group II Meeting, Chaired by Slater of UK, 17–18 October 1991. The preparatory work of the ECT also clarifies that the negotiators initially explored the possibility of including an explicit reference to GATT Article V in ECT Article 7 (then Article 8). The main reason they decided not to do so is because this would imply that other provisions of the ECT that do not refer to the GATT were intended to derogate from the latter. See European Energy Charter Conference Secretariat: Comments on Issues Assigned to the Legal Sub-group, No. 1396, 14 May 1993 at 3; and Organisation for Economic Co-operation and Development (OECD), 'Memorandum', 27 May 1993 at 3.

[74] Vienna Convention on the Law of Treaties of 1969, 1155 *United Nations Treaty Series* (1980) 331.

[75] Energy Charter Treaty, above n 63, emphasis added.

[76] See Chapter II, Section II.A and figure 2.

The important exception here is that the ECT allows its Members (who are not Members of the WTO within the meaning of Article 4) to exclude transit originating and destined for the 'Area' of one and the same Member from the scope of Article 7. Another particular feature of the ECT's definition of 'transit' is that the term 'Area' that it employs, at first glance, appears to be broader than the term 'territory' used in the GATT. The term 'Area' refers not only to the *territory* that includes land, internal waters and the territorial sea, but also to the sea, seabed and subsoil with regard to which a Member exercises jurisdiction (that is, continental shelf and EEZ).[77] Such a broad application of the Treaty is, however, tempered by a number of caveats prioritising the international law of the sea governing submarine cables and pipelines (established, inter alia, by the LOSC and customary international law) over ECT transit rules.[78] In addition, it is noteworthy that the text of the ECT's provision regulating capacity establishment rights, Article 7(4), does not mention the term 'Area', which implies that the rights/obligations under this provision do not extend beyond the *territory* of a transit state.[79] These caveats to the regulation of transit in the Area of ECT's Members can be explained by the fact that, whereas under the law of the sea transit via submarine pipelines on the continental shelf and EEZ is technically subject to very limited sovereign rights, transit via overland pipelines is subject to much broader sovereignty.[80]

One last point to mention with respect to the ECT's terminology is that, unlike WTO law, the ECT explicitly refers to transit via 'Energy Transport Facilities', which it defines, inter alia, as 'high-pressure gas

[77] ECT Article 1(10) defines the term 'Area' of a Contracting Party as '(a) the territory under its sovereignty, it being understood that territory includes land, internal waters and the territorial sea'; and '(b) subject to and in accordance with the international law of the sea: *the sea, sea-bed and its subsoil with regard to which that Contracting Party exercises sovereign rights and jurisdiction*'. Energy Charter Treaty, above n 63, emphasis added.

[78] See ECT Articles 1(10)(b), 7(8), and Declarations with respect to Article 7, ibid.

[79] ECT Articles 7(4), ibid. As the negotiating history of the ECT indicates, the drafters supported the proposal of Norway that Article 7(4) would only concern transit within the '*territory*' of Contracting Parties, leaving that on the continental shelf to the law of the sea. European Energy Charter Conference Secretariat, Room Document 26, Plenary Session, Brussels, 28 April 1993 at 6.

[80] See Belyi, Andrei, Nappert, Sophie and Pogoretskyy, Vitaliy, 'Modernizing the Energy Charter Process? The Energy Charter Conference Road Map and the Russian Draft Convention on Energy Security', 29(3) *Journal of Energy & Natural Resources Law* (2011) 383 at 393; Vinogradov, Sergei, 'Challenges of Nord Stream: Streamlining International Legal Frameworks and Regimes for Submarine Pipelines', 9 *Oil, Gas & Energy Law Intelligence* (2011), www.ogel.org, accessed 15 February 2014 at 40–47.

transmission pipelines', and transit of 'Energy Materials and Products' including natural gas.[81]

(ii) Third-party access The first provision of the ECT relevant to third-party access and capacity establishment rights is Article 7(1). It was noted earlier that this provision requires ECT Members 'to take the necessary measures to facilitate the Transit of Energy Materials and Products consistent with the principle of freedom of transit'.[82] There has been a debate among scholars on whether Article 7(1) establishes a weaker or stronger obligation than GATT Article V:2 (first sentence).[83] Roggenkamp, for example, contends that 'the wording used in the ECT is not as strong as in Article V GATT'.[84] Other analysts argue that Article 7 goes considerably beyond the GATT by requiring a positive action to 'take the necessary measures to facilitate the transit . . . consistent with the principle of freedom of transit'.[85] Given the fact that ECT transit rules must be interpreted in light of GATT Article V, it is unlikely that the obligation under ECT Article 7(1) could somehow diminish the freedom of transit obligation under the GATT, as it is applicable among WTO Members.

The next question to address is what type of 'measures' relating to third-party access (as well as to other aspects of transit) a transit state

[81] See ECT Articles 1(4), 7(4), 7(10)(b) and Annex EM, above n 63. As explained at the outset of this study, although in the course of the WTO Doha negotiations on trade facilitation there was a proposal to include the term 'pipeline' in the definition of 'traffic in transit' in GATT Article V:1 and the new Agreement on Trade Facilitation, this term was removed from the final text of the latter Agreement. See Chapter II, Section III.B.

[82] This provision reads as follows: 'Each Contracting Party shall take the necessary measures to facilitate the Transit of Energy Materials and Products consistent with the principle of freedom of transit and without distinction as to the origin, destination or ownership of such Energy Materials and Products or discrimination as to pricing on the basis of such distinctions, and without imposing any unreasonable delays, restrictions or charges.' Energy Charter Treaty, above n 63.

[83] GATT Article V:2 provides: '[t]here shall be freedom of transit through the territory of each contracting party, via the routes most convenient for international transit, for traffic in transit.' GATT 1994, *WTO Legal Texts*, above n 2.

[84] Roggenkamp, Martha M., 'Transit of Network-Bound Energy: The European Experience', in T. W. Wälde (ed.), *The Energy Charter Treaty: An East–West Gateway for Investment & Trade* (London: Kluwer Law International, 1996) 499 at 509. See a similar view in Azaria, Danae, *Treaties on Transit of Energy via Pipelines and Countermeasures* (Oxford: Oxford University Press, 2015) at 68–69.

[85] See Ehring and Selivanova, above n 64 at 83–84; Wälde and Gunst, above n 61 at 213.

must take to facilitate gas transit within the meaning of Article 7(1). These measures are set out in the relevant context of this provision.[86]

With respect to third-party access rights, the ECT, in Understanding IV:1(b)(i), states explicitly that Members do not have an obligation to introduce this right in a mandatory way.[87] Yet, if a Member (transit state) chooses to do so, inter alia, under its national law liberalising the gas market,[88] or if that Member grants a negotiated third-party access right to an individual transit-dependent state, it would have to comply with certain rules of the ECT.

The first such rule is the non-discrimination obligation with respect to Energy Materials and Products in transit. In particular, Article 7(1) of the ECT prohibits a 'distinction as to the origin, destination or ownership of such Energy Materials and Products or discrimination as to pricing on the basis of such distinctions'.[89] As the term 'distinction' indicates, Article 7(1) draws on a similar obligation under GATT Article V:2 (second sentence), which also uses this term.[90] The fact that the ECT drafters did not intend this treaty to depart from the GATT and the fact that the non-discrimination obligations under both Articles V:2 and 7(1) are similarly worded suggest that the normative content of these provisions should be substantially the same. Consequently, the interpretation of GATT Article V:2 (second sentence) provided in previous chapters should apply *mutatis mutandis* to Article 7(1). One important aspect of

[86] Ehring and Selivanova, above n 64 at 85; Liesen, Rainer, 'Transit under the 1994 Energy Charter Treaty', 17 *Journal of Energy and Natural Resources Law* (1999) 56 at 64.

[87] Energy Charter Treaty, above n 63. This Understanding accommodated the then concern of the EU and its gas industry (EUROGAS) that a mandatory third-party access right under ECT Article 7 would contradict the EU gas transit directive. See European Energy Charter Conference Secretariat, 'Basic Agreement for the European Energy Charter (Draft)', 31/92 (BA 13), 19 July 1992; European Energy Charter Conference Secretariat, 'Revised Draft', Annex, 15/93 (BA-35), Brussels, 9 February 1993 at 37, 39; and EUROGAS, 'Comment on Revised Draft Basic Agreement for the European Energy Charter', S/EUR/93/172, 8 February 1993 at 1.

[88] As noted earlier, during the negotiations on the ECT, it was also suggested that, in practice, a third party in the sending or receiving country could have third-party access to the transit state's pipeline on the basis of that transit state's legislation. See European Energy Charter Secretariat, 'Note for the Attention of S. Fremantle by C. Jones' 583 (CLJ/JI) (Brussels, 30 August 1993).

[89] Energy Charter Treaty, above n 63.

[90] To recall, GATT Article V:2 (second sentence) provides: 'No *distinction* shall be made which is based on the flag of vessels, the place of origin, departure, entry, exit or destination, or on any circumstances relating to the ownership of goods, of vessels or of other means of transport.' GATT 1994, *WTO Legal Texts*, above n 2, emphasis added.

this interpretation was that GATT Article V:2 (second sentence) covers both MFN and national treatment standards.[91]

According to Article 7(1), there must not be discrimination as to pricing *based on* the origin, destination or ownership of energy.[92] Arguably, Article 7(1) refers, inter alia, to the pricing of energy transportation services provided in the territory of a transit state. Logically, however, 'discrimination' resulting from distinctions of a purely commercial nature, which are origin-neutral and *are not based* on the previously listed factors (i.e. origin, destination or ownership of energy), must not fall afoul of Article 7(1). The same was argued earlier with respect to the meaning of a non-discrimination obligation under GATT Article V:2.[93]

Article 7(3) of the ECT establishes another non-discrimination obligation, which provides as follows:

> Each Contracting Party undertakes that its provisions relating to transport of Energy Materials and Products and the use of Energy Transport Facilities shall treat Energy Materials and Products in Transit in no less favourable a manner than its provisions treat such materials and products *originating in or destined for its own Area*, unless an existing international agreement provides otherwise.[94]

Article 7(3) appears to elaborate on the basic non-discrimination obligation under Article 7(1) by explicitly stating that Members' provisions relating to energy transport and the use of energy transport facilities (such as pipelines) must not discriminate against energy in transit. As Azaria suggests, the term 'provisions relating to transport' may refer to provisions regulating transportation tariffs.[95]

The text of Article 7(3), by comparing energy in transit with that 'originating in' or 'destined for' the Area of a transit state, clearly outlaws discrimination against transit as compared to the treatment of exported or imported energy.[96] Nevertheless, there have been different views in the literature as to whether this non-discrimination standard also extends to the comparison between energy in transit and domestic (internal) traffic. One author noted in this regard that: '[i]t is this ambiguity that makes all the difference for Russia, where the state-regulated internal gas

[91] See Chapter IV, Section III.B.1(i). [92] Energy Charter Treaty, above n 63.
[93] See Chapter IV, Section III.B.1(ii).
[94] Energy Charter Treaty, above n 63, emphasis added.
[95] Azaria, 'Energy Transit under the Energy Charter Treaty and the General Agreement on Tariffs and Trade', above n 68 at 581.
[96] Ehring and Selivanova, above n 64 at 87.

transportation tariffs are at the moment much lower than they should be in order to be cost-based, i.e., to cover actual expenses and guarantee the reasonable rate of return on investments.'[97] This situation could be one of the reasons why, in the negotiations on the ECT Transit Protocol, the Russian position has been that the obligation under Article 7(3) does not refer to domestic traffic.[98]

Some scholars have pointed out that the meaning of the obligation under Article 7(3) turns on whether the conjunction 'or', separating the phrases 'originating in' and 'destined for', is inclusive (that is, it captures energy that may both 'originate in' and be 'destined for' the Area of a transit state) or exclusive (that is, either 'originating in' or 'destined for').[99] Scholars who advocate for the inclusive meaning of the conjunction 'or' rely on its ordinary meaning and the structure of the provision in question. Ehring and Selivanova, for example, contend that 'the New Shorter Oxford Dictionary which the WTO Appellate Body uses habitually ... recognizes that the word "or" can have an inclusive as well as an exclusive meaning'; they note that the Appellate Body explicitly acknowledged this point in *US – Line Pipe*.[100] In the view of these authors, the text of the provision itself points to its following interpretation:

> [T]he two concepts 'originating in its own Area' and 'destined for its own Area' do not exclude each other. Rather, 'products originating in its own Area' conceptually *includes* 'products destined for its own Area', just as it includes products not destined for that Area that instead are export-bound. Likewise, 'products destined for its own area' conceptually *includes* 'products originating in its own Area', just as it includes products not originating in that Area that instead are imports.[101]

[97] Ibid. (citing T. Shtilkind).

[98] See Nychay, Nadiya and Shemelin, Dmitry, 'Interpretation of Article 7(3) of the Energy Charter Treaty', the Graduate Institute of International and Development Studies, Trade and Investment Law Clinic Papers (Geneva 2012), http://graduateinstitute.ch/home/research/centresandprogrammes/ctei/projects-1/trade-law-clinic.html, accessed 12 September 2013 at 3 (citing the Russian proposal on the new Understanding on Article 10 of the draft Transit Protocol).

[99] Ehring and Selivanova, above n 64 at 87.

[100] See Ibid. at 88, footnote 133; Oxford Dictionaries, http://oxforddictionaries.com, accessed 30 March 2012; and Appellate Body Report, *US – Line Pipe*, para. 163.

[101] Ehring and Selivanova, above n 64 at 88, emphasis in original. They give the following example of the inclusive meaning of the term 'or': 'If the dress code for a party is to wear a tuxedo or black, nobody would imagine that this could exclude black tuxedos and that men must wear either a tuxedo or black, but not a black tuxedo.' Ibid.

Ehring and Selivanova also state that, had the drafters intended that Article 7(3) prohibited discrimination only between traffic in transit and export or import, they would explicitly say so.[102] This point is supported by the negotiating history of the ECT, which shows that drafters insisted on specific and clear solutions to the problematic meaning of particular obligations in the draft texts of the ECT. As an example, an explicit recognition that the mandatory third-party access be excluded has already been mentioned. In addition, under Article 7(10)(a)(ii) and Annex N, certain ECT drafters expressly exempted themselves from the application of Article 7 to transit between two (as opposed to three) states. The explicit statement in Article 7(9) that ECT transit obligations must not be interpreted so as to oblige any Member who does not have a certain type of transit facilities to take any positive measure under Article 7 is also illustrative.[103] In light of these provisions, the fact that the language of Article 7(3) was left broad indicates that the drafters did not intend a restrictive interpretation of this provision.

Finally, Ehring and Selivanova argue that a broad (inclusive) reading of the term 'or' in Article 7(3) would be in line with the object and purpose of this Article to provide for a non-discriminatory treatment of traffic in transit; this purpose would be better achieved if domestic traffic were not to be excluded from the comparison.[104] A similar point was made earlier with respect to the meaning of the non-discrimination obligation under GATT Article V:2 (second sentence).[105] Some other scholars support this interpretation of Article 7(3).[106]

This view appears more convincing than the arguments in favour of the exclusive meaning of the term 'or' in Article 7(3) and, therefore, a narrow interpretation of this provision, covering only discrimination between energy transit and import or export. In particular, the proponents of such an interpretation, Nychay and Shemelin,[107] find support for the exclusive meaning of the term 'or' in the judgments of some national courts. These authors, however, conceded that the interpretation of this

[102] Ibid. at 89. [103] Energy Charter Treaty, above n 63.

[104] Ehring and Selivanova, above n 64 at 89. Nychay and Shemelin appear to agree with this argument, Nychay and Shemelin, above n 98 at 20.

[105] See Chapter IV, Section III.B.1(i).

[106] For example, see Azaria, 'Energy Transit under the Energy Charter Treaty and the General Agreement on Tariffs and Trade', above n 68 at 581; Nychay and Shemelin cite Cameron. Nychay and Shemelin, above n 98 at 3.

[107] Although these authors acknowledge that both inclusive and exclusive interpretations of the term 'or' are feasible, they seem to conclude that the latter is better supported. Nychay and Shemelin, above n 98 at 30–31.

term in national jurisdictions has not been consistent and was often based on the object and purpose of the provision in question (which in this case is to facilitate transit as well as to provide for a non-discriminatory treatment of traffic in transit).[108]

In addition, on the basis of certain provisions in regional trade agreements regulating the 'rules of origin' (such as NAFTA Article 401), Nychay and Shemelin argue that the term 'originating in' is the term of art which is used mainly in the context of cross-border transactions (i.e. export/import), as opposed to the internal movement of goods.[109] Admittedly, it is difficult to see how a provision, the key function of which is to assist customs officers in their determination of a customs duty that should be imposed on a particular imported product, is relevant for interpreting the meaning of transit obligations under ECT Article 7(3), especially given that traffic in transit must be exempted from such duties. Moreover, it is questionable that the term 'originating in' is in fact a term of art (that is, a term that has a 'special meaning'). According to VCLT Article 31(4), a party asserting that a particular term has a 'special meaning', displacing the ordinary one in the sense of Article 31(1) of the VCLT, bears the burden of substantiating this assertion.[110] The evidence Nychay and Shemelin provide in support of their proposition that the term 'originating in' is a special term – a term of art – is less than convincing. The 'special meaning' of terms is normally set out in definitional sections in treaties.[111] Article 1 of the ECT, which defines the key terms in this treaty, does not address the term 'originating in'.[112]

One important limitation that Article 7(3) imposes on its non-discrimination obligation is that it applies 'unless an existing international agreement provides otherwise'.[113] This provision appears to refer broadly to *any* international agreement, as opposed to an agreement between states involved in a particular traffic of energy in transit, or an

[108] Ibid. at 9–10. [109] Ibid. at 12–13.

[110] VCLT Article 31(4) reads: 'A special meaning shall be given to a term *if it is established* that the parties so intended.' Vienna Convention on the Law of Treaties, above n 74, emphasis added. Weeramantry states in this regard: 'The ILC considered whether Article 31(4) served any purpose because some members considered the special meaning would be apparent from the context of the terms subject to interpretation. Nevertheless, it was considered that there was a certain utility in including the rule "if only to emphasise that the burden of proof lies on the party invoking the special meaning of the term".' Weeramantry, Romesh, J., *Treaty Interpretation in Investment Arbitration* (Oxford: Oxford University Press, 2012) at 95 (citing the negotiating history of the VCLT).

[111] Weeramantry, above n 110 at 96 (citing inter alia the *Aguas del Tunari* tribunal).

[112] Energy Charter Treaty, above n 63. [113] Ibid.

agreement that preceded the conclusion of the ECT. Among such agreements could be regional or bilateral agreements (that is, transit agreements or general economic agreements) that may provide for a more favourable treatment of the internal traffic between the constituent territories than the treatment of traffic in transit. One such agreement appears to be the Treaty on the Eurasian Economic Union. As explained earlier, this Treaty establishes the obligation for its Members to ensure a negotiated third-party access right to gas pipeline infrastructure in their territories, on a non-discriminatory basis, which, however, is only applicable to entities of constituent Members and internal gas traffic.[114]

In other words, should a group of ECT Members conclude an international agreement establishing special rules on gas transportation that may be more favourable than those applicable to 'traffic in transit' originating in or destined for territories outside those Members, Article 7(3) would appear to exempt the aforementioned special rules from the general non-discrimination obligation under this provision. In a way, this exception is similar to that established by GATT Article XXIV (Customs Unions and Free-trade Areas), which exempts certain types of regional trade agreements formed by WTO Members from the general MFN obligation under the GATT. Importantly, however, this exception pursues greater policy goals (that is, a general expansion of trade) than merely providing for transit and is subject to a number of requirements, including that duties and other restrictive regulations of commerce be eliminated between the constituent territories of a regional trade agreement on substantially all trade.[115] By contrast, Article 7(3) only concerns transit and does not subject an 'international agreement' to similar requirements.

In addition to non-discrimination obligations under Articles 7(1) and 7(3), a transit state granting a third-party access right has an obligation, under Article 7(1), not to impose any unreasonable delays, restrictions or charges on energy in transit.[116] As the text of Article 7(1)

[114] See Article 83(4) and Annex 22 of the Treaty on the Eurasian Economic Union of 29 May 2014 (Договор 'О Евразийском Экономическом Союзе'), http://www.economy.gov.ru/wps/wcm/connect/economylib4/mer/about/structure/depsng/agreement-eurasian-economic-union, accessed 4 July 2014.

[115] See Van den Bossche, Peter and Zdouc, Werner, *The Law and Policy of the World Trade Organization*, 3rd ed. (New York: Cambridge University Press, 2013) at 651–664; Chapter IV, Section V.A.2.

[116] Energy Charter Treaty, above n 63.

indicates, it was likely modelled on GATT Articles V: 3 and V:4, and should therefore be interpreted in light of the relevant requirements in the GATT.[117]

(iii) Capacity establishment Article 7(4) is of key relevance to capacity establishment rights under the ECT. It provides that, in the event that gas transit cannot be achieved on commercial terms by means of energy transport facilities (such as pipelines), a transit state must not place obstacles in the way of new capacity being established.[118] This provision plays an important role in facilitating gas transit within the meaning of Article 7(1) of the Treaty. Otherwise, in the absence of a compulsory third-party access right, the establishment of pipeline gas transit via the Area of a transit state would depend on the discretion of the transit state, which would undermine the effectiveness of the principle of freedom of transit recognised by the ECT.

This provision, however, has qualifications. First of all, the second part of Article 7(4) stipulates that the obligation not to 'place obstacles' is subject to applicable national legislation which, however, must be consistent with Article 7(1) (that is, the obligation of a transit state to take the necessary measures to facilitate transit in line with the principles of freedom of transit and non-discrimination).[119] Articles 7(4) and 7(1), when read together, make clear that the right of a transit state to legislate within the meaning of Article 7(4) is not unlimited and must concern measures that aim at fulfilling certain legitimate objectives. In this regard, Understanding IV.8 with respect to Article 7(4) clarifies that '[t]he applicable legislation would *include* provisions on environmental protection, land use, safety, or technical standards'.[120] The examples of the regulatory areas mentioned in the Understanding confirm that the legislative acts of a transit state may only pursue legitimate goals. The word 'include' in the Understanding means that the list of such goals is not exhaustive.

Second, Article 7(5) of the ECT provides that a transit state is not required to permit the construction or modification of Energy Transport Facilities or

[117] See the discussion of GATT Articles V:3 and V:4 in Chapter IV, Sections III.A.3 and V.A.1

[118] Energy Charter Treaty, above n 63.

[119] Article 7(4) reads as follows: 'In the event that Transit of Energy Materials and Products cannot be achieved on commercial terms by means of Energy Transport Facilities the Contracting Parties shall not place obstacles in the way of new capacity being established, except as may be otherwise provided in applicable legislation which is consistent with paragraph (1).' Ibid.

[120] Ibid., emphasis added.

permit new or additional transit through the existing facilities 'which it demonstrates' to the other Member concerned 'would endanger the security or efficiency of its energy systems, including the security of energy supply'.[121] In this way, Article 7(5) appears to recognise that the integrity of the pipeline system can be considered a legitimate reason for denying the right to establish additional capacity in a transit state's territory. Importantly, the phrase 'it demonstrates' indicates that, as in the case of any exception, Article 7(5) places the burden of proof on the party invoking it.

Apart from these explicit qualifications to the obligation under Article 7(4), there are additional, subtle qualifications that define and delimit the scope of what a transit state is actually obliged to do (or more accurately obliged not to do) under this provision. It is noteworthy that, during the negotiations on Article 7(4), a number of parties (such as Japan, Poland and Kazakhstan) were concerned that, under ECT transit rules and in particular Articles 7(1) and 7(4), parties that do not have transit facilities could be compelled to take *positive* measures to build them.[122] This concern appears to be reflected in Article 7(9) of the Treaty, which states that Article 7 'shall not be so interpreted as to oblige any Contracting Party which does not have a certain type of Energy Transport Facilities used for Transit *to take any measure* under this Article with respect to that type of Energy Transport Facilities'.[123] Notably, in the following sentence, the same provision makes clear that a transit state must comply with its obligations under Article 7(4).[124]

This appears to mean that Article 7(9) leaves some discretion to a transit state as to how it would prefer to operationalise freedom of transit in its Area and emphasises the *negative* nature of the obligation under Article 7(4). The transit state certainly cannot be obliged to build a new pipeline, especially at its own expense, when there are other reasonably available ways to effect transit. It is also clear, however, that, pursuant to Article 7(4), a transit state is not released from its obligation to effect freedom of transit merely because it does not have certain facilities.[125]

[121] Ibid.

[122] See European Energy Charter Conference Secretariat, 'Room Document 10', Plenary Session, 24–26 March 1993 at 2; European Energy Charter Conference Secretariat, 'Room Document 8', Plenary Session, 26 March 1993 at 7; and OECD, 'Memorandum', above n 73 at 2.

[123] Energy Charter Treaty, above n 63, emphasis added.

[124] Ibid.

[125] Ehring and Selivanova also note that Article 7(9) takes precedence over other provisions in Article 7, with the exception of Article 7(4), which outlaws obstacles for the establishment of new transport capacity. Ehring and Selivanova, above n 64 at 86.

The obligation under Article 7(4) not to 'place obstacles' in the way of the establishment of a new pipeline capacity must be further distinguished from the obligation to allow new investment ('making of investment').[126] The obligation to allow new investment is regulated by the ECT in a rather soft manner.[127] Article 10(2) of the ECT (Promotion, Protection and Treatment of Investments) contains only a best-endeavour commitment of Members to accord to investors of other Members, as regards the 'making of investments', national treatment or MFN treatment, whichever is the most favourable.[128] While the 'fair and equitable treatment' principle under Article 10(1) applies to the 'making of investment', disputes relating to the pre-investment phase would appear to be outside the scope of the investor–state dispute settlement mechanism established by ECT Article 26.[129] Therefore, according to Article 10 of the ECT, it would be unlikely that a transit state would be obliged to allow investments for the purposes of establishing new pipeline capacity in its territory.

As with third-party access rights, should a transit state decide to operationalise its obligation under Article 7(1) of the ECT to take the necessary measures to facilitate transit by permitting the establishment of a new pipeline capacity, it will have to comply with other obligations under Article 7. Those would appear to include non-discrimination obligations under Articles 7(1) and 7(3) as well as the prohibition on imposing any unreasonable delays, restrictions or charges on traffic in transit.[130]

[126] The ECT negotiating history also indicates that the concepts of 'capacity establishment' and 'making of investment' were not perceived by negotiators as identical. See Official Preparatory Work of the ECT in the European Energy Charter Secretariat: European Energy Charter Conference Secretariat, 'Room Document 10', above n 122 at 3. The term 'making of investment' is defined in Article 1(8) of the ECT as 'establishing new Investments, acquiring all or part of existing Investments or moving into different fields of Investment activity'. Energy Charter Treaty, above n 63; see also Chapter III, Section III.C.2(i).

[127] Wälde, Thomas W., 'International Investment under the 1994 Energy Charter Treaty', in T. W. Wälde (ed.), *The Energy Charter Treaty: An East–West Gateway for Investment and Trade* (London: Kluwer Law International, 1996) 251 at 277–284.

[128] Energy Charter Treaty, above n 63.

[129] Ibid. As Wälde notes, ECT Article 26 regulates '[d]isputes between a Contracting Party and an Investor of another Contracting Party relating to an *Investment* of the latter [that is, an investment which is already made]'. See Wälde, 'International Investment under the 1994 Energy Charter Treaty', above n 127 at 278, 283, emphasis added.

[130] Recall that, under Article 7(4), the legislation of a transit state applicable to capacity establishment must be consistent with the overarching principles set out in ECT Article 7(1). Article 7(4) does not specifically mention Article 7(3). However, the text of Article

Finally, among other provisions of the ECT relevant to both capacity establishment and third-party access rights are those extending ECT obligations to the conduct of private entities. In Article 7(2), the Treaty requires that each Member encourage relevant entities to cooperate in, inter alia, modernisation, development and operation of cross-border transport facilities, and the facilitation of the interconnection of pipelines.[131] While it was observed that the obligation under Article 7(2) for the most part is of a soft law nature,[132] it is not deprived of any normative value. For example, if it was established that a transit state instructed a pipeline operator to block an interconnection or frustrate the cooperative operation of a cross-border pipeline (including third-party access rights), such actions would likely be inconsistent with Article 7(2). Similarly, in the same context, the reluctance of the competent authorities of a transit state to intervene in the private conduct of a pipeline operator that manifestly frustrates gas transit, especially if they possess such powers under national law, would hardly be in line with the requirement to *encourage* cooperation under Article 7(2).

This interpretation of Article 7(2) appears to be in line with other rules of the ECT governing the conduct of private or quasi-private entities. In particular, Article 22 (State and Privileged Enterprises), in paragraphs 2 and 4, prohibits Members from encouraging or requiring domestic state enterprises or entities, to which they grant exclusive or special privileges, to conduct their activities in a manner inconsistent with ECT obligations.[133] Article 22, paragraph 3, requires each Member to ensure that if it establishes or maintains an entity and entrusts this entity with regulatory, administrative or other governmental authority, such an entity will exercise that authority in an ECT-consistent manner.[134] These provisions, in their essence, appear to reflect the customary rules

7(3) is broad enough to cover transit requirements imposed in the context of capacity establishment to the extent that an existing international agreement does not provide otherwise. See Energy Charter Treaty, above n 63.

[131] Although the ECT does not define the term 'relevant entities', it is broad enough to cover a company operating or owning a pipeline. Such an interpretation is supported by the context of Article 7(2) – Article 22(5). This provision provides that the term 'entity' that it employs includes '*any* enterprise'. Ibid., emphasis added.

[132] See Clark, above n 67 at 4; Liesen, above n 86 at 64. As the negotiating history of the ECT clarifies, some drafters (such as the EU) were unwilling to interfere in private commercial practices and insisted on a soft law provision regulating the relationship between states and their domestic pipeline operators. See European Energy Charter Conference Secretariat, Working Group II, 1–5 June 1992.

[133] Energy Charter Treaty, above n 63. [134] Ibid.

on state responsibility for internationally wrongful acts, discussed in Chapter V.[135]

(iv) Summary and concluding remarks on the ECT's regulation of third-party access and capacity establishment rights As described in this section, in general, the ECT's regulation of gas transit, including third-party access and capacity establishment rights, does not depart from the basic principles of the GATT, such as freedom of transit and non-discrimination. The Treaty elaborates on those principles by establishing certain specific rules addressing particular features of energy transit and, in particular, pipeline gas transit. The overarching approach of the ECT to regulating third-party access and capacity establishment rights is that it leaves some discretion to a transit state as to how it prefers to operationalise freedom of gas transit via its territory in circumstances in which there is not adequate pipeline capacity or infrastructure.

Although the Treaty rules out a *mandatory* third-party access right, it does not exclude the possibility that a transit state may grant a *negotiated* third-party access right to a particular transit-dependent state, with a view to facilitating gas transit, or require a form of the third-party access right under its national law. In both cases, the transit state would have to comply with certain requirements of the ECT, including the prohibition on discriminatory treatment, as well as unreasonable delays, restrictions or charges imposed on transit. In the same context, the text of ECT Article 7(3) establishes an exception according to which the prohibition of discrimination applies 'unless an existing international agreement [such as bilateral or regional agreements] provides otherwise'. Currently, there are no equivalent exceptions in WTO law. This means that should a WTO Member challenge this form of discrimination in the WTO, this discrimination would likely be found inconsistent with WTO transit rules.

The question can be asked whether this exception would be desirable if WTO Members were to adopt the ECT approach to regulating energy transit in the WTO legal framework. On one hand, this exception may appear to encourage WTO Members to negotiate on bilateral or regional agreements with a view to operationalising WTO transit provisions in the context of trade in pipeline gas. The results of such negotiations, if set out in an international agreement, would be exempted from WTO non-discrimination obligations (MFN). On the other hand, this exception

[135] See Chapter V, Section III.B.

could significantly affect the current balance between third-party access and capacity establishment rights under WTO law and may provide a wide scope for circumventing a number of WTO transit rules, such as rules on transit regulations and transportation charges. Moreover, as explained in Chapter IV, in any event, the existing non-discrimination obligations in Article V, including Article V:2 (second sentence), appear to allow Members to afford formally different treatment to goods originating in different WTO Members, provided that such treatment is *based on* external, non-origin related factors, such as different conditions of transportation.[136] Finally, certain favourable conditions of transit could be made part and parcel of regional trade agreements, which are regulated by GATT Article XXIV (Customs Unions and Free-Trade Areas), although, as explained earlier, WTO jurisprudence has yet to provide clear guidance on which types of special and differential treatment of traffic in transit this provision would cover.[137]

In circumstances in which a third-party access right is not a viable or a technically feasible option for a transit state, the ECT transit rules allow the transit state to operationalise its transit obligations by permitting others to establish an additional pipeline capacity. Under the ECT, the notion of 'capacity establishment' means a *negative* obligation of a transit state not to 'place obstacles in the way of new capacity being established'. The ECT states explicitly that transit states do not have a *positive* obligation to build or expand transit facilities. In addition, the Treaty leaves some freedom to a transit state as to whether it wishes to allow new investments in its territory.

Nevertheless, this regulatory freedom of a transit state must be balanced against its obligation under the ECT to take measures to facilitate freedom of transit. This means, for example, that should a transit state, in cooperation with a transit-dependent state, choose the capacity establishment option as a mechanism to effect freedom of gas transit, the question will have to be answered as to who exactly will have the right or obligation to build the pipeline in question or provide for its extension. This could be a transit state (alone or together with a transit-dependent state) or national or foreign investors. On one hand, the pipeline project would have to comply with the transit state's legislation (including environmental protection, land use, safety or technical standards). On the other hand, this legislation, as well as other requirements and conditions of a transit state, must be consistent with the ECT, including

[136] See Chapter IV, Section III.B.1(ii). [137] See Chapter IV, Section V.A.2.

the principle of non-discrimination and the prohibition on imposing unreasonable delays, restrictions or charges.

The important obligations of transit states relevant to both third-party access and capacity establishment are to encourage relevant entities (such as pipeline operators within their jurisdiction) to cooperate in the modernisation, development and operation of cross-border transport facilities, as well as the facilitation of the interconnection of pipelines.

3. Analogues of third-party access and capacity establishment rights under the LOSC

(i) General background Throughout this study and in particular in Chapter III, it was explained that the LOSC is one of the multilateral treaties that recognises the principle of freedom of transit. It emerged from more than twenty years of negotiations on the codification and progressive development of the law of the sea, which culminated in 1982.[138] The LOSC devotes special attention to the issue of freedom of transit in Part X (Right of Access of Land-Locked States to and from the Sea and Freedom of Transit).

It will be demonstrated in the following sections that, in some respects, the LOSC appears to contain a more progressive and elaborate approach to the regulation of transit than the basic approach adopted under the GATT, back in 1947, which did not undergo changes during the Uruguay round of negotiations. Some of the GATT transit rules were further elaborated under the Agreement on Trade Facilitation. However, as explained in Chapter IV, the new amendments concerned third-party access and capacity establishment rights in an indirect manner, namely through best-endeavour cooperation and institutional provisions.

The important limitation of the LOSC is that it only addresses freedom of transit from the perspective of the right of access of land-locked states to and from the sea for the purpose of exercising other rights provided for in the Convention, such as those relating to the freedom of the high seas

[138] The parties to this Convention recognised in its preamble that the Convention achieved 'the codification and progressive development of the law of the sea'. The preamble links this achievement to other overarching goals of the LOSC, namely to strengthen 'peace, security, cooperation and friendly relations among all nations in conformity with the principles of justice and equal rights', 'the economic and social advancement of all peoples of the world, in accordance with the Purposes and Principles of the United Nations' and 'facilitat[ion] [of] international communication'. See preamble to the LOSC in United Nations Convention on the Law of the Sea, http://www.un.org/depts/los/convention_agreements/texts/unclos/unclos_e.pdf, accessed 30 December 2012.

and the common heritage of mankind.[139] This essentially means that, while recognising freedom of transit for land-locked states,[140] the LOSC does not extend this freedom to the wider international community. The latter includes geographically disadvantaged countries that have a very limited coastal corridor (like Iraq or the Democratic Republic of the Congo), which may not serve as a sufficient channel for the international communication and trade of those countries or as the most convenient route to the sea.

Furthermore, as stated earlier, freedom of transit under the LOSC is provided for the purpose of access to and from the sea. Thus, for example, a communication between two land-locked states, or a land-locked state and a coastal state through an intermediate transit country, which has nothing to do with the right of access to and from the sea, would appear to fall outside the scope of the Convention.[141] This, however, does not mean that freedom of transit is limited to transit through the territory of a neighbouring (coastal) state. It may also entail passage through as many states as necessary (including other land-locked states) with the ultimate goal of access to the sea.[142]

Finally, the purpose of transit under the LOSC should be linked to other rights established by the Convention, such as those relating to freedom of the high seas, including freedom of navigation and fishing.[143] Importantly, while, among other freedoms in areas outside the territory of a transit state (namely the high seas, EEZ and continental shelf), the LOSC recognises the freedom to lay submarine pipelines,[144] it does not directly establish a similar freedom in the *territory* of transit states. Article 124(2) of the Convention merely states that '[l]and-locked States and transit States may, *by agreement* between them, include as

[139] See the title of Part X and the key obligations in Article 125. Ibid.

[140] The term 'land-locked state' is defined in Article 124(1)(a) as 'a State which has no sea-coast'. Ibid.

[141] Similarly, Dupuy and Vignes argue that '[t]ransit for land-locked States consists of the passage of goods across the territory of a Contracting Party between a land-locked State *and the sea* . . . '. See Dupuy, René-Jean and Vignes, Daniel (eds.), *A Handbook on the New Law of the Sea*. vol. 1 (Boston: Hague Academy of International Law, Martinus Nijhoff Publishers, 1991) at 510–511, emphasis added.

[142] See Nandan, Satya N. and Rosenne, Shabtai (eds.), *United Nations Convention on the Law of the Sea 1982: A Commentary*, vol. 3 (The Hague: Centre for Oceans Law and Policy, Martinus Nijhoff Publishers, 1995) at 406.

[143] Ibid. at 371.

[144] See LOSC, Articles 58 (Rights and duties of other states in the exclusive economic zone), 79 (Submarine cables and pipelines on the continental shelf), and 87(1)(c) (Freedom of the high seas). United Nations Convention on the Law of the Sea, above n 138.

means of transport [identified by the Convention in Article 124(1)(d)][145] pipelines and gas lines'.[146] However, should such an agreement be concluded, the traffic in transit via gas pipelines would appear to fall within the scope of Part X of the LOSC as any other kind of traffic.

These limitations narrow the scope of freedom of transit under the LOSC. Nevertheless, as a general method of regulating transit, the LOSC can provide useful guidance on how to improve the clarity and effectiveness of WTO transit rules. In general, this method appears to be in line with the principles of international law discussed in this study. The following sections provide an overview of the provisions in the LOSC regulating freedom of transit of land-locked states in the territory of a transit state as well as discuss how the LOSC addresses the issues of the lack of adequate transit facilities and access to existing transit facilities. Various rights and obligations under the LOSC relating to submarine pipelines outside the territory of a transit state, which are regulated by different legal regimes and derived from a separate legal principle of the freedom of the high seas (rather than freedom of transit), are not discussed here.[147]

(ii) The key transit provisions under the LOSC It was explained earlier that Article 125 of the LOSC (Right of access to and from the sea and freedom of transit) strikes a delicate balance between the interests of land-locked and transit states.[148] In paragraph 1, this provision provides, inter alia, that land-locked states must enjoy freedom of transit through the territory of transit states by all means of transport identified by the Convention.[149] Article 125(3) recognises that transit states, in exercising sovereignty over their territory, must have the right to take

[145] These 'means' include: 'railway rolling stock, sea, lake and river craft and road vehicles [and] where local conditions so require, porters and pack animals'. Ibid.

[146] Ibid., emphasis added.

[147] See the latter issue discussed in Dupuy and Vignes, above n 141 at 976–988; and Vinogradov, 'Challenges of Nord Stream', above n 80 at 34–47.

[148] See Chapter III, Section III.B.2(i); Nandan and Rosenne (eds.), above n 142 at 409.

[149] The term 'means of transport' was defined in the previous section. In Article 124(1)(c), the LOSC defines the term 'traffic in transit' in a manner similar to how it is defined under GATT Article V, taking into account the particular features of the transit regulation under this Convention. This provision states as follows: '"traffic in transit" means transit of persons, baggage, goods and means of transport across the territory of one or more transit States, when the passage across such territory, with or without trans-shipment, warehousing, breaking bulk or change in the mode of transport, is only a portion of a complete journey which begins or terminates within the territory of the land-locked State'. United Nations Convention on the Law of the Sea, above n 138.

all measures necessary to ensure that the rights and facilities provided to land-locked states, in accordance with the LOSC, 'in no way infringe their *legitimate interests*'.[150] Paragraph 2 of Article 125 reconciles the interests of land-locked and transit states by establishing the obligation for those states to agree on '[t]he terms and modalities for exercising freedom of transit . . . through bilateral, sub-regional or regional agreements'.[151]

Importantly, the balance between the interests of land-locked and transit states struck by Article 125 does not mean that, according to this provision, freedom of transit *depends on* the conclusion of an agreement with a transit state and, therefore, on the will or caprice of the latter state. As explained earlier and in particular in Chapter V, a transit state may not deny 'freedom of transit', which is clearly established by international law, including the LOSC. Only the terms and modalities for exercising this freedom, such as the designation of a transit route, may be agreed on at a bilateral or regional level.[152]

It is clear that, in line with Article 125(3), a transit state may regulate transit via its territory by adopting necessary measures.[153] However, as this provision indicates, such measures should aim at protecting *legitimate* interests only and, therefore, must not amount to unjustified or disguised restrictions on transit.[154] Furthermore, other provisions of the LOSC impose additional requirements on transit states' regulations

[150] Ibid., emphasis added. [151] Ibid.

[152] See Chapter V, section II.B.3(iii); as well as Dupuy and Vignes, above n 141 at 517–518, 521; Melgar, Beatriz H., *The Transit of Goods in Public International Law* (Leiden: Brill Nijhoff, 2015) at 196–197; and Einhorn, Talia, 'Transit of Goods over Foreign Territory', in *Max Planck Encyclopaedia of Public International Law* (2010), www.mpepil.com, accessed 30 November 2011 at 11–12. Other scholars expressed concern as to whether the obligations of transit states under Article 125 could be circumvented if a transit state refused to conclude an agreement mentioned in paragraph 2 of this provision. See Caflisch, Lucius C., 'Land-Locked States and Their Access to and from the Sea', 49 *British Year Book of International Law* (1978) 71 at 96; Vasciannie, Stephen C., *Land-Locked and Geographically Disadvantaged States in the International Law of the Sea* (Oxford: Clarendon Press, 1990) at 188–189 (although the author accepts that Article 125(2) 'may be read as a *pactum de contrahendo*, with the result that the States concerned would be legally compelled to reach an agreement'); and Churchill, R. R. and Lowe, A. V., *The Law of the Sea*, 3rd ed. (Manchester: Manchester University Press, 1999) at 444. Churchill and Lowe, however, rightly observed that, under the LOSC as well as general international law principles of good faith and abuse of rights (*abus de droit*), 'a transit State must in good faith seek to conclude a transit agreement'. Ibid.

[153] See Dupuy and Vignes, above n 141 at 518.

[154] The meaning of the terms 'legitimate interests' are not elaborated on in the LOSC. While during the negotiations on the Convention the drafters discussed among those interests, public morality, public security, health and sanitation, they ultimately failed to reach an agreement on the precise list of legitimate interests. In Nandan and Rosenne's view, '[a]t

similar to those set out in WTO law and the ECT. Without delving into all the details of those requirements, it would suffice to mention a few relevant provisions.

Article 127(1) prohibits the imposition of 'any customs duties, taxes or other charges [on traffic in transit] except charges levied for specific services rendered in connection with such traffic'.[155] Paragraph 2 of this Article imposes a non-discrimination obligation on transit states, pursuant to which '[m]eans of transport in transit and other facilities provided for and used by land-locked States shall not be subject to taxes or charges higher than those levied for the use of means of transport of the transit State'.[156] Some aspects of this provision are discussed in the next section.

Article 130 establishes a twofold obligation on a transit state. On one hand, it requires that a transit state 'take all appropriate measures to avoid delays or other difficulties of a technical nature in traffic in transit'.[157] On the other hand, it states that '[s]hould such delays or difficulties occur, the competent authorities of the transit States and land-locked States concerned *shall* cooperate towards their expeditious elimination'.[158] Thus, this provision requires cooperation as a mandatory solution to any delays or technical difficulties impeding transit.

Finally, outside Part X, Article 300 of the LOSC incorporates the principles of good faith and abuse of rights (*abus de droit*), which provide additional protection against unjustifiable restrictions of transit. This provision establishes that 'States Parties shall fulfil in good faith the obligations assumed under this Convention and shall exercise the rights, jurisdiction and freedoms recognized in this Convention in a manner which would not constitute an abuse of right'.[159] As explained in the previous chapter, these principles are the balancing tools that can assist in determining whether a particular transit regulation is 'reasonable' (that is, inter alia, appropriate and necessary to fulfil a legitimate policy objective and compatible with international law).[160]

(iii) The lack of adequate transit facilities and access to existing transit facilities under the LOSC Unlike the GATT, the LOSC directly addresses the issues of the lack of adequate transit facilities in some

a minimum, they [i.e. legitimate interests] should be consistent with generally accepted rules of international law.' See Nandan and Rosenne (eds.), above n 142 at 417, 422.

155 United Nations Convention on the Law of the Sea, above n 138. 156 Ibid.

157 Ibid. 158 Ibid., emphasis added. 159 Ibid. 160 See Chapter V, section II.B.3(ii).

transit states and access to existing transit facilities. This Convention, therefore, explicitly recognises the relevance of these issues to effective freedom of transit. As discussed later, the LOSC regulates these issues mainly by requiring cooperation between transit and land-locked states in improving and constructing additional transit facilities, which in effect is analogous to the capacity establishment right. In addition, the Convention imposes certain obligations that are relevant for the exploitation of existing transit facilities, which have certain implications for third-party access rights.

With respect to the use of existing transit facilities ('third-party access' rights), it was mentioned earlier that Article 127 imposes a non-discrimination obligation relating to charges levied on the use of transit facilities provided to land-locked states. In light of the context of this provision – Articles 124(1)(c) and 129 mentioning warehouses and port installations – it becomes clear that these facilities may include fixed infrastructure.[161] Furthermore, it was explained earlier that, by a special agreement, transit and land-locked states may include pipelines within the scope of the 'means of transport' listed in Article 124(1)(d). In this case, the non-discrimination obligation under Article 127 would extend to the latter type of facilities.

Notably, Article 126 provides that 'special agreements relating to the exercise of the right of access to and from the sea, establishing rights and facilities on account of the special geographical position of land-locked States, are excluded from the application of the most-favoured-nation clause'.[162] In this way, Article 126 excludes any facilities or rights granted pursuant to a special agreement between a transit state and a land-locked state, such as agreements governing the 'terms and modalities' of transit within the meaning of Article 125(2) of the Convention, from an MFN treatment.

[161] United Nations Convention on the Law of the Sea, above n 138.

[162] Ibid. The ITO Charter, which was never adopted, also addressed the issue of granting land-locked countries additional facilities, albeit in a different manner. In particular, Ad Note to Article 33(6) reads as follows: 'If, as a result of negotiations in accordance with paragraph 6, a Member grants to a country which has no direct access to the sea more ample facilities than those already provided for in other paragraphs of Article 33, such special facilities may be limited to the land-locked country concerned unless the Organization finds, on the complaint of any other Member, that the withholding of the special facilities from the complaining Member contravenes the most-favoured-nation provisions of this Charter.' This provision was not included in the GATT. See International Trade Organization Charter, http://www.wto.org/english/docs_e/legal_e/havana_e.pdf, accessed 10 January 2014.

As Vasciannie noted, however, it is not entirely clear whether Article 126 refers to an MFN treatment of land-locked states as a *sui generis* group vis-à-vis other states or also to the treatment of one land-locked state vis-à-vis another. Put differently, if a transit state, under a special bilateral agreement, provides certain favourable treatment to a land-locked state relating to the exploitation of its pipeline, would Article 126 of the LOSC allow that transit state not to extend this treatment to other land-locked states?[163] If so, as with a similar exception to a non-discrimination obligation under ECT Article 7(3), the LOSC would appear to deviate from the GATT transit rules. To recall, under Article V:2 (second sentence), a transit state may not make distinctions solely on the basis of the origin of goods.[164]

Article 129 of the LOSC sets out the key obligation relating to cooperation in the construction and improvement of means of transport ('capacity establishment'). It provides as follows:

> Where there are no means of transport in transit States to give effect to the freedom of transit or where the existing means, including the port installations and equipment, are inadequate in any respect, the transit States and land-locked States concerned *may* cooperate in constructing or improving them.[165]

It is noteworthy that Article 129 uses the word 'may'. Yet, in some scholars' opinions, this provision does not leave much discretion to transit states as to whether they should engage in the good faith cooperation on the construction or improvement of transit facilities, especially in circumstances where the existing facilities are inadequate to effect freedom of transit. In this regard, Dupuy and Vignes state as follows:

> The absence or inadequacy of means of transport in the transit State can constitute a material obstacle to the effective exercise of the freedom of transit. For such cases, the 1982 Convention has established a *duty* of co-operation between the land-locked and the transit State for the construction of means of transport or their improvement.[166]

This view appears to be supported by the fact that Article 300 of the LOSC requires states to exercise their rights under the Convention in good faith and avoid any abuse of these rights.[167]

[163] See the discussion of this issue in Vasciannie, above n 152 at 193–194.
[164] See Chapter IV, Section III.B.1.
[165] United Nations Convention on the Law of the Sea, above n 138, emphasis added.
[166] Dupuy and Vignes, above n 141 at 521, emphasis added.
[167] United Nations Convention on the Law of the Sea, above n 138.

This cooperation in constructing additional facilities could result in the conclusion of a bilateral or regional agreement, mentioned in Article 125(2), where the 'terms and modalities' of transit would concern rights analogous to capacity establishment rights discussed in this study. If the negotiating parties so wish, this could also be an agreement governing the construction or improvement of a gas pipeline.

(iv) Summary and concluding remarks on the LOSC's regulation of third-party access and capacity establishment rights The LOSC's regulation of transit is not free from certain pitfalls and may not be transposed 'lock, stock and barrel' into the WTO legal framework. In particular, the LOSC does not directly regulate pipeline gas transit and establishes transit rules with a view to addressing particular concerns of land-locked states, derived from their right of access to and from the sea. Furthermore, like the ECT, the Convention exempts from a non-discriminatory (MFN) treatment 'special agreements' concluded between transit states and land-locked states regulating the exercise of the land-locked state's right of access to and from the sea.

Nonetheless, some aspects of the LOSC's regulation of transit could provide useful guidance to WTO Members on how to improve the effectiveness of WTO transit rules when there is no adequate transit infrastructure or capacity. In this respect, the LOSC appears to accurately delimit the scope of the different layers of transit governance (that is, multilateral and regional, or bilateral). Its overarching approach to transit regulation is: (i) through the recognition of the basic principles relating to transit, such as freedom of transit and sovereignty; and (ii) striking the balance between those competing principles through the requirement of cooperation between transit states and transit-dependent (land-locked) states. As the LOSC provides, this cooperation can take the form of bilateral or regional agreements determining the terms and modalities of transit. Importantly, the LOSC explicitly extends this requirement of cooperation to circumstances where transit facilities, which are otherwise inadequate or absent, must be constructed or improved.

The LOSC does not expressly refer to rights analogous to third-party access. However, as with transit in general, such rights would be subject to the aforementioned requirement of cooperation. The Convention also imposes certain specific limitations on the sovereign right of transit states to regulate transit. In particular, it establishes that this right must be used exclusively for the purposes of protecting the legitimate interests of a transit state and exercised in good faith in accordance with the principle of the prohibition of the abuse of rights (*abus de droit*). Other such limitations include the non-discrimination obligation under Article 127 of the LOSC. It is noteworthy, that, unlike the ECT, the LOSC does not appear to directly address the conduct of private companies operating transit facilities.

B. Progressive Development Approach: Multilateralising Ukraine's Additional Commitments on Pipeline Transport or a Reference Paper on Energy Services?

1. What role could the GATS play in the regulation of pipeline gas transit?

It was explained in Chapter IV that the GATT and the GATS regulate different aspects of trade in energy: namely, trade in goods and services, respectively.[168] These Agreements may not always overlap in the context of gas transit. For example, in the hypothetical scenario discussed in Chapter I, a gas seller X, located in the WTO Member A, had an interest in selling gas to consumers located in the Member C. This required the transportation of X's gas through the pipeline network operated in the territory of the transit Member B by company Y. Absent transit barriers imposed by B, X could conclude a private gas transportation contract with Y according to which Y would undertake to ship X's gas through the territory of B until the gas is handed over to a third party or even to a subsidiary of X at the border between B and C. On the basis of similar contracts, for decades, Russian gas was shipped to the EU market through the territory of Ukraine by the Ukrainian national company Naftogaz Ukrayini.[169] In other words, to organise the delivery of its gas to the market of C, X does not have to rely on B's specific commitments under the GATS with respect to pipeline transportation of fuels or obtain the permission of B to supply energy transportation services in B's territory.

[168] Chapter IV, Section II.

[169] See Contract between Gazprom and Naftogaz Ukrayini on the Volumes and Conditions of Transit of Natural Gas through the Territory of Ukraine for the Period of 2009–2019, signed in Moscow on 19 January 2009, available at Ukrayinska Pravda, 'The Contract on Transit of Russian Gas + Additional Agreement on the Advance Payment to Gazprom' ('Контракт о транзите российского газа + Допсоглашение об авансе "Газпрома"'), http://www.pravda.com.ua/rus/articles/2009/01/22/4462733/, accessed 10 October 2012.

However, even if B did liberalise its market for pipeline transportation services by undertaking, inter alia, market access and national treatment commitments, these commitments alone would not be sufficient to establish a right for X to ship its gas through the territory of B to C. As noted in Chapter IV, there are important limitations in the GATS' regulation of gas transit. In particular, if B were to make market access and national treatment commitments with respect to modes 1 and 3 of the supply of pipeline transportation services, these commitments would prevent B from imposing restrictions or discriminatory measures on the supply by X of these services from the territory of A to consumers located in B or on the supply of these services within the territory of B. These commitments would, however, not require B to allow X to ship its gas *through* its territory to a third-country market. Likewise, these commitments would not require B to compel a private company Y to provide third-party access for X to the pipeline network that Y operates. Finally, if X is a vertically integrated company selling and transporting its own gas to consumers, it is questionable whether the services component of this transaction would be covered by the GATS.[170]

These limitations in the GATS' regulation of transit barriers affecting third-party access and capacity establishment rights could, in principle, be corrected through a legislative reform. This reform would, however, require extensive additional commitments by Members on various energy services relating to gas transit. Depending on how ambitious Members are, they could:

- commit themselves to take positive measures against anti-competitive conducts of their domestic pipeline operators hindering third-party access to gas networks, drawing on the model of Ukraine's additional commitments on pipeline transport discussed in Chapter IV; or
- liberalise their markets for energy services, based on the model of the Telecom Reference Paper.

Both approaches are discussed in the following sections.

The advantage of such additional commitments over the existing transit obligations under GATT Article V will be the improved access to 'natural monopolies'/'essential facilities'[171] under Members' national jurisdiction, which, from an economic viewpoint, cannot be easily duplicated. Under current transit rules, WTO Members do not

[170] See these issues discussed in detail in Chapter IV, Sections IV.B and IV.C.
[171] See these terms explained in Chapter II, Section III.C.1.

have to ensure this access and can operationalise their transit obligations through alternative means, such as by negotiating the terms and modalities of the establishment of new pipeline capacity. The latter option may not be the most attractive one for a transit-dependent state. Therefore, the legislative reform under the GATS discussed in the following sections will have a positive effect on the regulation of third-party access and capacity establishment rights under WTO law.

2. The model of Ukraine's additional commitments on pipeline transport

One way to improve the regulation of gas transit under the GATS would be if all Members undertook additional commitments similar to those Ukraine made in its Schedule of Specific Commitments under the GATS with respect to pipeline transport. To recall, Ukraine agreed:

> to ensure adherence to the principles of non-discriminatory treatment in access to and use of pipeline networks under its jurisdiction, within the technical capacities of these networks, with regard to the origin, destination, or ownership of product transported, without imposing any unjustified delays, restrictions, or charges, as well as without discriminatory pricing based on the differences in origin, destination or ownership.[172]

In addition, Ukraine committed itself to 'to provide full transparency in the formulation, adoption, and application of measures affecting access to and trade in services of pipeline transportation'.[173]

These commitments mean that, pursuant to its Schedule, Ukraine must take positive measures to ensure that non-discriminatory third-party access is provided to foreign gas shippers by its domestic pipeline operators. This commitment is not limited to gas transportation services rendered or terminated within Ukraine's territory, since discrimination based on the *destination* of foreign gas is also prohibited. Consequently, the commitment has practical implications for the ability of foreign gas suppliers to transit gas through the territory of Ukraine by

[172] World Trade Organization, 'Schedule of Specific Commitments of Ukraine', GATS/SC/144 (10 March 2008) at 34.

[173] World Trade Organization, 'Schedule of Specific Commitments of Ukraine', ibid.; Chapter IV, Section IV.A.

obtaining third-party access to pipeline networks under Ukraine's jurisdiction.

3. The model of the Telecom Reference Paper

A more ambitious approach to improving the regulation of gas transit under the GATS, which could remove restrictions on third-party access and capacity establishment rights, would be the liberalisation of the international market for energy services by developing a 'reference paper on energy services' in line with the model of the Telecom Reference Paper. A similar proposal was made by Norway and the United States during the WTO on-going negotiations on energy services, although these Members did not refer to energy transit as such. The key elements of the 'reference paper' proposed by Norway and the United States included: (i) rules on transparency; (ii) non-discriminatory access to energy networks and interconnection; (iii) an independent regulator; (iv) non-discriminatory, objective and timely procedures for the transportation and transmission of energy; and (iv) rules preventing anti-competitive practices in the area of energy services. It can be inferred from these proposals that the 'reference paper' would constitute a separate agreement regulating particular features of trade in energy services, converted into WTO law through additional commitments undertaken by all Members or at least Members that are active in this area.[174]

The Telecom Reference Paper is the first legal instrument in the WTO directly regulating competition, albeit at a plurilateral rather than multilateral level.[175] Some of its provisions contain rights analogous to third-party access and capacity establishment in gas trade. The Telecom Reference Paper aims to create a competitive market for telecommunications services by establishing the key principles of the functioning of this market that must be observed by Members as well as by their 'major suppliers' of such services. The 'major suppliers' are defined as 'a supplier which has the ability to materially affect the terms of participation ... in the relevant market for basic telecommunications services as a result of:

[174] See World Trade Organization, Council for Trade in Services Special Session, Communication from Norway, 'The Negotiations on Trade in Services', S/CSS/W/59, 21 March 2001 at 10; World Trade Organization, Council for Trade in Services Special Session, Communication from the United States, 'Energy Services', S/CSC/W/24, 18 December 2000, para. 16. These proposals appear to be in line with the position of the EU in the negotiations on energy services. Poretti, Pietro and Rios-Herran, Roberto, 'A Reference Paper on Energy Services', above n 61, section III.B.

[175] As explained, it is incorporated into some WTO Members' GATS Schedules of Specific Commitments. See Chapter IV, Section IV.B.

(a) control over essential facilities; or (b) use of its position in the market'.[176] The term 'essential facility' is, in turn, defined as 'facilities of a public telecommunications transport network or service that (a) are exclusively or predominantly provided by a single or limited number of suppliers; and (b) cannot feasibly be economically or technically substituted in order to provide a service'.[177]

In simple terms, the Telecom Reference Paper disciplines the conduct of services suppliers that can materially affect the terms of competition in the 'relevant market' for telecommunications services by either 'abusing their dominant position' or hampering access to 'essential facilities' that they control. The terms 'relevant market', 'dominant position' and 'essential facilities' are defined differently in national competition laws.[178] However, as commentators have noted, they may have a separate meaning in WTO law.[179] Importantly, gas pipelines have been regarded as 'essential facilities' in the EU and US legal systems, which indicates that their operators could fall within the definition of a 'major supplier'.[180]

The Telecom Reference Paper establishes a number of regulatory requirements relating to a third-party access right. In Section 2 (Interconnection), it requires Members to ensure the interconnection with their 'major suppliers' at any technically feasible point in the network to allow the users of one supplier to communicate with users of another supplier and to access

[176] See 'Definitions' in World Trade Organization, Reference Paper on Telecommunications Services, www.wto.org/english/tratop_e/serv_e/telecom_e/tel23_e.htm, accessed 30 March 2012.

[177] Ibid.

[178] See Talus, Kim, 'Just What Is the Scope of the Essential Facilities Doctrine in the Energy Sector?: Third Party Access-Friendly Interpretation in the EU v. Contractual Freedom in the US', 48 *Common Market Law Review* (2011) 1571; Janssens, Thomas and Wessely, Thomas (eds.), 'Dominance: The Regulation of Dominant Firm Conduct in 40 Jurisdictions Worldwide' (Global Competition Review, 2010). The term 'essential facilities' was also discussed in Chapter II, Section III.C.1.

[179] Bronckers, Marco and Larouche, Pierre, 'A Review of the WTO Regime for Telecommunications Services', in K. Alexander and M. Andenas (eds.), *The World Trade Organization and Trade in Services* (The Netherlands: Koninklijke Brill NV, 2008) 319 at 332–334. In *Mexico – Telecoms*, the Panel found that Mexico's Telmex was the 'major supplier' in the market for certain types of telecommunications services supplied cross-border from the United States to Mexico by applying the principles of 'demand substitution' and analysing Telmex's 'ability to materially affect the terms of participation' in the market at issue. Panel Report, *Mexico – Telecoms*, paras. 7.146–7.159. See this case discussed in Chapter V, Section II.B.3(iii).

[180] See, inter alia, *American Central Eastern Texas Gas Company v. Union Pacific Resources Group Inc.* and *GDF Suez/E.ON* cases discussed in Talus, 'Just What Is the Scope of the Essential Facilities Doctrine in the Energy Sector?', above n 178 at 1580–1581, 1584–1586.

services provided by another supplier. The interconnection must be provided under: (i) non-discriminatory terms, conditions and rates; and (ii) in a timely fashion, on terms, conditions and cost-oriented rates that are transparent, reasonable, having regard to economic feasibility.[181] In essence, the requirement of interconnection establishes the third-party access right of a supplier to the foreign network, on non-discriminatory and commercially reasonable conditions.

Some requirements in the Telecom Reference Paper appear to be relevant to capacity establishment. For example, Section 2.2(c) requires Members to ensure interconnection upon request 'at points in addition to the network termination points offered to the majority of users, subject to charges that reflect the cost of construction of necessary additional facilities'.[182] As Mathew states, this provision was created to ensure the possibility of interconnection beyond standard interconnection points as long as an applicant is willing to pay the additional cost (including the cost of the construction of necessary additional facilities).[183]

Among rules relevant to both third-party access and capacity establishment rights, Section 1 (Competitive Safeguards) establishes an obligation for Members to maintain appropriate measures preventing their 'major suppliers' from engaging in or continuing 'anti-competitive practices'. The latter term is defined loosely by listing three illustrative examples: (i) anti-competitive cross-subsidisation; (ii) using information obtained from competitors for anti-competitive purposes; and (iii) withholding information, including technical information about essential facilities.[184] This list, however, does not represent all 'anti-competitive practices' covered by Section 1.[185] Furthermore, pursuant to Section 5 (Independent Regulators), Members are required to set up a regulatory body separate from, and not accountable to, any supplier of basic telecommunications services, which may be charged with the task of resolving disputes on the conditions of interconnection.[186] In national gas markets, such bodies fix or approve

[181] See Section 2.2, in particular subparagraphs (a) and (b). World Trade Organization, Reference Paper on Telecommunications Services, above n 176.

[182] Ibid. Emphasis added.

[183] See Mathew, Bobjoseph, *The WTO Agreements on Telecommunications* (Bern: Peter Lang, 2003) at 169–170.

[184] World Trade Organization, Reference Paper on Telecommunications Services, above n 176.

[185] Panel Report, *Mexico – Telecoms*, para. 7.232; Bronckers and Larouche, above n 179 at 336.

[186] See Sections 5 and 2.5 in World Trade Organization, Reference Paper on Telecommunications Services, above n 176.

transmission or distribution tariffs or their methodologies; ensure that there are no cross-subsidies between different energy supply activities; monitor investment plans of pipeline operating companies; and fulfil other functions relating to third-party access and capacity establishment rights.[187] Finally, Section 6 (Allocation and Use of Scarce Resources) provides that '[a]ny procedures for the allocation and use of scarce resources, including frequencies, numbers and rights of way, will be carried out in an objective, timely, transparent and non-discriminatory manner'.[188]

In the academic literature, commentators suggested that the 'reference paper on energy services' must also include the requirement of 'vertical unbundling' (namely separation of pipeline operation from other activities, such as supply and production). According to Poretti and Rios-Herran, this mechanism 'remains the most efficacious competition safeguard' preventing cross-subsidisation and enhancing the third-party access to essential facilities.[189] Nartova argues that the failure to impose this requirement on domestic monopolies 'would largely reduce the incentives for other suppliers to enter a given market and compete in the supply'.[190]

The liberalisation of energy services based on these principles would remove obstacles to gas transit by creating competitive opportunities in the market for pipeline transportation services for both domestic and foreign gas shippers. In such a competitive environment, a pipeline operator (unbundled from domestic gas producers and shippers) would not have a commercial incentive to discriminate against foreign gas shippers in allocating third-party access rights among different applicants. Moreover, it would not resist the expansion of its pipeline network if this is commercially justified. If the additional commitments discussed in this section are not limited to cross-border supply or the supply through commercial presence in the importing Members, they will expand Members' transit obligations.

The aforementioned approach proposed by some Members and scholars appears to be inspired by the liberalisation and privatisation of energy markets in a number of Western states, namely the EU and the United

[187] See Article 41 of the Directive 2009/73/EC of the European Parliament and of the Council of 13 July 2009 concerning Common Rules for the Internal Market in Natural Gas and Repealing Directive No. 2003/55/EC, OJ L 211, 14.8.2009, 94.

[188] World Trade Organization, Reference Paper on Telecommunications Services, above n 176.

[189] See Poretti and Rios-Herran, 'A Reference Paper on Energy Services', above n 61, section III.A.

[190] Nartova, Olga, *Energy Services and Competition Policies under WTO Law*, above n 61 at 249.

States. It is based on the assumption that such reforms can create 'new business opportunities and more cross-border trade, so as to achieve efficiency gains, competitive prices, and higher standards of service, ... [and will] contribute to security of supply and sustainability'.[191] In national energy markets, these reforms have normally been triggered by: the fact that the amount of investment required exceeded the capacities of governments and multinational institutions; the lack of public resources to manage the development of domestic infrastructure; and the belief that private companies operating in a competitive environment are more efficient than public companies.[192] It is thus clear that, apart from improving the conditions of gas transit, the model of liberalisation of energy services outlined in this section can produce overall positive effects on the international and national gas markets, stimulating investment, increasing the quality of services and reducing prices.

Nevertheless, as explained in the following section, in reality, the idea to introduce competition in the market for pipeline transportation services, implemented through any of these approaches, will unlikely be readily acceptable to all major stakeholders in the international gas market. This situation poses serious challenges to the progressive development of WTO transit rules through the additional commitments under the GATS described earlier.

4. The improvement of the regulation of gas transit under the GATS – key challenges

In its proposal on the liberalisation of trade in energy services, Norway aptly noted that 'Members are in various phases of regulatory development,' 'competition for energy services varies significantly from one country to another' and that '[d]ifferences in the level of commitments undertaken should ... be expected'.[193] Even the liberalised gas markets of developed Members such as the EU and the United States are based on different rationales and principles. Whereas in the United States the pipeline networks were from the outset built in a competitive environment, the EU pipeline infrastructure was constructed through state funding or under special rights.[194] Even the current 'third energy package' of the EU contains various exceptions to the obligation of

[191] See preamble of the Directive 2009/73/EC, above n 187.

[192] Poretti and Rios-Herran, 'A Reference Paper on Energy Services', above n 61, section I.C.

[193] World Trade Organization, Communication from Norway, above n 174, para. 55.

[194] Talus, 'Just What Is the Scope of the Essential Facilities Doctrine in the Energy Sector?', above n 178 at 1577–1578.

pipeline operators to provide third-party access, including those relating to investment in major new infrastructure.[195]

The vision of the international gas market of other major stakeholders, such as Russia, is based on principles very different to those promoted by the EU and the United States. The vast gap between the visions of the EU and Russia became obvious when, following the 2009 Russo–Ukrainian gas crisis,[196] Russia withdrew its provisional application of the ECT and set forth its own legal framework for East–West energy cooperation, embodied in the Draft Convention on Energy Security. This Convention focuses heavily on energy security (in the sense of the security of gas supply from Russia to its European counterparts), achieved through almost unfettered domestic regulation of the energy-chain operations with very limited interference from the outside.[197] In line with Russia's approach to energy security, Russia's internal energy market is dominated by vertically integrated state-controlled companies that own gas transportation systems, production, and distribution assets, such as Gazprom.[198]

In light of the significant discrepancies between Members' approaches to the development of an international gas market, the ambitious liberalisation reforms in the area of the GATS' regulation of energy services will inevitably face challenges. It is, therefore, unclear whether the 'critical mass' needed for reaching the deal on the improved regulation of gas transit under the GATS is possible. Among ways out here could be a plurilateral agreement on energy services or Members' individual or differentiated commitments in this area. However, if major energy players, such as Russia, were excluded from the negotiations or their outcome, the practical utility of these negotiations for the development of an international gas market would be diminished.

IV. Summary and Concluding Remarks

This chapter discussed two roles the WTO could play to regulate pipeline gas transit more actively and effectively: (i) a dispute settlement forum

[195] See Article 30 in Regulation (EC) No. 715/2009 of the European Parliament and the Council of 13 July 2009 on Conditions for Access to the Natural Gas Transmission Networks and Repealing Regulation, OJ 2009 L 211, 36.

[196] On this crisis, see Chapter II, Section III.A.1.

[197] On this convention, see Belyi et al., above n 80 at 385.

[198] Mitrova, Tatiana, 'Natural Gas in Transition: Systemic Reform Issues', in S. Pirani (ed.), *Russian and CIS Gas Markets and Their Impact on Europe* (New York: Oxford Institute for Energy Studies, Oxford University Press, 2009) 13 at 21–23.

where restrictions on gas transit in the form of a denial of third-party access or capacity establishment rights can be challenged; and (ii) a diplomatic platform for the development and further improvement of the international regulation of particular features of gas transit, including third-party access and capacity establishment rights.

As regards the first role, no formal procedural obstacles were identified that could prevent a Member from challenging restrictions on third-party access and/or capacity establishment rights as inconsistent with WTO transit rules in a WTO dispute settlement proceeding. Such a challenge could have important practical implications for a transit state and transit-dependent states. It could help the parties involved in the dispute to crystallise the scope of their disagreement as well as provide WTO-consistent means to exert both political and economic pressure on the infringer of WTO law, including in the form of retaliation. In this regard, certain practical aspects of the application of WTO rules on retaliation were discussed.

However, should WTO Members wish to improve the WTO's regulation of pipeline gas transit through a legislative reform of WTO transit provisions, this chapter analysed two approaches to this reform. The first approach was the codification of the already existing principles of general public international law in the WTO legal system, namely the principles of effective and integrated rights and economic cooperation, the obligation to negotiate in good faith with a view to implementing freedom of transit and *abus de droit*. These principles, when balanced against the principle of territorial sovereignty in various factual contexts, may point to different ways to implement the obligation to provide a non-discriminatory freedom of gas transit. These ways may include different forms of third-party access and/or capacity establishment rights. The codification would result in a precise articulation of these rights under WTO law, facilitating Members' implementation of their transit obligations and providing a stronger market signal for investors in gas production and supply chains.

Given the relative indeterminacy of principles in general, this study did not attempt to suggest specific transit rules based on these particular principles of general international law. Instead, two legal frameworks were analysed – the ECT and the LOSC – as the examples of a possible regulation of third-party access and capacity establishment rights under WTO law. Both treaties have certain pitfalls. However, some aspects of their regulation of these rights could provide guidance to WTO Members on how to update and improve WTO transit rules in light of Members' evolving concerns pertinent to pipeline gas. The following aspects of the regulation of transit under the ECT and the LOSC are of particular relevance to this study:

a. Both treaties appear to leave sufficient policy space to transit states in regulating transit as well as the access to or use of transit infrastructure.
b. These treaties, however, recognise the importance of access to and use of the infrastructure for operationalising freedom of transit established by these treaties.
c. To ensure effective freedom of transit, the ECT requires that transit states 'take the necessary measures' either in the form of a negative obligation to allow capacity establishment or, if a transit state so wishes, a third-party access right.
d. The ECT neither imposes an obligation on transit states to guarantee a mandatory third-party access right to pipelines nor requires these states to take positive measures to build transit facilities – pipelines.
e. The LOSC addresses the problem of access to and use of transit infrastructure through the prism of good faith cooperation between transit states and transit-dependent states, as well as stipulates that the conditions and modalities of this cooperation must be set out in a bilateral or regional agreement.

As demonstrated throughout this study, the fundamental aspects of the transit regulation under the ECT and the LOSC appear to reflect the common practice of states in the area of transit cooperation as well as principles of general international law applicable in this area.

The second approach to the reform is the progressive development of WTO law through negotiations on the additional commitments under the GATS relevant to gas transit, including third-party access and/or capacity establishment rights. These commitments could be modelled on the additional commitments of Ukraine on pipeline transport or the Telecom Reference Paper. The key challenge these negotiations will likely face lies in a huge gap between Members' visions of how the international market for energy services must be structured. Whereas some Members have advocated the liberalised model based on the principles of competition law, others have resisted this model and have historically relied on state-controlled, vertically integrated enterprises in energy trade. Whether Members would be able to bridge the gap between their different visions of the international energy market remains to be seen.

VII

General Summary and Conclusions

The general question this study aimed to explore is how the WTO could promote the development of an international gas market and ensure energy security and the sustainable development of WTO Members by playing a more prominent role in regulating rights essential to effective pipeline gas transit. Gas transit is network-dependent and, because of its particular physical features, cannot be transported by means other than pipelines. Even an LNG shipment must first be delivered to a vessel via these means, before it is liquefied and sent on to other markets overseas.

Nevertheless, at an inter-regional level, there is no sufficient pipeline infrastructure that would allow gas to travel freely from a supplier to the most lucrative markets. In many instances, this infrastructure has yet to be built. The existing infrastructure, for various reasons, including commercial reasons, often does not contain enough capacity to handle all possible gas supplies.

This, in turn, impedes the shift of WTO Members' economies to cleaner energy sources, such as natural gas, from coal and oil-based energy and raises concerns regarding energy security. Energy security in this context should be understood as access, availability and acceptability of energy sources, such as gas. The important role international trade in energy plays in promoting energy security and sustainable development (encompassing economic, social and environmental concerns of the international community at large and its members) has long been recognised on the political level.

From a supplier perspective, especially when it is a developing and land-locked country, the ability to export gas, including via transit, could make a significant contribution to attracting foreign direct investments as well as economic growth. Without this ability, clearly established by international law, it is unlikely that investors will receive a sufficient market signal to invest in the upstream and downstream segments of energy markets of those states, such as energy production and transportation.

In addition to technical problems that hinder the development of the international gas market and effective gas transit, there are political problems. These political problems result from a commercial reality in which allowing energy transit may not be in the best interest of a transit state. One good example of such a situation is when a transit state is an energy exporter and, for this state, freedom of transit means more competition and loosening its own bargaining power in global trade, as well as foreign politics. However, even if, at a political level, a transit state were to allow gas transit through its territory, the ability to access the existing pipeline infrastructure in its territory could be undermined by the anti-competitive conduct of private pipeline owners or operators, which strive to maintain their monopoly position in relevant markets by foreclosing new entries. As the practice of the EU competition authority demonstrates, this is often the case even in liberalised gas markets, such as the EU.

From a more general standpoint, transit states have naturally been interested in maintaining their absolute territorial sovereignty, which in the view of some scholars and policy makers is not easily reconcilable with the idea of freedom of transit. The principle of freedom of transit has been promoted throughout centuries in different branches of international law, including network-bound energy trade. However, the apparent conflict between the principles of freedom of transit and territorial sovereignty, discussed in Chapters I to III of this study, has been restraining the development of the international regulation of transit, even for land-locked developing countries. For such countries the right to pass through the territory of a transit state is an important prerequisite for their independence vis-à-vis other states.

Nevertheless, despite this restraint, a number of principles of general international law relating to transit appear to have crystallised in state practice, as reflected in a significant number of multilateral, regional and bilateral agreements, including the ECT, the LOSC and the WTO Agreement (namely the GATT 1994). These principles prohibit levying customs duties on traffic in transit and imposing on this traffic charges and regulations that are unreasonable and discriminatory. Importantly, the principle of freedom of transit itself appears to have entered the corpus of general international law as an 'imperfect right' to pass through foreign territory. This right is called 'imperfect' because its practical implementation may depend on the conclusion of a bilateral or regional treaty with a transit state, stipulating the terms and modalities of transit. This practice of giving effect to freedom of transit is widely accepted, especially in the areas of trade that are technically complex, such as trade in network-bound energy. That said, if the terms and conditions of transit, as set out in the implementation agreement, are fulfilled, the right to transit per se cannot be denied, subject to limited exceptions.

It is hardly controversial that WTO law regulates freedom of energy transit (including pipeline gas) under the GATT. Moreover, some aspects of energy transit are also addressed in the GATS and other WTO agreements, such as the new Agreement on Trade Facilitation. The question that was not adequately addressed in the academic literature prior to this study is whether and, if so, how WTO law regulates particular aspects of pipeline gas transit, such as: (i) the right of access to pipeline capacity, and (ii) the right to expand or build new pipelines. It has been explained throughout this study that without these two fundamental rights, which naturally complement each other depending on how well the pipeline infrastructure is developed in a transit state, the problem of effective gas transit cannot be resolved in practice. This was clearly demonstrated during the negotiations on the ECT – so far, the only multilateral treaty specifically devoted to energy cooperation – and in some regional regimes, such as the EU. These rights have been broadly termed in this study as 'third-party access' and 'capacity establishment' rights, respectively.

This study analysed two main questions. The first question examined is whether and how WTO law regulates third-party access and capacity establishment rights in the context of pipeline gas transit. To answer this question, Chapter IV examined the relevant provisions of the GATT, in particular: (i) the principle of freedom of transit in Article V:2 (first sentence); (ii) the prohibition of discrimination under Articles V:2 (second sentence) and V:5; and (iii) the conditions for filing a non-violation complaint established by Article XXIII:1(b). Other provisions in the GATT as well as other WTO agreements were also analysed to the extent that they were directly or indirectly relevant to transit obligations.

It was found in Chapter IV that WTO transit rules are relatively vague and do not explicitly address the relationship between relevant primary rights established by WTO law, such as freedom of transit, and inherent ancillary rights, such as third-party access and capacity establishment. There are different views among scholars on the scope of WTO obligations with respect to gas transit. However, one common limitation in the existing scholarly works, which appears to undermine their conclusions significantly, is that, while focusing on the textual interpretation of WTO law, scholars have generally been reluctant to walk through all the interpretative steps required by the principles of treaty interpretation, including resorting to the principle of systemic integration. Yet, when the treaty itself does not provide a clear answer to a question, the principle of systemic integration is of particular importance, since it allows the interpreter to fall back on the external legal environment encompassing, inter alia, the principles of general international law and other relevant rules applicable among WTO Members. This fall-back fulfils a number of important functions. First, it may clarify the meaning of the interpreted treaty term. Second, it is likely that the systemic approach to treaty interpretation will ensure the coherent application and development of international law sources.

The broad wording of WTO transit rules, however, does not in and of itself indicate that these rules do not regulate third-party access and/or capacity establishment rights. On the contrary, it was argued in this study that certain aspects of these rights fall within the scope of the obligations to provide freedom transit, on a non-discriminatory basis, under GATT Article V:2 (first and second sentences), to the extent that these rights are inherent in the practical implementation of the GATT transit obligations in the context of trade in pipeline gas. Such an interpretation of WTO transit rules is consistent with the ordinary meaning of the generic term 'freedom' of transit, as it is understood in the light of the contemporary concerns of WTO Members in the area of network-bound energy trade, the relevant context and the object and purpose of the WTO Agreement, and is in line with the principle of freedom of transit in general international law. Third-party access and capacity establishment rights could also be enforced in the WTO through a 'non-violation complaint' under GATT Article XXIII:1(b), based on Members' reasonable expectation that WTO transit obligations would not be circumvented through restrictions on rights essential to effective gas transit. As explained in Chapter IV, this provision can be invoked regardless of whether a particular measure *violates* WTO law and could, therefore, serve as a complementary or alternative cause of action for claims regarding restrictions on third-party access and/or capacity establishment rights.

Chapter V analysed a number of additional legal principles that appear to have crystallised in general international law and support the interpretation of WTO transit rules as encompassing certain aspects of third-party access and/or capacity establishment rights. These principles are: 'effective or integrated rights', 'economic cooperation', the principle that 'negotiations (on the terms and modalities of gas transit) must be conducted in good faith' (derived from the '*pactum de contrahendo*' nature of

GATT Article V:2), and 'abuse of rights (*abus de droit*)'. They are the balancing tools that may assist WTO Members or a panel in striking a reasonable balance between effective freedom of transit, on one hand, and legitimate sovereign concerns of transit states, on the other hand.

The principle of effective rights requires that when the implementation of a primary obligation established by international law (inter alia, the obligation to provide freedom of transit under GATT Article V:2) depends on the recognition of certain inherent ancillary rights, potentially including third-party access and/or capacity establishment rights, both the primary obligation and the inherent ancillary rights must be given effect. This principle finds support in the commonly recognised national law doctrines of servitudes and easements, judgements of international courts and tribunals analysing the relationship between primary and inherent ancillary rights in different areas of international law, in particular fluvial law, as well as treaty regimes regulating trade in energy, such as the ECT and regional economic agreements.

Furthermore, to the extent that a transit state intends to give effect to its obligations under GATT Article V, it must be guided by the principle of economic cooperation and a general duty to negotiate in good faith with a view to achieving a non-discriminatory freedom of gas transit. Pursuant to these legal instruments, when a particular obligation established by international law (such as freedom of transit under the GATT) contains an element of uncertainty as to how it must be implemented in a particular practical situation, the terms and modalities of the implementation of this obligation must be devised through cooperation and good faith negotiations, including at the level of bilateral or regional agreements. Such negotiations may involve international institutions or organisations promoting the development of cross-border infrastructure and willing to finance a particular pipeline project. WTO Members may formally raise issues relating to the implementation of transit obligations in the Committee on Trade Facilitation, which is contemplated under the Agreement on Trade Facilitation. In the context of network-bound energy trade, the cooperative approach to opening energy markets and developing cross-border pipeline infrastructure is commonly endorsed by the members of the international community in different international declarations and UN resolutions.

It is possible that in certain cases the negotiations on the terms and modalities of gas transit may not be fruitful and the parties involved could legitimately disagree on particular commercial terms. In such cases, a transit state cannot be held liable for the failure. Nevertheless, the abuse of rights in the course of the negotiations on matters of a highly political nature such as energy cannot be excluded. With a view to delimiting the boundary between a genuine disagreement on the conditions of gas transit from an abusive conduct of a transit state, Chapter V analysed a number of principles controlling the process of the negotiations. Furthermore, it discussed the principle of *abus de droit*, which may assist international adjudicators in assessing the 'reasonableness' of regulatory requirements or demands of a transit state imposed as a precondition to giving the permission to transit. Pursuant to the principle of *abus de droit*, a reasonable regulation must be compatible with the letter and spirit of the transit state's international obligations (including obligations under both WTO law and other relevant rules and principles of international law), appropriate and necessary for achieving a rational public policy objective, and must not confer an unfair advantage on the relevant stakeholders in the transit state. A transit regulation will likely be found compatible with the transit state's international obligations if it addresses the conditions of transit and not the right to transit, which has already been established under international law as an 'imperfect right'. It has been posited in this study that a refusal to negotiate on the practical implementation of WTO transit rules or the frustration of the negotiations, whether in an open or disguised form, can lead to a violation of WTO law, as well as other relevant rules and principles of public international law.

The final terms and modalities of gas transit agreed between the negotiators may vary depending on particular facts, including geography and commercial considerations. Certainly, no gas transit project is the same as others, due to the differences in the projects' regulatory structures and the stakeholders involved. Normally, these terms and modalities are set out in a bilateral or regional agreement. The approach of dividing the regulation of complex transit matters, such as energy transit, into different layers of international governance, whereby the fundamental principles and rules would be laid down at a multilateral level (such as the WTO), whereas the specifics would be addressed at a bilateral or regional level, has a long history. This history stretches from the transit doctrine postulated by the classical exponents of natural law to the Treaty of Versailles and then the LOSC. This approach appears to be commonly recognised by states and applied by the gas industry.

The cooperation-based approach to the regulation of gas transit supported by the systemic interpretation of WTO transit obligations in the light of principles of general international law, suggested in this study, provides transit states with the regulatory autonomy necessary to achieve their legitimate policy objectives, such as environmental protection or the safe operation of the pipeline system. According to this approach, a transit state does not have an obligation to provide compulsory third-party access to its pipeline system if the system is not open to general use and is operated for private purposes. Likewise, it is not obliged to allow investments in its territory to construct new pipeline facilities. Thus, the transit state appears to have certain discretion as to how it prefers to operationalise its WTO transit obligations, provided that the terms and modalities of gas transit that it accepts or proposes are consistent with international law. Given, however, that there are only a few technical mechanisms through which freedom of transit can be practically effected in the context of pipeline gas, these terms and modalities would necessarily have to include some form of a third-party access right or a capacity establishment right. For example, a transit state, independently or jointly with other states or their investors, may agree to construct a new pipeline, allocate a certain portion of its existing pipeline capacity for transit purposes, or merely not place obstacles in the way of new pipeline capacity being established in its territory by domestic or foreign private entities, consistently with its national legislation governing, inter alia, easements or licensing procedures. The question of which entities would be entitled to operate the constructed gas pipeline in the transit state's territory (i.e. render pipeline transportation services) is addressed separately under the GATS. As was explained in Chapters IV and VI, in light of the current GATS' obligations, this question has limited relevance to transit rights per se, established by the GATT.

In Chapter VI, two roles the WTO could play in regulating gas transit, including third-party access and capacity establishment rights, were explored. These roles include: (i) a dispute settlement forum for the enforcement of existing transit obligations under the GATT; and (ii) a diplomatic platform for further improvement of the international regulation of particular features of gas transit. This study, in its largest part, focused on the first of these roles. While Chapters III to V analysed rules and principles of international law relevant to the WTO's regulation of gas transit, Chapter VI assessed whether and, if so, how these legal sources can be applied in the WTO dispute settlement proceedings to enforce third-party access and capacity establishment rights. The main conclusion here is that, in principle and subject to particular facts, these rights are enforceable as any other rights under the covered agreements – their violation can, therefore, result in concrete practical consequences for a transit state, including retaliatory measures.

In addition, Chapter VI emphasised certain advantages of reforming WTO transit rules. This study does not suggest that the WTO is the only appropriate forum to regulate all aspects of gas transit. It was explained in Chapters I and III that there are other international regimes, such as the ECT and various regional and bilateral treaties, dealing with energy-specific issues. The WTO and bilateral/regional modes of regulating energy transit are complementary and fulfil different functions. The WTO, as a multilateral economic treaty, can provide general principles and rules regulating this matter, acceptable to all 162 Members, while other agreements can be properly regarded as 'implementation agreements' setting out the detailed terms and modalities of gas transit relevant to a particular region or a particular cross-border gas transit project. The reason this study discussed the improvement of transit rules within the WTO specifically is that, unlike other bilateral or regional economic frameworks, the WTO is a rule-based regime that has a robust dispute settlement mechanism, with an impressive record of compliance with panel and the Appellate Body's decisions.

Chapter VI analysed two approaches to the legislative reform of WTO transit rules: (i) the codification of the relevant principles of general international law discussed in this study in the WTO legal system; and (ii) the progressive development of these rules by expanding Members' additional commitments on energy services relevant to third-party access and/or capacity establishment rights under the GATS. The codification (i.e. the first approach) could improve the clarity of WTO transit obligations relevant to pipeline gas and build on the consensus among all WTO Members regarding the application of these obligations in particular contexts. Although there could be different ways to codify these principles, the ECT and the LOSC were discussed as illustrative examples of how third-party access and capacity establishment rights or their analogues, established by these treaties, could resolve the problem of the lack of adequate transit infrastructure or capacity and consequently enhance the effectiveness of freedom of transit. Some aspects of the regulation of these rights under the ECT and the LOSC could, therefore, guide WTO Members in their efforts to codify principles of general international law relevant to transit into WTO law.

The progressive development of transit rules through the negotiations on the additional commitments under the GATS is the second approach.

As explained in Chapter VI, these additional commitments could be modelled on the additional commitments of Ukraine on pipeline transport, or, as proposed by some Members and scholars, on the Telecom Reference Paper. Both models contain certain elements of competition rules. The question that can be asked here is whether these negotiations would likely be successful, as significant differences exist between Members' regulatory approaches to trade in energy services and given the fact that competition matters have traditionally been regulated outside WTO law. In light of these factors, the negotiations on additional commitments on energy services based on these models will have to overcome serious political challenges.

By analysing these approaches to the legislative reform of WTO transit rules, Chapter VI answered the second question raised in this study: how could WTO transit rules be improved through a legislative reform to regulate particular aspects of pipeline gas transit better – namely third-party access and capacity establishment rights?

None of the roles the WTO could play in regulating gas transit is new to this organisation. The WTO has played these roles successfully throughout its history, with more than 300 disputes resolved and new legal instruments concluded at the Bali Ministerial Conference coming into force. The questions that remain to be addressed in the context of the WTO's regulation of pipeline gas transit are of a political nature: (i) whether this matter is ripe enough to be regarded seriously in the WTO; and (ii) how the panels and/or the Appellate Body would approach the interpretation of WTO transit rules when faced with restrictions imposed on WTO Members' right to gas transit (including third-party access and capacity establishment rights). This study made a modest attempt to lay out a legal framework that would, in the author's view, strike a reasonable balance between the effective freedom of gas transit, on one hand, and legitimate sovereign concerns of transit states, on the other hand. It is, of course, at the discretion of WTO Members, or the WTO DSB, whether and how to apply this legal framework to the particular facts of each case or in the process of law making.

APPENDIX 1

SELECTED LEGAL MATERIALS

1. Vienna Convention on the Law of Treaties, Vienna, 23 May 1969

Article 31: General rule of interpretation

1. A treaty shall be interpreted in good faith in accordance with the ordinary meaning to be given to the terms of the treaty in their context and in the light of its object and purpose.

2. The context for the purpose of the interpretation of a treaty shall comprise, in addition to the text, including its preamble and annexes:

(a) any agreement relating to the treaty which was made between all the parties in connexion with the conclusion of the treaty;
(b) any instrument which was made by one or more parties in connexion with the conclusion of the treaty and accepted by the other parties as an instrument related to the treaty.

3. There shall be taken into account, together with the context:

(a) any subsequent agreement between the parties regarding the interpretation of the treaty or the application of its provisions;
(b) any subsequent practice in the application of the treaty which establishes the agreement of the parties regarding its interpretation;
(c) any relevant rules of international law applicable in the relations between the parties.

4. A special meaning shall be given to a term if it is established that the parties so intended.

Article 32: Supplementary means of interpretation

Recourse may be had to supplementary means of interpretation, including the preparatory work of the treaty and the circumstances of its conclusion, in order to confirm the meaning resulting from the application of article 31, or to determine the meaning when the interpretation according to article 31:

(a) leaves the meaning ambiguous or obscure; or
(b) leads to a result which is manifestly absurd or unreasonable.

2. Convention and Statute on Freedom of Transit Barcelona, 20 April 1921

. . .

Desirous of making provision to secure and maintain freedom of communications and of transit,

Being of opinion that in such matters general conventions to which other Powers may accede at a later date constitute the best method of realising the purpose of Article 23(e) of the Covenant of the League of Nations,

Recognizing that it is well to proclaim the right of free transit and to make regulations thereon as being one of the best means of developing co-operation between States without prejudice to their rights of sovereignty or authority over routes available for transit,

. . .

Statute: Article 1

Persons, baggage and goods, and also vessels, coaching and good stock, and other means of transport, shall be deemed to be in transit across territory under the sovereignty or authority of one of the Contracting States, when the passage across such territory, with or without trans-shipment, warehousing, breaking bulk, or change in the mode of transport, is only a portion of a complete journey, beginning and terminating beyond the frontier of the State across whose territory the transit takes place.

Traffic of this nature is termed in this Statute "traffic in transit".

Article 2

Subject to the other provisions of this Statute, the measures taken by Contracting States for regulating and forwarding traffic across territory under their sovereignty or authority shall facilitate free transit by rail or waterway on routes in use convenient for international transit. No distinction shall be made which is based on the nationality of persons, the flag of vessels, the place of origin, departure, entry, exit or destination, or on any circumstances relating to the ownership of goods or of vessels, coaching or goods stock or other means of transport.

In order to ensure the application of the provisions of this Article, Contracting States will allow transit in accordance with the customary conditions and reserves across their territorial waters.

Article 3

Traffic in transit shall not be subject to any special dues in respect of transit (including entry and exit). Nevertheless, on such traffic in transit there may be levied dues intended solely to defray expenses of supervision and administration entailed by such transit. The rate of any such dues must correspond as nearly as possible with the expenses which they are intended to cover, and the dues must be imposed under the conditions of equality laid down in the preceding Article, except that on certain routes, such dues may be reduced or even abolished on account of differences in the cost of supervision.

Article 4

The Contracting States undertake to apply to traffic in transit on routes operated or administered by the State or under concession, whatever may be the place of departure or destination of the traffic, tariffs which, having regard to the conditions of the traffic and to considerations of commercial competition between routes, are reasonable as regards both their rates and the method of their application. These tariffs shall be so fixed as to facilitate international traffic as much as possible. No charges, facilities or restrictions shall depend, directly or indirectly, on the nationality or ownership of the vessel or other means of transport on which any part of the complete journey has been or is to be accomplished.

...

3. Relevant WTO Instruments: The General Agreement on Tariffs and Trade Marrakesh, 12–15 April 1994

Article V: Freedom of transit

1. Goods (including baggage), and also vessels and other means of transport, shall be deemed to be in transit across the territory of a contracting party when the passage across such territory, with or without trans-shipment, warehousing, breaking bulk, or change in the mode of transport, is only a portion of a complete journey beginning and terminating beyond the frontier of the contracting party across whose territory the traffic passes. Traffic of this nature is termed in this article "traffic in transit".

2. There shall be freedom of transit through the territory of each contracting party, via the routes most convenient for international transit, for traffic in transit to or from the territory of other contracting parties. No distinction shall be made which is based on the flag of vessels, the place of origin, departure,

entry, exit or destination, or on any circumstances relating to the ownership of goods, of vessels or of other means of transport.

3. Any contracting party may require that traffic in transit through its territory be entered at the proper custom house, but, except in cases of failure to comply with applicable customs laws and regulations, such traffic coming from or going to the territory of other contracting parties shall not be subject to any unnecessary delays or restrictions and shall be exempt from customs duties and from all transit duties or other charges imposed in respect of transit, except charges for transportation or those commensurate with administrative expenses entailed by transit or with the cost of services rendered.

4. All charges and regulations imposed by contracting parties on traffic in transit to or from the territories of other contracting parties shall be reasonable, having regard to the conditions of the traffic.

5. With respect to all charges, regulations and formalities in connection with transit, each contracting party shall accord to traffic in transit to or from the territory of any other contracting party treatment no less favourable than the treatment accorded to traffic in transit to or from any third country.*

6. Each contracting party shall accord to products which have been in transit through the territory of any other contracting party treatment no less favourable than that which would have been accorded to such products had they been transported from their place of origin to their destination without going through the territory of such other contracting party. Any contracting party shall, however, be free to maintain its requirements of direct consignment existing on the date of this Agreement, in respect of any goods in regard to which such direct consignment is a requisite condition of eligibility for entry of the goods at preferential rates of duty or has relation to the contracting party's prescribed method of valuation for duty purposes.

7. The provisions of this Article shall not apply to the operation of aircraft in transit, but shall apply to air transit of goods (including baggage).

**Ad Article V, paragraph 5*

With regard to transportation charges, the principle laid down in paragraph 5 refers to like products being transported on the same route under like conditions.

Article XXIII: Nullification or impairment

1. If any contracting party should consider that any benefit accruing to it directly or indirectly under this Agreement is being nullified or impaired or that the attainment of any objective of the Agreement is being impeded as the result of

(a) the failure of another contracting party to carry out its obligations under this Agreement, or
(b) the application by another contracting party of any measure, whether or not it conflicts with the provisions of this Agreement, or
(c) the existence of any other situation . . .

Telecommunications Services: Reference Paper, 24 April 1996

Definitions

Users mean service consumers and service suppliers.

Essential facilities mean facilities of a public telecommunications transport network or service that

(a) are exclusively or predominantly provided by a single or limited number of suppliers; and
(b) cannot feasibly be economically or technically substituted in order to provide a service.

A major supplier is a supplier which has the ability to materially affect the terms of participation (having regard to price and supply) in the relevant market for basic telecommunications services as a result of:

(a) control over essential facilities; or
(b) use of its position in the market.

1. Competitive safeguards

1.1 Prevention of anti-competitive practices in telecommunications

Appropriate measures shall be maintained for the purpose of preventing suppliers who, alone or together, are a major supplier from engaging in or continuing anti-competitive practices.

1.2 Safeguards

The anti-competitive practices referred to above shall include in particular:

(a) engaging in anti-competitive cross-subsidization;
(b) using information obtained from competitors with anti-competitive results; and
(c) not making available to other services suppliers on a timely basis technical information about essential facilities and commercially relevant information which are necessary for them to provide services.

2. *Interconnection*

2.1 This section applies to linking with suppliers providing public telecommunications transport networks or services in order to allow the users of one supplier to communicate with users of another supplier and to access services provided by another supplier, where specific commitments are undertaken.

2.2 Interconnection to be ensured

Interconnection with a major supplier will be ensured at any technically feasible point in the network. Such interconnection is provided.

(a) under non-discriminatory terms, conditions (including technical standards and specifications) and rates and of a quality no less favourable than that provided for its own like services or for like services of non-affiliated service suppliers or for its subsidiaries or other affiliates;
(b) in a timely fashion, on terms, conditions (including technical standards and specifications) and cost-oriented rates that are transparent, reasonable, having regard to economic feasibility, and sufficiently unbundled so that the supplier need not pay for network components or facilities that it does not require for the service to be provided; and
(c) upon request, at points in addition to the network termination points offered to the majority of users, subject to charges that reflect the cost of construction of necessary additional facilities.

2.3 Public availability of the procedures for interconnection negotiations

The procedures applicable for interconnection to a major supplier will be made publicly available.

2.4 Transparency of interconnection arrangements

It is ensured that a major supplier will make publicly available either its interconnection agreements or a reference interconnection offer.

2.5 Interconnection: dispute settlement

A service supplier requesting interconnection with a major supplier will have recourse, either:

(a) at any time or
(b) after a reasonable period of time which has been made publicly known

to an independent domestic body, which may be a regulatory body as referred to in paragraph 5 below, to resolve disputes regarding appropriate terms, conditions and rates for interconnection within a reasonable period of time, to the extent that these have not been established previously.

3. *Universal service*

Any Member has the right to define the kind of universal service obligation it wishes to maintain. Such obligations will not be regarded as anti-competitive per se, provided they are administered in a transparent, non-discriminatory and competitively neutral manner and are not more burdensome than necessary for the kind of universal service defined by the Member.

4. *Public availability of licensing criteria*

Where a licence is required, the following will be made publicly available:

(a) all the licensing criteria and the period of time normally required to reach a decision concerning an application for a licence and
(b) the terms and conditions of individual licences.

The reasons for the denial of a licence will be made known to the applicant upon request.

5. *Independent regulators*

The regulatory body is separate from, and not accountable to, any supplier of basic telecommunications services. The decisions of and the procedures used by regulators shall be impartial with respect to all market participants.

6. *Allocation and use of scarce resources*

Any procedures for the allocation and use of scarce resources, including frequencies, numbers and rights of way, will be carried out in an objective, timely, transparent and non-discriminatory manner. The current state of allocated frequency bands will be made publicly available, but detailed identification of frequencies allocated for specific government uses is not required.

Bali Ministerial Declaration
Bali, 7 December 2013

Part II – Doha Development Agenda Trade Facilitation

- Agreement on Trade Facilitation – Ministerial Decision – WT/MIN(13)/36 – WT/L/911

In this regard, we reaffirm that the non-discrimination principle of Article V of GATT 1994 remains valid.

Agreement on Trade Facilitation

Article 11: Freedom of transit

1. Any regulations or formalities in connection with traffic in transit imposed by a Member shall not be:

a. maintained if the circumstances or objectives giving rise to their adoption no longer exist or if the changed circumstances or objectives can be addressed in a reasonably available less trade-restrictive manner,
b. applied in a manner that would constitute a disguised restriction on traffic in transit.

2. Traffic in transit shall not be conditioned upon collection of any fees or charges imposed in respect of transit, except the charges for transportation or those commensurate with administrative expenses entailed by transit or with the cost of services rendered.

3. Members shall not seek, take, or maintain any voluntary restraints or any other similar measures on traffic in transit. This is without prejudice to existing and future national regulations, bilateral or multilateral arrangements related to regulating transport, consistent with WTO rules.

4. Each Member shall accord to products which will be in transit through the territory of any other Member treatment no less favourable than that which would be accorded to such products if they were being transported from their place of origin to their destination without going through the territory of such other Member.

5. Members are encouraged to make available, where practicable, physically separate infrastructure (such as lanes, berths and similar) for traffic in transit.

6. Formalities, documentation requirements and customs controls in connection with traffic in transit shall not be more burdensome than necessary to:

(a) identify the goods; and
(b) ensure fulfilment of transit requirements.

7. Once goods have been put under a transit procedure and have been authorized to proceed from the point of origination in a Member's territory, they will not be subject to any customs charges nor unnecessary delays or restrictions until they conclude their transit at the point of destination within the Member's territory.

8. Members shall not apply technical regulations and conformity assessment procedures within the meaning of the Agreement on Technical Barriers to Trade to goods in transit.

9. Members shall allow and provide for advance filing and processing of transit documentation and data prior to the arrival of goods.

10. Once traffic in transit has reached the customs office where it exits the territory of a Member, that office shall promptly terminate the transit operation if transit requirements have been met.

11. Where a Member requires a guarantee in the form of a surety, deposit or other appropriate monetary or non-monetary[1] instrument for traffic in transit, such guarantee shall be limited to ensuring that requirements arising from such traffic in transit are fulfilled.

12. Once the Member has determined that its transit requirements have been satisfied, the guarantee shall be discharged without delay.

13. Each Member shall, in a manner consistent with its laws and regulations, allow comprehensive guarantees which include multiple transactions for same operators or renewal of guarantees without discharge for subsequent consignments.

14. Each Member shall make publicly available the relevant information it uses to set the guarantee, including single transaction and, where applicable, multiple transaction guarantee.

15. Each Member may require the use of customs convoys or customs escorts for traffic in transit only in circumstances presenting high risks or when compliance with customs laws and regulations cannot be ensured through the use of guarantees. General rules applicable to customs convoys or customs escorts shall be published in accordance with Article 1.

16. Members shall endeavour to cooperate and coordinate with one another with a view to enhancing freedom of transit. Such cooperation and coordination may include, but is not limited to, an understanding on:

(a) charges;
(b) formalities and legal requirements; and
(c) the practical operation of transit regimes.

17. Each Member shall endeavour to appoint a national transit coordinator to which all enquiries and proposals by other Members relating to the good functioning of transit operations can be addressed.

4. United Nations Convention on the Law of the Sea, Montego Bay, 10 December 1982

Part X: Right of access of land-locked states to and from the sea and freedom of transit

[1] Nothing in this provision shall preclude a Member from maintaining existing procedures whereby the means of transport can be used as a guarantee for traffic in transit.

Article 124: Use of terms

1. For the purposes of this Convention:

(a) "land-locked State" means a State which has no sea-coast;
(b) "transit State" means a State, with or without a sea-coast, situated between a land-locked State and the sea, through whose territory traffic in transit passes;
(c) "traffic in transit" means transit of persons, baggage, goods and means of transport across the territory of one or more transit States, when the passage across such territory, with or without trans-shipment, warehousing, breaking bulk or change in the mode of transport, is only a portion of a complete journey which begins or terminates within the territory of the land-locked State;
(d) "means of transport" means:
   (i) railway rolling stock, sea, lake and river craft and road vehicles;
   (ii) where local conditions so require, porters and pack animals.

2. Land-locked States and transit States may, by agreement between them, include as means of transport pipelines and gas lines and means of transport other than those included in paragraph 1.

Article 125: Right of access to and from the sea and freedom of transit

1. Land-locked States shall have the right of access to and from the sea for the purpose of exercising the rights provided for in this Convention including those relating to the freedom of the high seas and the common heritage of mankind. To this end, land-locked States shall enjoy freedom of transit through the territory of transit States by all means of transport.

2. The terms and modalities for exercising freedom of transit shall be agreed between the land-locked States and transit States concerned through bilateral, subregional or regional agreements.

3.Transit States, in the exercise of their full sovereignty over their territory, shall have the right to take all measures necessary to ensure that the rights and facilities provided for in this Part for land-locked States shall in no way infringe their legitimate interests.

Article 126: Exclusion of application of the most-favoured-nation clause

The provisions of this Convention, as well as special agreements relating to the exercise of the right of access to and from the sea, establishing rights and facilities on account of the special geographical position of land-locked States, are excluded from the application of the most-favoured-nation clause.

Article 127: Customs duties, taxes and other charges

1. Traffic in transit shall not be subject to any customs duties, taxes or other charges except charges levied for specific services rendered in connection with such traffic.

2. Means of transport in transit and other facilities provided for and used by land-locked States shall not be subject to taxes or charges higher than those levied for the use of means of transport of the transit State.

Article 128: Free zones and other customs facilities

For the convenience of traffic in transit, free zones or other customs facilities may be provided at the ports of entry and exit in the transit States, by agreement between those States and the land-locked States.

Article 129: Cooperation in the construction and improvement of means of transport

Where there are no means of transport in transit States to give effect to the freedom of transit or where the existing means, including the port installations and equipment, are inadequate in any respect, the transit States and land-locked States concerned may cooperate in constructing or improving them.

Article 130: Measures to avoid or eliminate delays or other difficulties of a technical nature in traffic in transit

1. Transit States shall take all appropriate measures to avoid delays or other difficulties of a technical nature in traffic in transit.

2. Should such delays or difficulties occur, the competent authorities of the transit States and land-locked States concerned shall cooperate towards their expeditious elimination.

Article 131: Equal treatment in maritime ports

Ships flying the flag of land-locked States shall enjoy treatment equal to that accorded to other foreign ships in maritime ports.

Article 132: Grant of greater transit facilities

This Convention does not entail in any way the withdrawal of transit facilities which are greater than those provided for in this Convention and which are agreed between States Parties to this Convention or granted by a State Party. This Convention also does not preclude such grant of greater facilities in the future.

...

Part XVI: General Provisions Article 300: Good faith and abuse of rights

States Parties shall fulfil in good faith the obligations assumed under this Convention and shall exercise the rights, jurisdiction and freedoms recognized in this Convention in a manner which would not constitute an abuse of right.

5. The Energy Charter Treaty Lisbon, 17 December 1994

Understanding IV:1(b)(i)

1. With respect to the Treaty as a whole ... (b) The provisions of the Treaty do not:

(i) oblige any Contracting Party to introduce mandatory third party access.

Article 7: Transit[2]

(1) Each Contracting Party shall take the necessary measures to facilitate the Transit of Energy Materials and Products consistent with the principle of freedom of transit and without distinction as to the origin, destination or ownership of such Energy Materials and Products or discrimination as to pricing on the basis of such distinctions, and without imposing any unreasonable delays, restrictions or charges.
(2) Contracting Parties shall encourage relevant entities to cooperate in:
    (a) modernising Energy Transport Facilities necessary to the Transit of Energy Materials and Products;
    (b) the development and operation of Energy Transport Facilities serving the Areas of more than one Contracting Party;

[2] See Final Act of the European Energy Charter Conference, Declarations, n. 3 with respect to Article 7, p. 31.

(c) measures to mitigate the effects of interruptions in the supply of Energy Materials and Products;
(d) facilitating the interconnection of Energy Transport Facilities.

(3) Each Contracting Party undertakes that its provisions relating to transport of Energy Materials and Products and the use of Energy Transport Facilities shall treat Energy Materials and Products in Transit in no less favourable a manner than its provisions treat such materials and products originating in or destined for its own Area, unless an existing international agreement provides otherwise.

(4) In the event that Transit of Energy Materials and Products cannot be achieved on commercial terms by means of Energy Transport Facilities the Contracting Parties shall not place obstacles in the way of new capacity being established, except as may be otherwise provided in applicable legislation which is consistent with paragraph (1).[3]

(5) A Contracting Party through whose Area Energy Materials and Products may transit shall not be obliged to

(a) permit the construction or modification of Energy Transport Facilities; or
(b) permit new or additional Transit through existing Energy Transport Facilities,
which it demonstrates to the other Contracting Parties concerned would endanger the security or efficiency of its energy systems, including the security of supply.
Contracting Parties shall, subject to paragraphs (6) and (7), secure established flows of Energy Materials and Products to, from or between the Areas of other Contracting Parties.

(6) A Contracting Party through whose Area Energy Materials and Products transit shall not, in the event of a dispute over any matter arising from that Transit, interrupt or reduce, permit any entity subject to its control to interrupt or reduce, or require any entity subject to its jurisdiction to interrupt or reduce the existing flow of Energy Materials and Products prior to the conclusion of the dispute resolution procedures set out in paragraph (7), except where this is specifically provided for in a contract or other agreement governing such Transit or permitted in accordance with the conciliator's decision.

(7) The following provisions shall apply to a dispute described in paragraph (6), but only following the exhaustion of all relevant contractual or other dispute resolution remedies previously agreed between the Contracting Parties party to the dispute or between any entity referred to in paragraph (6) and an entity of another Contracting Party party to the dispute:

[3] See Article 32(1), p. 79 and Annex T, pp. 113 and 122.

(a) A Contracting Party party to the dispute may refer it to the Secretary-General by a notification summarizing the matters in dispute. The Secretary-General shall notify all Contracting Parties of any such referral.
(b) Within 30 days of receipt of such a notification, the Secretary-General, in consultation with the parties to the dispute and the other Contracting Parties concerned, shall appoint a conciliator. Such a conciliator shall have experience in the matters subject to dispute and shall not be a national or citizen of or permanently resident in a party to the dispute or one of the other Contracting Parties concerned.
(c) The conciliator shall seek the agreement of the parties to the dispute to a resolution thereof or upon a procedure to achieve such resolution. If within 90 days of his appointment he has failed to secure such agreement, he shall recommend a resolution to the dispute or a procedure to achieve such resolution and shall decide the interim tariffs and other terms and conditions to be observed for Transit from a date which he shall specify until the dispute is resolved.
(d) The Contracting Parties undertake to observe and ensure that the entities under their control or jurisdiction observe any interim decision under subparagraph (c) on tariffs, terms and conditions for 12 months following the conciliator's decision or until resolution of the dispute, whichever is earlier.
(e) Notwithstanding subparagraph (b) the Secretary-General may elect not to appoint a conciliator if in his judgement the dispute concerns transit that is or has been the subject of the dispute resolution procedures set out in subparagraphs (a) to (d) and those proceedings have not resulted in a resolution of the dispute.
(f) The Charter Conference shall adopt standard provisions concerning the conduct of conciliation and the compensation of conciliators.

(8) Nothing in this Article shall derogate from a Contracting Party's rights and obligations under international law including customary international law, existing bilateral or multilateral agreements, including rules concerning submarine cables and pipelines.
(9) This Article shall not be so interpreted as to oblige any Contracting Party which does not have a certain type of Energy Transport Facilities used for Transit to take any measure under this Article with respect to that type of Energy Transport Facilities. Such a Contracting Party is, however, obliged to comply with paragraph (4).

(10) For the purposes of this Article:

(a) 'Transit' means:

(i) the carriage through the Area of a Contracting Party, or to or from port facilities in its Area for loading or unloading, of Energy Materials and Products originating in the Area of another state and destined for the Area of a third state, so long as either the other state or the third state is a Contracting Party; or

(ii) the carriage through the Area of a Contracting Party of Energy Materials and Products originating in the Area of another Contracting Party and destined for the Area of that other Contracting Party, unless the two Contracting Parties concerned decide otherwise and record their decision by a joint entry in Annex N. The two Contracting Parties may delete their listing in Annex N by delivering a joint written notification of their intentions to the Secretariat, which shall transmit that notification to all other Contracting Parties. The deletion shall take effect four weeks after such former notification.

(b) 'Energy Transport Facilities' consist of high-pressure gas transmission pipelines, high-voltage electricity transmission grids and lines, crude oil transmission pipelines, coal slurry pipelines, oil product pipelines, and other fixed facilities specifically for handling Energy Materials and Products.

APPENDIX 2

REGULATION OF TRANSIT IN SELECTED MULTILATERAL TREATIES

Treaty:	Barcelona Convention on Freedom of Transit 1921	United Nations Convention on the Law of the Sea 1982	GATT 1994	Energy Charter Treaty 1994
Number of Contracting Parties:[1]	36 (original) and 20 (subsequent)	166	162	54
Common Transit Principles:				
Freedom of transit	. . . the measures taken by Contracting States for regulating and forwarding traffic across territory under their sovereignty or authority shall facilitate free transit by rail or waterway on routes in use convenient for international transit. (Statute on Freedom of Transit, Article 2)	Land-locked States shall have the right of access to and from the sea for the purpose of exercising the rights provided for in this Convention . . . To this end, land-locked States shall enjoy freedom of transit through the territory of transit States by all means of transport (Article 125(1)).	There shall be freedom of transit through the territory of each contracting party, via the routes most convenient for international transit, for traffic in transit to or from the territory of other contracting parties (Article V:2).	Each Contracting Party shall take the necessary measures to facilitate the Transit of Energy Materials and Products consistent with the principle of freedom of transit . . . (Article 7(1))

[1] Based on official data from the United Nations, World Trade Organization and Energy Charter's websites.

Prohibition of distinctions	No distinction shall be made which is based on . . . the flag of vessels, the place of origin, departure, entry, exit or destination, or on any circumstances relating to the ownership of goods or of vessels, coaching or goods stock or other means of transport (Statute on Freedom of Transit, Article 2).	Means of transport in transit and other facilities provided for and used by land-locked States shall not be subject to taxes or charges higher than those levied for the use of means of transport of the transit State (Article 127(2)). Ships flying the flag of land-locked States shall enjoy treatment equal to that accorded to other foreign ships in maritime ports (Article 131).	No distinction shall be made which is based on the flag of vessels, the place of origin, departure, entry, exit or destination, or on any circumstances relating to the ownership of goods, of vessels or of other means of transport (Article V:2).	. . . without distinction as to the origin, destination or ownership of such Energy Materials and Products or discrimination as to pricing on the basis of such distinctions . . . (Article 7(1))
Exemption from customs duties and	Traffic in transit shall not be subject to any	Traffic in transit shall not be subject to any	. . . traffic coming from or going to the	Note that, pursuant to the Trade

(*cont*).

Treaty:	Barcelona Convention on Freedom of Transit 1921	United Nations Convention on the Law of the Sea 1982	GATT 1994	Energy Charter Treaty 1994
other charges linked to transit	special dues in respect of transit (including entry and exit). Nevertheless, on such traffic in transit there may be levied dues intended solely to defray expenses of supervision and administration entailed by such transit. (Statute on Freedom of Transit, Article 3).	customs duties, taxes or other charges except charges levied for specific services rendered in connection with such traffic (Article 127(1)).	territory of other contracting parties . . . shall be exempt from customs duties and from all transit duties or other charges imposed in respect of transit, except charges for transportation or those commensurate with administrative expenses entailed by transit or with the cost of services rendered (Article V:3).	Amendment, the ECT incorporates by reference Article V of the GATT.

Transit charges and regulations must be reasonable Unnecessary delays and restrictions must be avoided	The Contracting States undertake to apply to traffic in transit on routes operated or administered by the State or under concession . . . tariffs which, having regard to the conditions of the traffic and to considerations of commercial competition between routes, are reasonable as regards both their rates and the method of their application. These tariffs shall be so fixed as to facilitate international traffic as much as possible (Statute on Freedom of Transit, Article 4).	Transit States shall take all appropriate measures to avoid delays or other difficulties of a technical nature in traffic in transit (Article 130(1)).	. . . traffic coming from or going to the territory of other contracting parties shall not be subject to any unnecessary delays or restrictions (Article V:3) All charges and regulations imposed by contracting parties on traffic in transit to or from the territories of other contracting parties shall be reasonable, having regard to the conditions of the traffic (Article V:4).	. . . and without imposing any unreasonable delays, restrictions or charges (Article 7(1)).

APPENDIX 3

EXAMPLES OF TRANSIT GAS PIPELINES AND FLOWS, 2014

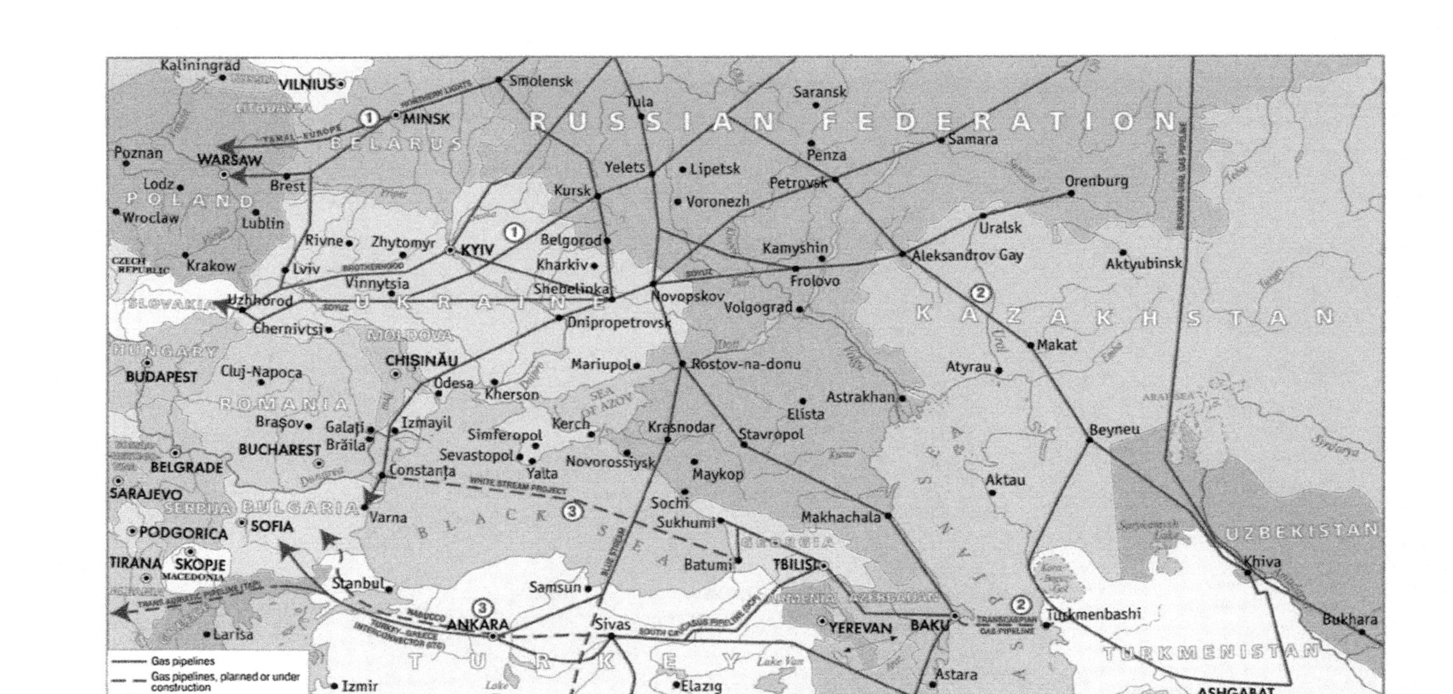

MAP 1: EAST TO WEST

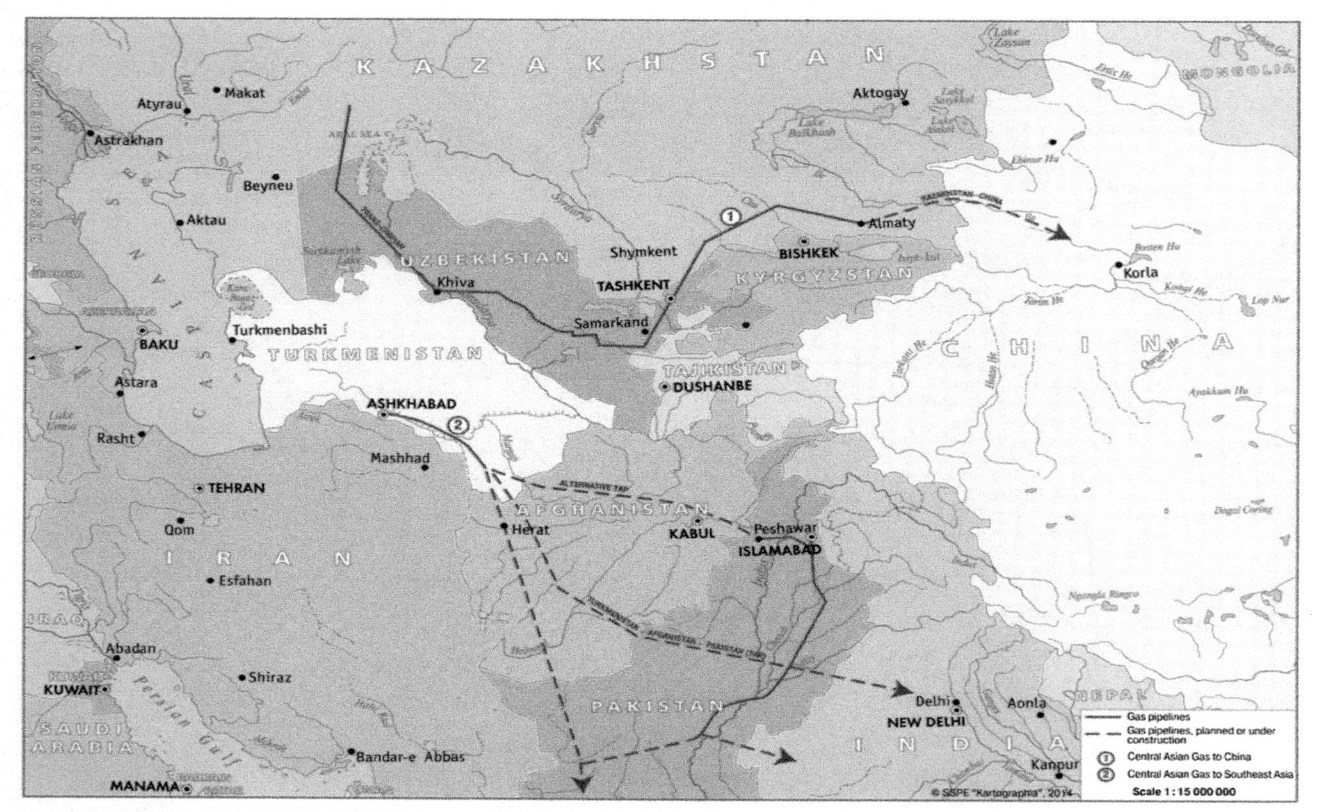

MAP 2: CENTRAL ASIA TO THE FAR EAST AND SOUTH EAST

BIBLIOGRAPHY

1. Books, Articles, Research Papers and Studies

Alexy, Robert, *A Theory of Constitutional Rights*, J. Rivers (trans.) (New York: Oxford University Press, 2010)

Alford, Roger P., 'The Self-Judging WTO Security Exception', 3 *Utah Law Review* (2011) 697

Armstrong, S. W., 'The Doctrine of the Legal Equality of Nations in International Law and the Relations of the Doctrine to the Treaty of Versailles', 14 *American Journal of International Law* (1920) 540

Azaria, Danae, 'Energy Transit under the Energy Charter Treaty and the General Agreement on Tariffs and Trade', 27 *Journal of Energy & Natural Resources Law* (2009) 559

Azaria, Danae, *Treaties on Transit of Energy via Pipelines and Countermeasures* (Oxford: Oxford University Press, 2015)

Ballem, John B., 'International Pipelines: Canada – United States', 18 *The Canadian Yearbook of International Law* (1980) 148

Barnidge, Robert P., 'The International Law of Negotiations as a Means of Dispute Settlement', 36 *Fordham International Law Journal* (2013) 545

Barrera-Hernández, Lila, 'South American Energy Networks Integration: Mission Possible?', in M. M. Roggenkamp et al. (eds.), *Energy Networks and the Law: Innovative Solutions in Changing Markets* (New York: Oxford University Press, 2012) 61

Bedjaoui, Mohammed, 'The Right to Development', in M. Bedjaoui (ed.), *International Law: Achievements and Prospects* (Dordrecht: Martinus Nijhoff Publishers, 1991) 1177

Belyi, Andrei, 'International Energy Security Viewed by Russia', 5 *Oil, Gas & Energy Law Intelligence* (2007), www.ogel.org, accessed 15 February 2014

Belyi, Andrei, Nappert, Sophie and Pogoretskyy, Vitaliy, 'Modernizing the Energy Charter Process? The Energy Charter Conference Road Map and the Russian Draft Convention on Energy Security', 29(3) *Journal of Energy & Natural Resources Law* (2011) 383

Benvenisti, Eyal, 'Sovereigns as Trustees of Humanity: On the Accountability of States to Foreign Stakeholders', 107 *American Journal of International Law* (2013) 295

Beukenkamp, Annelieke, 'Pipeline-to-Pipeline Competition: An EU Assessment', 27 *Journal of Energy & Natural Resources Law* (2009) 5

Bigos, Bradley J., 'Contemplating GATS Article XVIII on Additional Commitments', 42 *Journal of World Trade* (2008) 723

Botchway, Francis N. 'Can the Law Compel Business Parties to Negotiate?', 3 *Journal of World Energy Law & Business* (2010) 286

Boute, Anatole, 'The Good Neighbourliness Principle in EU External Energy Relations: The Case of Energy Transit', in D. Kochenov and E. Basheska (eds.), *The Principle of Good Neighbourliness in the European Legal Context* (Leiden: Brill Nijhoff, 2015), 354

Bown, Chad P. and Pauwelyn, Joost (eds.), *The Law, Economics and Politics of Retaliation in WTO Dispute Settlement* (New York: Cambridge University Press, 2010)

Brölmann, Catherine, 'Law-Making Treaties: Form and Function in International Law', 74 *Nordic Journal of International Law* (2005) 383

Bronckers, Marco and Baetens, Freya, 'Reconsidering Financial Remedies in WTO Dispute Settlement', 16(2) *Journal of International Economic Law* (2013) 281

Bronckers, Marco and Larouche, Pierre, 'A Review of the WTO Regime for Telecommunications Services', in K. Alexander and M. Andenas (eds.), *The World Trade Organization and Trade in Services* (The Netherlands: Koninklijke Brill NV, 2008) 319

Broom, Herbert, *A Selection of Legal Maxims: Classified and Illustrated*, 10th ed., R. H. Kersley (ed.) (Universal Law Publishing, 2006, reprinted)

Broude, Tomer, 'Principles of Normative Integration and the Allocation of International Authority: The WTO, the Vienna Convention on the Law of Treaties, and the Rio Declaration', The Hebrew University of Jerusalem, Research Paper No. 07–08 (August 2008)

Brownlie, Ian, *Principles of Public International Law*, 7th ed. (New York: Oxford University Press, 2008)

Buckland, W. W., *A Text-Book of Roman Law from Augustus to Justinian*, P. Stein (ed.) (New York: Cambridge University Press, 1963)

Buckland, W. W. and McNair, Arnold, D., *Roman Law & Common Law: A Comparison in Outline*, 2nd ed., F. H. Lawson (ed.) (Cambridge: University Press, 1952)

Byers, Michael, 'Abuse of Rights: An Old Principle, A New Age', 47 *McGill Law Journal* (2002) 390

Caflisch, Lucius, C., 'Land-Locked States and Their Access to and from the Sea', 49 *British Year Book of International Law* (1978) 71

Cameron, Peter D., 'Completing the Internal Market in Energy: An Introduction to the New Legislation', in P. Cameron (ed.), *Legal Aspects of EU Energy Regulation: Implementing the New Directives on Electricity and Gas across Europe* (New York: Oxford University Press, 2005) 7

Cameron, Peter D., 'The Energy Charter Treaty and East-West Transit', in G. Coop (ed.), *Energy Dispute Resolution: Investment Protection, Transit and the Energy Charter Treaty* (New York: Juris, 2011) 297

Cameron, Peter D., 'The EU and Energy Security: A Critical Review of the Legal Issues', in A. Antoniadis, R. Shultze and E. Spaventa (eds.), *The European Union and Global Emergencies: A Law and Policy Analysis* (Oxford: Hart Publishing, 2011) 125

Campbell, David, and Thomas, Philip (eds.), *Fundamental Legal Conceptions as Applied in Judicial Reasoning by Wesley Newcomb Hohfeld* (Aldershot: Ashgate Darmouth, 2001)

Casas de las Peñas del Corral, Amalia, 'Regional Energy Integration: A Wide and Worthy Challenge for South America', 5 *Journal of World Energy Law & Business* (2012), 166

Chapman, Simon, 'Multi-tiered Dispute Resolution Clauses: Enforcing Obligations to Negotiate in Good Faith', 27 *Journal of International Arbitration* (2010) 89

Cheng, Bin, *General Principles of Law as Applied by International Courts and Tribunals* (New York: Cambridge University Press, 2006)

Chua, A. 'Reasonable Expectations and Non-violation Complaints in GATT/WTO Jurisprudence', 32 *Journal of World Trade* (1998) 27

Churchill, R. R. and Lowe, A. V., *The Law of the Sea*, 3rd ed. (Manchester: Manchester University Press, 1999)

Clark, Bryan, 'Transit and the Energy Charter Treaty: Rhetoric and Reality', 5 *Web Journal of Current Legal Issues* (1998)

Condorelli, Luigi and Kress, Claus, 'The Rules of Attribution: General Considerations', in J. Crawford et al. (eds.), *The Law of International Responsibility* (New York: Oxford University Press, 2010) 221

Cook, Graham, *A Digest of WTO Jurisprudence on Public International Law Concepts and Principle* (Cambridge: Cambridge University Press, 2015)

Cossy, Mireille, 'Energy Services under the General Agreement on Trade in Services', in Y. Selivanova (ed.), *Regulation of Energy in International Trade Law: WTO, NAFTA and Energy Charter* (The Netherlands: Kluwer Law International, 2011) 149

Cossy, Mireille, 'Energy Trade and WTO Rules: Reflexions on Sovereignty over Natural Resources, Export Restrictions and Freedom of Transit', in C. Herrmann and J. P. Terhechte (eds.), 3 *European Yearbook of International Economic Law* (2012) 281

Cottier, Thomas and Foltea, Marina, 'Constitutional Functions of the WTO and Regional Trade Agreements', in L. Bartels and F. Ortino (eds.), *Regional Trade*

Agreements and the WTO Legal System (New York: Oxford University Press, 2006) 43

Cottier, Thomas and Nadakavukaren, Schefer K., 'Non-violation Complaints in WTO/GATT Dispute Settlement: Past, Present and Future', in E-U. Petersmann (ed.), *International Trade Law and the GATT/WTO Dispute Settlement System* (London: Kluwer Law International, 1997) 143

Cottier, Thomas et al., 'Energy in WTO Law and Policy', NCCR, Working Paper No 2009/25 (May 2009), http://phase1.nccr-trade.org/images/stories/projects/ip6/IP6%20Working%20paper.pdf (last visited 30 November 2013)

D'Amato, Anthony A., *The Concept of Custom in International Law* (London: Cornell University Press, 1971)

Davey, William J. 'Dispute Settlement in the WTO and RTAs: A Comment', in L. Bartels and F. Ortino (eds.), *Regional Trade Agreements and the WTO Legal System* (New York: Oxford University Press, 2006) 343

Dawidowicz, Martin, 'The Effect of the Passage of Time on the Interpretation of Treaties: Some Reflections on Costa Rica v. Nicaragua', 24 *Leiden Journal of International Law* (2011) 201

de Vattel, Emer, *The Law of Nations, Or Principles of the Law of Nature, Applied to the Conduct and Affairs of Nations and Sovereigns, with Three Early Essays on the Origin and Nature of Natural Law and on Luxury*, vol. 2, Kapossy, B. and Whatmore, R. (eds.) (Indianapolis, IN: Liberty Fund, 2008)

de Visscher, Charles, *Le Droit International des Communications* (Paris: Université de Gand, 1921)

Degan, Vladimir D., *Sources of International Law* (The Netherlands: Kluwer Law International, 1997)

Desta, Melaku G., 'GATT/WTO Jurisprudence in the Energy Sector and Movements in the Marketplace', *Oil, Gas & Energy Law Intelligence* (2003), www.ogel.org, accessed 15 February 2014

Desta, Melaku G., 'The GATT/WTO System and International Trade in Petroleum: An Overview', 21 *Journal of Energy and Natural Resources Law* (2003) 385

Desta, Melaku G., 'Soft Law in International Law: An Overview', in A. K. Bjorklund and A. Reinisch (eds.), *International Investment Law and Soft Law* (Cheltenham: Edward Elgar Publishing, 2012) 39

Doré, Julia, 'Negotiating the Energy Charter Treaty', in T. W. Wälde (ed.), *The Energy Charter Treaty: An East–West Gateway for Investment & Trade* (London: Kluwer Law International, 1996) 137

Dow, Stephen, Siddiky, Ishrak A. and Ahmmad, Yadgar K., 'Cross-border Oil and Gas Pipelines and Cross-border Waterways: A Comparison between the Two Legal Regimes', 6 *Journal of World Energy Law and Business* (2013) 107

Dupuy, René-Jean and Vignes, Daniel (eds.), *A Handbook on the New Law of the Sea*, 2 vols. (Boston: Hague Academy of International Law, Martinus Nijhoff Publishers, 1991)

Durling, James P., and Lester, Simon N., 'Original Meanings and the Film Dispute: The Drafting History, Textual Evolution, and Application of the Non-violation Nullification or Impairment Remedy', 32 *George Washington Journal of International Law and Economics* (1999) 211

Dworkin, Ronald, *Taking Rights Seriously* (London: Duckworth, 1977)

Ehlermann, Claus-Dieter, 'Six Years on the Bench of the "World Trade Court": Some Personal Experiences as Member of the Appellate Body of the World Trade Organization', 36 *Journal of World Trade* (2002) 605

Ehring, Lothar and Selivanova, Yulia, 'Energy Transit', in Y. Selivanova (ed.), *Regulation of Energy in International Trade Law: WTO, NAFTA and Energy Charter* (The Netherlands: Kluwer Law International, 2011) 49

Einhorn, Talia, 'Transit of Goods over Foreign Territory', in *Max Planck Encyclopaedia of Public International Law* (2010), www.mpepil.com, accessed 30 November 2011

Eljuri, Elisabeth and Johnston, Daniel, 'Mexico's Energy Sector Reform', 17 *Journal of World Energy Law and Business* (2014) 168

Energy Charter Secretariat, 'Bringing Gas to the Market. Gas Transit and Transmission Tariffs in Energy Charter Treaty Countries: Regulatory Aspects and Tariff Methodologies' (Brussels: Energy Charter Secretariat, 2012)

Energy Charter Secretariat, 'Gas Transit Tariffs in Selected Energy Charter Treaty Countries' (Brussels: Energy Charter Secretariat, 2006)

Energy Charter Secretariat, 'Model Intergovernmental and Host Government Agreements for Cross-border Pipelines', 2nd ed. (Brussels: Energy Charter Secretariat, 2007)

Energy Charter Secretariat, 'Putting a Price on Energy: International Pricing Mechanisms for Oil and Gas' (Brussels: Energy Charter Secretariat, 2007)

Energy Charter Secretariat, 'Trade in Energy: WTO Rules Applying under the Energy Charter Treaty' (Brussels: Energy Charter Secretariat, 2001)

Evans, Peter C., 'Strengthening WTO Member Commitments in Energy Services: Problems and Prospects', in P. Sauvé and A. Mattoo (eds.), *Domestic Regulation and Services Trade Liberalization* (Washington, DC: World Bank Publications, 2003) 167

Farran d'Olivier, C., 'International Enclaves and the Question of State Servitudes', 4 *International and Comparative Law Quarterly* (1955) 294

Fastenrath, Ulrich, 'Relative Normativity in International Law', 4 *European Journal of International Law* (1993) 305

Fawcett, J., 'Trade and Finance in International Law', 123 *Recueil des Cours* (1968) 214

Flower, Andy, 'Natural Gas from the Middle East', in J. Stern (ed.), *Natural Gas in Asia: The Challenges of Growth in China, India, Japan, and Korea* (New York: Oxford Institute for Energy Studies, Oxford University Press, 2008) 330

Footer, Mary, 'The Return to "Soft Law" in Reconciling the Antinomies in WTO Law', 11 *Melbourne Journal of International Law* (2010) 241

French, Duncan, 'Treaty Interpretation and the Incorporation of Extraneous Legal Rules', 55 *International and Comparative Law Quarterly* (2006) 281

Fridley, David, 'Natural Gas in China', in J. Stern (ed.), *Natural Gas in Asia: The Challenges of Growth in China, India, Japan, and Korea* (New York: Oxford Institute for Energy Studies, Oxford University Press, 2008) 7

Friedrich, H. M., 'Legal Aspects of Transit Carriage in Gas Transmission Systems: State Obligations and Private Ownership' (thesis on file at the Centre for Energy, Petroleum and Mineral Law and Policy, 1989)

Gaffney, John P., 'The GATT and the GATS: Should They Be Mutually Exclusive Agreements?', 12 *Leiden Journal of International Law* (1999) 135

García-Castrillón, Carmen O., 'Private Parties under the Present WTO (Bilateralist) Competition Regime', 35 *Journal of World Trade* (2001) 99

Gardiner, Richard, *Treaty Interpretation* (New York: Oxford University Press, 2008)

Gentili, Alberico, *De Iure Belli Libri Tres*, vol. 2, J. B. Scott (ed.), J. C. Rolfe (trans.) (London: Clarendon Press, 1933, reprinted)

Gómez-Palacio, Ignacio and Muchlinski, Peter, 'Admission and Establishment', in P. Muchlinski et al. (eds.), *The Oxford Handbook of International Investment Law* (New York: Oxford University Press, 2008) 227

González, Anabel, 'The Rationale for Bringing Investment into the WTO', in S. J. Evenett and A. Jara, *Building on Bali: A Work Programme for the WTO* (A VoxEU.org eBook, Centre for Economic Policy Research, 2013) 67

Gordley, James and Von Mehren, Arthur T., *An Introduction to the Comparative Study of Private Law: Readings, Cases, Materials* (New York: Cambridge University Press, 2006)

Grewlich, Klaus W., 'International Regulatory Governance of the Caspian Pipeline Policy Game', 29 *Journal of Energy & Natural Resources Law* (2011) 87

Grotius, Hugo, *The Rights of War and Peace*, vol. 2, Tuck, R. and Barbeyrac, J. (eds.) (Indianapolis, IN: Liberty Fund, 2005)

Haghighi, Sanam S., *Energy Security: The External Legal Relations of the European Union with Major Oil- and Gas-Supplying Countries* (Portland: Hart Publishing, 2007)

Hall, William E., *A Treatise on International Law*, 3rd ed. (Oxford: Clarendon Press, 1890)

Hestermeyer, Holger P., 'Article XXI Security Exceptions', in R. Wolfrum et al. (eds.), *WTO – Trade in Goods* (Leiden: Martinus Nijhoff Publishers, 2011) 569

Higgins, Rosalyn, *Problems and Process: International Law and How We Use It* (Oxford: Clarendon Press, 1994)

Hohfeld, Wesley, N., 'Some Fundamental Legal Conceptions as Applied in Judicial Reasoning', 23 *Yale Law Journal* (1913–1914) 16

Hudec, Robert E., *The GATT Legal System and World Trade Diplomacy* (New York: Praeger Publishers, Inc., 1975) at 4–5 (citing Clair Wilcox)

Hufbauer, Gary C. and Schott, Jeffrey J., *NAFTA Revisited: Achievements and Challenges* (Washington, DC: Institute for International Economics, 2005)

Hyde, Charles C., *International Law Chiefly as Interpreted and Applied by the United States*, vol. 1, 2nd ed. (Boston: Little Brown and Company, 1951)

International Energy Agency, 'Natural Gas Market Review' (Paris: IEA Publications, 2009)

International Energy Agency, 'Tracking Industrial Energy Efficiency and CO2 Emissions' (Paris: IEA Publications, 2007)

Jackson, John H., 'Comment on Chapter 2', in C. Bown and J. Pauwelyn (eds.), *The Law, Economics and Politics of Retaliation in WTO Dispute Settlement* (New York: Cambridge University Press, 2010) 66

Jackson, John H., 'History of the General Agreement on Tariffs and Trade', in R. Wolfrum et al. (eds.), *WTO – Trade in Goods* (Leiden: Martinus Nijhoff Publishers, 2011) 1

Jackson, John H., *Sovereignty, the WTO and Changing Fundamentals of International Law* (New York: Cambridge University Press, 2006)

Jackson, John H., *World Trade and the Law of GATT* (Indianapolis, IN: The Bobbs-Merill Company, Inc., 1969)

Jackson, John H., 'The WTO Dispute Settlement System after Ten Years: The First Decade's Promises and Challenges', in Y. Taniguchi, A. Yanovich and J. Bohanes, *The WTO in the Twenty-First Century: Dispute Settlement, Negotiations, and Regionalism in Asia* (New York: Cambridge University Press, 2007)

Janssens, Thomas and Wessely, Thomas (eds.), 'Dominance: The Regulation of Dominant Firm Conduct in 40 Jurisdictions Worldwide' (Global Competition Review, 2010)

Jenkins, David, 'An Oil and Gas Industry Perspective', in T. W. Wälde (ed.), *The Energy Charter Treaty: An East–West Gateway for Investment & Trade* (London: Kluwer Law International, 1996) 187

Jia, Bing B., *The Regime of Straits in International Law* (Oxford: Clarendon Press, 1998)

Jia, Bing B., 'The Relations between Treaties and Custom', 9 *Chinese Journal of International Law* (2010) 81

Kiss, Alexandre, 'Abuse of Rights', in *Max Planck Encyclopaedia of Public International Law* (2010), www.mpepil.com, accessed 30 March 2012

Kojima, Chie and Vereshchetin, Vladlen S., 'Implementation Agreements', in *Max Planck Encyclopaedia of Public International Law* (2010), www.mpepil.com, accessed 30 March 2012

Konoplyanik, Andrei, A., 'A Common Russia–EU Energy Space (The New EU–Russia Partnership Agreement, *Acquis Communautaire*, the Energy

Charter and the New Russian Initiative)', 7 *Oil, Gas & Energy Law Intelligence* (2009), www.ogel.org, accessed 15 February 2014

Konoplyanik, Andrei, A., 'Energy Security and the Development of International Energy Markets', in B. Barton et al. (eds.), *Energy Security: Managing Risk in a Dynamic Legal and Regulatory Environment* (New York: Oxford University Press, 2004) 47

Konoplyanik, Andrei A., 'Russia–EU Summit: WTO, the Energy Charter Treaty and the Issue of Energy Transit', 2 *International Energy Law and Taxation Review* (2005) 30

Konoplyanik, Andrei A., 'Russian–Ukrainian Gas Dispute: Prices, Pricing and ECT', 4 (4) *Oil, Gas & Energy Law Intelligence* (2006), www.ogel.org, accessed 15 February 2014

Konoplyanik, Andrei A. and Von Halem, Freidrich, 'The Energy Charter Treaty: A Russian Perspective', in T. W. Wälde (ed.), *The Energy Charter Treaty: An East–West Gateway for Investment & Trade* (London: Kluwer Law International, 1996) 156

Koskenniemi, Martti, *From Apology to Utopia: The Structure of International Legal Argument* (New York: Cambridge University Press, 2009, reissue)

Koskenniemi, Martti, 'General Principles: Reflexions on Constructivist Thinking in International Law', in M. Koskenniemi (ed.), *Sources of International Law* (Aldershot: Ashgate Darmouth, 2000) 360

Kunoy, Bjørn, 'The Ambit of *Pactum de Negotiatum* in the Management of Shared Fish Stock: A Rumble in the Jungle', 11 *Chinese Journal of International Law* (2012) 689

Kurmanov, Baurzhan, 'Transit of Energy Resources under GATT Article V', *Transnational Dispute Management* (provisional issue, January 2013)

Lamy, Pascal, *The Geneva Consensus: Making Trade Work for All* (New York: Cambridge University Press)

Lamy, Pascal, 'The Place of the WTO and Its Law in the International Legal Order', 17 *European Journal of International Law* (2006) 969

Latty, Franck, 'Actions and Omissions', in J. Crawford et al. (eds.), *The Law of International Responsibility* (New York: Oxford University Press, 2010) 355

Lauterpacht, Elihu, 'Freedom of Transit in International Law', 44 *Transactions of the Grotius Society* (1958–1959) 313

Lauterpacht, Elihu, 'River Boundaries: Legal Aspects of the Shatt-Al-Arab Frontier', 9 *International and Comparative Law Quarterly* (1960) 208

Lauterpacht, Hersch, 'Restrictive Interpretation and the Principle of Effectiveness in the Interpretation of Treaties', 26 *British Year Book of International Law* (1949) 48

Lauterpacht, Hersch, *Private Law Sources and Analogies of International Law* (USA: Archon Books, 1970, reprinted)

Lauterpacht, Hersch, *The Function of Law in the International Community* (Hamden: Archon Books, 1966)

Liesen, Rainer, 'Transit under the 1994 Energy Charter Treaty', 17 *Journal of Energy and Natural Resources Law* (1999) 56

Mahmoudi, Said, 'Transit Passage', in *Max Planck Encyclopaedia of Public International Law* (2010), www.mpepil.com, accessed 15 May 2011

Makil, A., 'Transit Rights of Land-Locked Countries: An Appraisal of International Conventions', 4 *Journal of World Trade Law* (1970) 35

Mankiw, Gregory N. and Taylor, Mark P., *Economics* (USA: Thomson, 2006)

Marceau, Gabrielle, 'The New TBT Jurisprudence in *US – Clove Cigarettes, WTO US – Tuna II* and *US – COOL*', 8 *Asian Journal of WTO & International Health Law and Policy* (2013) 1

Marceau, Gabrielle, 'The WTO in the Emerging Energy Governance Debate', in J. Pauwelyn (ed.), *Global Challenges at the Intersection of Trade, Energy and the Environment* (Geneva: The Graduate Institute, Centre for Trade and Economic Integration, 2010) 25

Marceau, Gabrielle, 'WTO Dispute Settlement and Human Rights', 13 *European Journal of International Law* (2002) 753

Marceau, Gabrielle, 'The WTO in the Emerging Energy Governance Debate', 5 *Global Trade and Customs Journal* (2010) 83

Marceau, Gabrielle and Morosini, Fabio C., 'The Status of Sustainable Development in the Law of the World Trade Organization' (8 November 2011), http://ssrn.com/abstract=2547282, accessed 22 December 2015

Mathew, Bobjoseph, *The WTO Agreements on Telecommunications* (Bern: Peter Lang, 2003)

McCaffrey, Stephen C, 'International Watercourses', in *Max Planck Encyclopaedia of Public International Law* (2014), www.mpepil.com, accessed 30 May 2014

McGrady, Benn, 'Fragmentation of International Law or "Systemic Integration" of Treaty Regimes: EC-Biotech Products and the Proper Interpretation of Article 31(3)(c) of the Vienna Convention of the Law of Treaties', 42(3) *Journal of World Trade* (2008) 589

McLachlan, Campbell, 'The Principle of Systemic Integration and Article 31(3)(c) of the Vienna Convention', 54 *International and Comparative Law Quarterly* (2005) 279

McNair, Arnold D., 'The Functions and Differing Legal Character of Treaties', 11 *British Year Book of International Law* (1930) 100

McNair, Arnold D., *The Law of Treaties* (Oxford: Clarendon Press, 1961)

McNair, Arnold D. 'So-Called State Servitudes', 6 *British Year Book of International Law* (1925) 111

Meggiolaro, Francesco and Vergano, Paolo R., 'Energy Services in the Current Round of WTO Negotiations', 11 *International Trade Law & Regulation* (2005) 97

Melgar, Beatriz H., *The Transit of Goods in Public International Law* (Leiden: Brill Nijhoff, 2015)

Merkouris, P. 'Keep Calm and Call (no, not Batman but . . .) Articles 31–32 VCLT: A Comment on Istrefi's Recent Post on R.M.T. v. The UK', *EJIL: Talk!* (19 June 2014), http://www.ejiltalk.org/keep-calm-and-call-no-not-batman-but-articles-31-32-vclt-a-comment-on-istrefis-recent-post-on-r-m-t-v-the-uk/, accessed 20 June 2014

Mitchell, Andrew D., *Legal Principles in WTO Disputes* (New York: Cambridge University Press, 2011)

Mitrova, Tatiana, 'Natural Gas in Transition: Systemic Reform Issues', in S. Pirani (ed.), *Russian and CIS Gas Markets and Their Impact on Europe* (New York: Oxford Institute for Energy Studies, Oxford University Press, 2009) 13

Mommsen, T. et al. (eds.), *The Digest of Justinian*, vol. 1 (Pennsylvania: University of Pennsylvania Press, 1985)

Momtaz, Djamchid, 'Attribution of Conduct to the State: State Organs and Entities Empowered to Exercise Elements of Governmental Authority', in J. Crawford et al. (eds.), *The Law of International Responsibility* (New York: Oxford University Press, 2010) 237

Mowat, R. B., *The Concert of Europe* (London: Macmillan and Co., 1930)

Musselli, Irene and Zarrilli, Simonetta, 'Oil and Gas Services: Market Liberalization and the Ongoing GATS Negotiations', 8 *Journal of International Economic Law* (2005) 551

Nandan, Satya N. and Rosenne, Shabtai (eds.), *United Nations Convention on the Law of the Sea 1982: A Commentary*, vol. 3 (The Hague: Centre for Oceans Law and Policy, Martinus Nijhoff Publishers, 1995)

Nartova, Olga, *Energy Services and Competition Policies under WTO Law* (Moscow: Infra-M, 2010)

Neumann, Jan and Türk, Elisabeth, 'Necessity Revisited: Proportionality in World Trade Organization Law after *Korea-Beef, EC-Asbestos* and *EC-Sardines*', 37 *Journal of World Trade* (2003) 199

Neven, Damien J. and Mavroidis Petros, C., 'El Mess in TELMEX: A Comment on Mexico – Measures Affecting Telecommunications Services', 5 *World Trade Review* (2006) 271

Nottage, Hunter and Sebastian, Thomas, 'Giving Legal Effect to the Results of WTO Trade Negotiations: An Analysis of the Methods of Changing WTO Law', 9 *Journal of International Economic Law* (2006) 989

Nussbaum, Arthur, *A Concise History of the Law of Nations* (New York: The Macmillan Company, 1954, revised edition)

Nychay, Nadiya and Shemelin, Dmitry, 'Interpretation of Article 7(3)of the Energy Charter Treaty', the Graduate Institute of International and Development Studies, Trade and Investment Law Clinic Papers (Geneva 2012), http://gradua

teinstitute.ch/home/research/centresandprogrammes/ctei/projects-1/trade-law-clinic.html, accessed 12 September 2013

O'Connell, D. P., 'A Re-consideration of the Doctrine of International Servitude', 30 *The Canadian Bar Review* (1952) 807

Owada, Hisashi, 'Pactum de Contrahendo, Pactum de Negotiando', in *Max Planck Encyclopaedia of Public International Law* (2010), www.mpepil.com, accessed 30 March 2012

Özen, Erdinç, 'Turkey's Natural Gas Market Expectations and Developments' (Deloitte, April 2012)

Panizzon, Marion, *Good Faith in the Jurisprudence of the WTO: The Protection of Legitimate Expectations, Good Faith Interpretation and Fair Dispute Settlement* (Oxford: Hart Publishing, 2006)

Pauwelyn, Joost, 'The Calculation and Design of Trade Retaliation in Context: What Is the Goal of Suspending WTO Obligations?', in C. P. Bown and J. Pauwelyn (eds.), *The Law, Economics and Politics of Retaliation in WTO Dispute Settlement* (New York: Cambridge University Press, 2010) 34

Pauwelyn, Joost, *Conflict of Norms in Public International Law: How WTO Law Relates to Other Rules of International Law* (New York: Cambridge University Press, 2003)

Phillipson, Coleman, *The International Law and Custom of Ancient Greece and Rome*, vol. 2 (London: The Macmillan Company, 1911)

Pirani, Simon, Stern, Jonathan and Yafimava, Katja, 'The Russo–Ukrainian Gas Dispute of January 2009: A Comprehensive Assessment', Fall (XVIII) *Energy Politics* (2009) 4

Pogoretskyy, Vitaliy, 'Energy Dual Pricing in International Trade: Subsidies and Anti-dumping Perspectives', in Y. Selivanova (ed.), *Regulation of Energy in International Trade Law: WTO, NAFTA and Energy Charter* (The Netherlands: Kluwer Law International, 2011) 181

Pogoretskyy, Vitaliy, 'The Transit Role of Turkey in the European Union's Southern Gas Corridor Initiative: An Assessment from the Perspective of the World Trade Organization and the Energy Charter Treaty', 2(2) *Journal of International Trade and Arbitration Law* (Istanbul, 2013) 121 (also available at www.ogel.org, accessed 15 February 2014)

Pogoretskyy, Vitaliy and Beketov, Sergey, 'Bridging the Abyss? Lessons from Global and Regional Integration of Ukraine', 46(2) *Journal of World Trade* (2012) 457

Pogoretskyy, Vitaliy and Melnyk, Sergii, 'Energy Security, Climate Change and Trade: Does the WTO Provide for a Viable Framework for Sustainable Energy Security?', in P. Delimatsis (ed.), *Research Handbook on Climate Change and Trade Law* (Cheltenham: Edward Elgar Publishing Ltd., 2016) 233

Poretti, Pietro and Rios-Herran, Roberto, 'A Reference Paper on Energy Services: The Best Way Forward?', 4 *Oil, Gas & Energy Law Intelligence* (2006), www.ogel.org, accessed 15 February 2014

Prichard, A. M., *League's Roman Private Law Founded on the Institutes of Gaius and Justinian*, 3rd ed. (London: Macmillan & Co Ltd, 1964)

Rakhmanin, Vladimir, 'Transportation and Transit of Energy and Multilateral Trade Rules: WTO and Energy Charter', in J. Pauwelyn (ed.), *Global Challenges at the Intersection of Trade, Energy and the Environment* (Geneva: The Graduate Institute, Centre for Trade and Economic Integration, 2010) 123

Reid, Helen D., *International Servitudes in Law and Practice* (Chicago, IL: University of Chicago Press, 1932)

Renouf, Yves, 'From Bananas to Byrd: Damage Calculation Coming Age?', in C. P. Bown and J. Pauwelyn (eds.), *The Law, Economics and Politics of Retaliation in WTO Dispute Settlement* (New York: Cambridge University Press, 2010) 135

Rheinstein, Max, 'Observations et Conclusions du Gouvernement de la République Portugaise sur les Exceptions Préliminaire du Gouvernement de l'inde (Annexe 20), Étude Comparative sur le Droit D'Accès aux Domaines Enclaves par le Professeur Rheinstein', in *Case Concerning Right of Passage over Indian Territory*, Judgment of 12 April 1960: I.C.J. Reports 1960, pp. 6, 714

Riley, Alan, 'Can Nordstream and Southstream Survive in a Changing European Gas Market?', 7 *Oil, Gas & Energy Law Intelligence* (2009), www.ogel.org, accessed 15 February 2014

Rivier, Alphonse, *Principes du Droit des Gens*, vol. 1, A. Rousseau (ed.) (Paris: Librairie Nouvelle de Droit et de Jurisprudence, 1896)

Roberts, Peter, *Gas Sales and Gas Transportation Agreements: Principles and Practice*, 2nd ed. (London: Sweet & Maxwell, 2008)

Roessler, Frieder, *The Legal Structure, Functions and Limits of the World Trade Order: A Collection of Essays* (London: Cameron May, 2000)

Roessler, Frieder, 'Should Principles of Competition Policy Be Incorporated into WTO Law through Non-violation Complaints?', 2 *Journal of International Economic Law* (1999) 413

Roggenkamp, Martha M., 'Implications of GATT and EEC on Network-Bound Energy Trade in Europe', 12 *Journal of Energy & Natural Resources Law* (1994) 59

Roggenkamp, Martha M., 'Transit of Network-Bound Energy: The European Experience', in T. W. Wälde (ed.), *The Energy Charter Treaty: An East–West Gateway for Investment & Trade* (London: Kluwer Law International, 1996) 499

Romero, Hugo and Piérola, Fernando, 'Unwritten Measures: Reflections on the Panel Reports in *Argentina – Measures Affecting the Importation of Goods*', 10(1) *Global Trade and Customs Journal* (2015) 54

Root, Elihu, *North Atlantic Fisheries Arbitration at the Hague: Argument on Behalf of the United States*, R. Bacon, and J. B. Scott, (eds.) (Cambridge, MA: Harvard University Press, 1917)

Rosenne, Shabtai, *Developments in the Law of Treaties 1945–1986* (New York: Cambridge University Press, 1989)

Rousseau, Charles, *Droit International Public*, Tome III (Paris: Sirey, 1977)

Sands, Philippe, 'Treaty, Custom and Cross-fertilization of International Law', 1 *Yale Human Rights and Development Law Journal* (1998) 85

Sartori, Nicolò, 'The European Commission vs. Gazprom: An Issue of Fair Competition or a Foreign Policy Quarrel?', Istituto Affari Internazionali Working Papers (3 January 2013)

Sauser-Hall, G., 'Les Droits Des Etats Sur La Force Motrice Des Cours D'Eau Internationaux', 83 *Recueil des Cours* (1953) 539

Sauvé, Pierre, 'Multilateral Rules on Investment: Is Forward Movement Possible?', 9 *Journal of International Economic Law* (2006) 325

Schachter, Oscar, *International Law in Theory and Practice* (Boston: Martinus Nijhoff Publishers, 1991)

Schindler, Dietrich, 'The Administration of Justice in the Swiss Federal Court in Intercantonal Disputes', 15 *American Journal of International Law* (1921) 149

Schwarzenberger, Georg, *International Law*, vol. 1, 3rd ed. (London: Stevens & Sons Limited, 1957)

Schwebel, Stephen M., 'Investor–State Disputes and the Development of International Law: The Influence of Bilateral Investment Treaties on Customary International Law', 98 *ASIL Proceedings* (2004) 27

Sebastian, Thomas, 'The Law of Permissible WTO Retaliation', in C. P. Bown and J. Pauwelyn (eds.), *The Law, Economics and Politics of Retaliation in WTO Dispute Settlement* (New York: Cambridge University Press, 2010) 89

Shaw, Malcolm N., *International Law*, 6th ed. (New York: Cambridge University Press, 2008)

Shelton, Dinah, 'Normative Hierarchy in International Law', 100 *American Journal of International Law* (2006) 291

Simma, Bruno and Paulus, Andreas L., 'The "International Community": Facing the Challenge of Globalization', 9 *European Journal of International Law* (1998) 266

Sinjela, Mpazi A., 'Freedom of Transit and the Right of Access for Land-Locked States: The Evolution of Principle and Law', 12 *Georgia Journal of International and Comparative Law* (1982) 31

Slotboom, Marco, 'Recent Developments of Competition Law and the Impact of the Sector Inquiry', in M. M. Roggenkamp and U. Hammer (eds.), *European Energy Law Report VII* (Antwerp: Intersentia, 2010) 97

Stern, Jonathan, *The Future of Russian Gas and Gazprom* (New York: Oxford Institute for Energy Studies, Oxford University Press, 2005)

Stern, Jonathan and Bradshaw, Michael, 'Russian and Central Asian Gas Supply for Asia', in J. Stern (ed.), *Natural Gas in Asia: Challenges of Growth in China, India, Japan, and Korea*, 2nd ed. (New York: Oxford Institute for Energy Studies, Oxford University Press, 2008) 220

Stevens, Paul, 'Cross-border Oil and Gas Pipelines: Problems and Prospects' (UNDP & World Bank Energy Sector Management Assistance Programme, 2003)

Stratakis, Alexandros, 'Comparative Analysis of the US and EU Approach and Enforcement of the Essential Facilities Doctrine', 27 *European Competition Law Review* (2006) 434

Talus, Kim, 'Just What Is the Scope of the Essential Facilities Doctrine in the Energy Sector?: Third Party Access-Friendly Interpretation in the EU v. Contractual Freedom in the US', 48 *Common Market Law Review* (2011) 1571

Talus, Kim, *Vertical Natural Gas Transportation Capacity, Upstream Commodity Contracts and EU Competition Law* (The Netherlands: Kluwer Law International, 2011)

Tiroch, Katrin, 'Rhine River', in *Max Planck Encyclopaedia of Public International Law* (2014), www.mpepil.com, accessed 30 May 2014

Toulmin, G. E., 'The Barcelona Conference on Communications and Transit and the Danube Statute', 3 *British Year Book of International Law* (1922–1923) 167

Tunkin, Grigory, 'Is General International Law Customary Law Only?', 4 *European Journal of International Law* (1993) 53

United Kingdom's Law Commission, 'Easements, Covenants and Profits à Prendre', Consultation Paper No 186 (2008)

Uprety, Kishor, *The Transit Regime for Landlocked States: International Law and Development Perspectives* (Washington, DC: The World Bank, 2006)

Váli, F. A., *Servitudes of International Law: A Study of Rights in Foreign Territory*, 2nd ed. (London: Stevens & Sons Limited, 1958)

Valles, Cherise, 'Article V Freedom of Transit', in R. Wolfrum et al. (eds.), *WTO – Trade in Goods* (Leiden: Martinus Nijhoff Publishers, 2011) 183

Van den Bossche, Peter and Zdouc, Werner, *The Law and Policy of the World Trade Organization*, 3rd ed. (New York: Cambridge University Press, 2013)

Vasciannie, Stephen C., *Land-Locked and Geographically Disadvantaged States in the International Law of the Sea* (Oxford: Clarendon Press, 1990)

Villiger, Mark E., *Commentary on the 1969 Vienna Convention on the Law of Treaties* (Leiden: Martinus Nijhoff Publishers, 2009)

Villiger, Mark E., *Customary International Law and Treaties: A Manual on the Theory and Practice of the Interrelation of Sources*, 2nd ed. (The Hague: Kluwer Law International, 1997)

Vinogradov, Sergei, 'Challenges of Nord Stream: Streamlining International Legal Frameworks and Regimes for Submarine Pipelines', 9 *Oil, Gas & Energy Law Intelligence* (2011), www.ogel.org, accessed 15 February 2014

Vinogradov, Sergei, 'The Legal Status of the Caspian Sea and Its Hydrocarbon Resources', in G. Blake et al. (eds.), *Boundaries and Energy: Problems and Prospects* (London: Kluwer Law International, 1998) 137

Voigt, Christina, *Sustainable Development as a Principle of International Law: Resolving Conflicts between Climate Measures and WTO Law* (Leiden: Martinus Nijhoff Publishers, 2009)

Von Pufendorf, Samuel, *De Jure Naturae Et Gentium Libri Octo*, vol. 2, J. B. Scott (ed.), C. H. Oldfather and W. A. Oldfather (trans.) (Oxford: Clarendon Press, 1934)

Wälde, Thomas W., 'International Investment under the 1994 Energy Charter Treaty', in T. W. Wälde (ed.), *The Energy Charter Treaty: An East–West Gateway for Investment and Trade* (London: Kluwer Law International, 1996) 251

Wälde, Thomas W., 'Interpreting Investment Treaties: Experiences and Examples', in C. Binder et al. (eds.), *International Investment Law for the 21st Century: Essays in Honour of Christoph Schreuer* (New York: Oxford University Press, 2009) 724

Watkins, G. C., 'Constitutional Imperatives and the Treatment of Energy in the NAFTA', 7 *Oil and Gas Law and Taxation Review* (1994) 199

Weeramantry, Romesh J., *Treaty Interpretation in Investment Arbitration* (Oxford: Oxford University Press, 2012)

Westlake, John, *International Law*, Part I (Cambridge: Cambridge University Press, 1904)

Willems, Arnoud R. and Li, Qing, 'Using WTO Rules to Enforce Energy Transit and Influence the Transit Fee', 4 *European Energy Journal* (2014) 34

Winrow, Gareth, M., 'Problems and Prospects for the "Fourth Corridor": The Positions and Role of Turkey in Gas Transit to Europe' (Oxford Institute for Energy Studies, June 2009)

WTO Secretariat, 'World Trade Report 2008: Trade in a Globalizing World' (Geneva, 2008)

WTO Secretariat, 'World Trade Report 2010: Trade in Natural Resources' (Geneva, 2010)

WTO Secretariat, 'World Trade Report 2011: The WTO and Preferential Trade Agreements: From Co-existence to Coherence' (Geneva, 2011)

Yafimava, Katja, *The Transit Dimension of EU Energy Security: Russian Gas Transit across Ukraine, Belarus, and Moldova* (New York: Oxford Institute for Energy Studies, Oxford University Press, 2011)

Yergin, Daniel, 'Ensuring Energy Security', 85 *Foreign Affairs* (2006) 69

Yergin, Daniel, *The Quest: Energy, Security, and the Remaking of the Modern World* (New York: Penguin Books, 2012)

Zedalis, Rex J., 'When Do the Activities of Private Parties Trigger WTO Rules?', 10 *Journal of International Economic Law* (2007) 335

Zeilinger, Anton F., 'Danube River', in *Max Planck Encyclopaedia of Public International Law* (2014), www.mpepil.com, accessed 30 May 2014

Zillman, Donald N. et al. (eds.), *Human Rights in Natural Resource Development: Public Participation in the Sustainable Development of Mining and Energy Resources* (New York: Oxford University Press, 2002)

Zimmermann, Reinhard, Visser, Daniel and Reid, Kenneth, *Mixed Legal Systems in Comparative Perspective: Property and Obligations in Scotland and South Africa* (New York: Oxford University Press, 2004)

2. Legal Sources

2.1 Treaties and Contracts

Agreement between Belarus, Russia and Kazakhstan of 2010 'On the Rules of Access to Services of Natural Monopolies in the Area of Gas Transportation through Gas Transportation Systems, including the Foundations of Pricing and Tariff Policy' (Соглашение от 9 декабря 2010 'О правилах доступа к услугам субъектов естественных монополий в сфере транспортировки газа по газотранспортным системам, включая основы ценообразования и тарифной политики'), http://www.economy.gov.ru/minec/activity/sections/formuep/agreement/doc20110404_08, accessed 5 September 2013

Agreement of the Southern African Customs Union, http://www.sacu.int, accessed 30 November 2011

Agreements between Azerbaijan, Georgia and Turkey on the South Caucasus Pipeline (SCP) System, in British Petroleum, *Legal Agreements*, http://www.bp.com/en_az/caspian/aboutus/legalagreements.html, accessed 10 December 2012

Association of Southeast Asian Nations, Framework Agreement on the Facilitation of Goods in Transit, http://www.asean.org, accessed 30 November 2011

Association of Southeast Asian Nations, Memorandum of Understanding on the Trans-ASEAN Gas at http://www.asean.org/communities/asean-economic-community/item/the-asean-memorandum-of-understanding-mou-on-the-trans-asean-gas, accessed 23 March 2014

Bilateral Investment Treaty between the Netherlands and the Czech Republic, http://www.unctadxi.org/templates/DocSearch.aspx?id=779, accessed 3 July 2013

Charter of the International Trade Organization, http://www.wto.org/english/docs_e/legal_e/havana_e.pdf, accessed 10 January 2014

Charter of the United Nations of 1945, http://www.un.org/en/documents/charter/, accessed 10 January 2014

Commonwealth of Independent States, Free Trade Agreement, http://www.cis.minsk.by/, accessed 30 November 2011

Contract between Gazprom and Naftogaz Ukrayini (Ukrainian national gas company) on the Volumes and Conditions of Transit of Natural Gas through the

Territory of Ukraine for the Period of 2009–2019, signed in Moscow on 19 January 2009, available at Ukrayinska Pravda, 'The Contract on Transit of Russian Gas + Additional Agreement on the Advance Payment to Gazprom' ('Контракт о транзите российского газа + Допсоглашение об авансе "Газпрома"'), http://www.pravda.com.ua/rus/articles/2009/01/22/4462733/, accessed 10 October 2012

Contract between Gazprom and Naftogaz Ukrayini on the Sales of Natural Gas, signed in Moscow on 19 January 2009, available at Ukrayinska Pravda, 'Gas Deal of Timoshenko-Putin – Full Text' ('Газова угода Тимошенко-Путіна. Повний текст'), http://www.pravda.com.ua/articles/2009/01/22/3686613, accessed 10 October 2012

Convention and Statute of the International Regime of Maritime Ports of 1923, *League of Nations Treaty Series*, No. 1379 (1926–1927) 287

Convention and Statute on Freedom of Transit of 1921, 8 *American Journal of International Law, Supplement* (1924) 118

Convention and Statute on the International Regime of Railways of 1923, *League of Nations Treaty Series*, No. 1129 (1926) 57

Convention and Statute on the Regime of Navigable Waterways of International Concern of 1921, http://www.fao.org/docrep/005/w9549e/w9549e02.htm, visited 5 September 2013

Convention relative to the Transmission in Transit of Electric Power of 1923, 22 *American Journal of International Law Supplement* (1928) 83

Customs Code of the Customs Union (Eurasian Economic Union), http://www.tsouz.ru/Docs/kodeks/Documents/TRANSLATION%20CUC.pdf, accessed 30 November 2015

Energy Charter Secretariat, 'The Energy Charter Treaty and Related Documents: A Legal Framework for International Energy Co-operation' (Brussels: Energy Charter Secretariat, 2004)

Energy Charter Secretariat, Transit Protocol, http://www.encharter.org/index.php?id=37, accessed 23 December 2013

Energy Community Secretariat, Energy Community Treaty, http://www.energy-community.org, accessed 12 September 2013

Final Act of the Congress of Vienna of 1815, http://www.fao.org/docrep/005/w9549e/w9549e02.htm, accessed 5 September 2013

General Act of the Berlin Conference in *Protocoles et Acts Général de la Conférence de Berlin* (1884–1885), Annexe au Protocole No. 10, available at the Peace Palace Library, http://www.peacepalacelibrary.nl/, accessed 10 March 2014

North American Free Trade Agreement, http://www.worldtradelaw.net/fta/agreements/nafta.pdf, accessed 5 September 2013

Protocol on Trade of the Southern African Development Community, http://www.sadc.int, accessed 30 November 2011

Protocol on Transit Trade and Transit Facilities of the Common Market for Eastern and Southern Africa, http://www.comesa.int/, accessed 30 November 2011

Statute of the International Court of Justice of 1945, http://www.icj-cij.org, accessed 10 January 2012

Statute of the International Law Commission, http://legal.un.org/ilc/texts/instruments/english/statute/statute_e.pdf, accessed 15 September 2013

Treaty of the East African Community, http://www.eac.int/, accessed 30 November 2011

Treaty of the Economic Community of West African States, http://www.ecowas.int/, accessed 30 November 2011

Treaty of Versailles of 1919, http://avalon.law.yale.edu/subject_menus/versailles_menu.asp, accessed 30 May 2010

Treaty of Westphalia of 1648, http://avalon.law.yale.edu/17th_century/westphal.asp, accessed 30 September 2011

Treaty on European Union and the Treaty on the Functioning of the European Union and the Charter of Fundamental Rights of the European Union (EU Treaty) at http://europa.eu/lisbon_treaty/full_text/index_en.htm, accessed 5 September 2013

Treaty on the Eurasian Economic Union of 29 May 2014 (Договор 'О Евразийском Экономическом Союзе'), http://www.economy.gov.ru/wps/wcm/connect/economylib4/mer/about/structure/depsng/agreement-eurasian-economic-union, accessed 4 July 2014

United Nations Convention on the Law of the Sea of 1982, http://www.un.org/depts/los/convention_agreements/texts/unclos/unclos_e.pdf, accessed 30 December 2012

Vienna Convention on the Law of Treaties of 1969, 1155 *United Nations Treaty Series* (1980) 331

World Customs Organization, Harmonized Commodity Description and Coding System, http://www.wcoomd.org/en/topics/nomenclature/instrument-and-tools/hs_nomenclature_2012/hs_nomenclature_table_2012.aspx, accessed 30 March 2014

World Trade Organization, 'Schedule of Specific Commitments of Ukraine', GATS/SC/144 (10 March 2008)

World Trade Organization, Ministerial Conference, Agreement on Trade Facilitation, Ministerial Decision of 7 December 2013, WT/MIN(13)/36, WT/L/911 (11 December 2013)/ Preparatory Committee on Trade Facilitation, Agreement on Trade Facilitation, WT/L/931 (15 July 2014)

World Trade Organization, Reference Paper on Telecommunications Services (incorporated into some WTO Members' GATS Schedules of Specific Commitments), www.wto.org/english/tratop_e/serv_e/telecom_e/tel23_e.htm, accessed 30 March 2012

WTO Agreement (Marrakesh Agreement Establishing the World Trade Organization) including other covered agreements, inter alia: Dispute Settlement Understanding, GATT 1994, GATS and TBT Agreement, *WTO*

Legal Texts, http://www.wto.org/english/docs_e/legal_e/legal_e.htm, accessed 15 February 2014

2.2 International Resolutions and Declarations

G8 Summit on Global Energy Security, St. Petersburg (16 July 2006), http://www.g8.utoronto.ca/summit/2006stpetersburg/energy.html, accessed 30 September 2014

Resolution of the Institute of International Law of 1894, 'Règles sur la Définition et le Régime de la Mer Territoriale', Session de Paris, 1894, http://www.idi-iil.org, accessed 30 September 2011

'Rio Declaration on Environment and Development', Rio de Janeiro (3–14 June 1992), http://www.unep.org/documents.multilingual/default.asp?documentid=78&articleid=1163, accessed 15 November 2015

The European Energy Charter, Energy Charter Secretariat, 'The Energy Charter Treaty and Related Documents: A Legal Framework for International Energy Co-operation' (Brussels: Energy Charter Secretariat, 2004)

The International Energy Charter, http://www.energycharter.org/process/international-energy-charter-2015/, accessed 15 September 2015

United Nations, 'Millennium Declaration', adopted by the GA Resolution at 8th plenary meeting, A/55/L.2 (8 September 2000)

United Nations, 'The UN Report of the International Ministerial Conference on Land-Locked and Transit Developing Countries and Transit Transport Co-operation', and the attached Almaty Declaration, A/CONF.202/3 (28 and 29 August 2003)

United Nations, GA Resolution 2625 of 24 October 1970, 'Declaration on Principles of International Law Concerning Friendly Relations and Co-operation among States in Accordance with the Charter of the United Nations', http://www.un.org/documents/resga.htm, accessed 15 November 2015

United Nations, GA Resolution 53/101 'Principles and Guidelines for International Negotiations', http://www.un.org/documents/resga.htm, accessed 15 November 2015

United Nations, GA Resolution 63/210 of 19 December 2008, 'Reliable and Stable Transit of Energy and Its Role in Ensuring Sustainable Development and International Cooperation', http://www.un.org/documents/resga.htm, accessed 15 November 2015

United Nations, GA Resolution 997(ES-I) of 2 November 1956, http://www.un.org/documents/resga.htm, accessed 15 November 2015

United Nations, Security Council Resolution 118 of 13 October 1956, http://www.un.org/en/sc/documents/resolutions/1956.shtml, accessed 15 November 2015

World Trade Organization, Ministerial Conference, Bali Ministerial Declaration, WT/MIN(13)/DEC (11 December 2013)

2.3 *Decisions and Miscellaneous Documents of International Organisations*

Application of Bolivia Instituting Proceedings before the International Court of Justice, http://www.icj-cij.org, last visited 30 August 2013

Crawford, J., Special Rapporteur, 'Second Report on State responsibility', A/CN.4/498 and Add.1–4 (17 March, 1 and 30 April, 19 July 1999)

Decision of the Energy Charter Secretariat, 'Road Map for the Modernisation of the Energy Charter Process' (24 November 2010)

GATT Council, 'Decision Concerning Article XXI of the General Agreement', L/5426 (30 November 1982)

GATT, Consultative Group of Eighteen, 'Restrictive Business Practices', Note by the Secretariat, CG.18/W/44 (10 October 1980)

International Court of Justice, Order of 18 June 2013, http://www.icj-cij.org/docket/files/153/17392.pdf, visited 30 August 2013

International Law Commission, 'Fragmentation of International Law: Difficulties Arising from the Diversification and Expansion of International Law', A/CN.4/L.682 (13 April 2006)

International Law Commission, 'Report of the International Law Commission Covering the Work of Its Twelfth Session', *Yearbook of the International Law Commission*, vol. 2 (1960)

International Law Commission, 'Responsibility of States for Internationally Wrongful Acts with Commentaries', *Yearbook of the International Law Commission*, vol. 2, Part 2 (2001)

International Law Commission, 'Third Report on Identification of Customary International Law, by Michael Wood, Special Rapporteur', A/CN.4/682 (27 March 2015)

International Law Commission, 'Ways and Means for Making the Evidence of Customary International Law More Readily Available', *Yearbook of the International Law Commission*, vol. 2 (1950)

United Nations Secretariat, 'Question of Free Access to the Sea of Land-Locked Countries', A/CONF.13/29 (14 January 1958)

World Trade Organization, Council for Trade in Services, 'Developments in the Air Transport Sector since the Conclusion of the Uruguay Round', Background Note by the Secretariat, S/C/W/163/Add.3 (13 August 2001)

World Trade Organization, DSB Minutes, WT/DSB/M/80, 26 June 2000

World Trade Organization, Request for Consultations Submitted to the WTO by the Russia Federation in *European Union and Its Member States – Certain Measures Relating to the Energy Sector (DS476)*

World Trade Organization, 'Services Classification List', MTN.GNS/W/120 (10 July 1991)
World Trade Organization, Working Party on the Accession of Montenegro, 'Report of the Working Party on the Accession of Montenegro – Addendum, Part II (Schedule of Specific Commitments on Services)', WT/ACC/CGR/38/Add.2/WT/MIN(11)/7/Add.2 (5 December 2011)
World Trade Organization, Working Party on the Accession of the Russian Federation, 'Draft Report of the Working Party on the Accession of the Russian Federation to the World Trade Organization', WT/ACC/SPEC/RUS/25/Rev.2 (27 May 2003)
World Trade Organization, Working Party on the Accession of the Russian Federation, 'Report of the Working Party on the Accession of the Russian Federation to the World Trade Organization', WT/ACC/RUS/70 (17 November 2011)
World Trade Organization, Working Party on the Accession of Ukraine, 'Report of the Working Party on the Accession of Ukraine to the World Trade Organization', WT/ACC/UKR/152 (25 January 2008)

2.4 Negotiating History of Selected Treaties

2.4.1 Energy Charter Treaty (on File at the European Energy Charter Secretariat)

EUROGAS, 'Comment on Revised Draft Basic Agreement for the European Energy Charter', S/EUR/93/172 (8 February 1993)
European Energy Charter Conference Secretariat, 'Basic Agreement for the European Energy Charter (Draft)', 31/92 (BA 13) (19 July 1992)
European Energy Charter Conference Secretariat, 'Memorandum to Legal Sub-group, Agenda for 21–25 June Legal Sub-group Meeting', 47/93 (LEG-9) (Brussels, 9 June 1993)
European Energy Charter Secretariat, 'Note for the Attention of S. Fremantle by C. Jones' 583 (CLJ/JI) (Brussels, 30 August 1993)
European Energy Charter Conference Secretariat, 'Note from the Chairman of Working Group II', 36/93 (BA-40) (Brussels, 6 May 1993)
European Energy Charter Conference Secretariat, 'Note from the Chairman of Working Group II', 14/91 (BP 3) (Brussels, 11 October 1991)
European Energy Charter Conference Secretariat, 'Note from the Chairman of Working Group II', Annex II, 8/91 (BP 2) (Brussels, 12 September 1991)
European Energy Charter Conference Secretariat, 'Revised Draft', Annex, 15/93 (BA-35) (Brussels, 9 February 1993)
European Energy Charter Conference Secretariat, 'Room Document 8', Plenary Session (26 March 1993)

European Energy Charter Conference Secretariat, 'Room Document 10', Plenary Session (24–26 March 1993)

European Energy Charter Conference Secretariat, 'Room Document 2 (Attached to the Basic Agreement)', Working Group II (Brussels, 1 June 1992)

European Energy Charter Conference Secretariat, 'Room Document 6', Working Group II (Brussels, 22–23 April 1993)

European Energy Charter Conference Secretariat, 'Submission from European Communities' (Brussels, 1 April 1992)

European Energy Charter Conference Secretariat, Room Document 26, Plenary Session (Brussels, 28 April 1993)

European Energy Charter Conference Secretariat, Working Group II Meeting, Chaired by Slater of UK (17–18 October 1991)

European Energy Charter Conference Secretariat, Working Group II (1–5 June 1992)

European Energy Charter Conference Secretariat: Comments on Issues Assigned to the Legal Sub-group, No. 1396 (14 May 1993)

General Secretariat of the Council, 'European Energy Charter of Negotiations: Meeting of EC Ad Hoc Group (High Level) on 19 May 1993', Note to Delegations by the Presidency (17 May 1993)

Organisation for Economic Co-operation and Development (OECD), 'Memorandum' (27 May 1993)

2.4.2 General Agreement on Tariffs and Trade of 1947, the Barcelona Convention and the Statute on Freedom of Transit of 1921

Economic and Social Council, 'Preparatory Committee of the International Conference on Trade and Employment', E/PC/T/C.II/54/Rev.1 (28 November 1946)

Economic and Social Council, Preparatory Committee of the International Conference on Trade and Employment (Committee II), 'Report of the Technical Sub-committee', E/PC/T/C.II/54 (16 November 1946)

League of Nations, 'General Transport Situation in 1921. Statements submitted by the States which took part in the First General Conference on Communications and Transit, held in March–April 1921, with an Introduction by Prof. Tajani', vol. 1 (Geneva, 1922)

League of Nations, 'Verbatim Reports and Texts Relating to the Convention on Freedom of Transit', Barcelona Conference (Geneva, 1921)

2.4.3 *Agreement on Trade Facilitation*

United Nations Conference on Trade and Development, 'Trust Fund for Trade Facilitation Negotiations', 'Freedom of Transit', Technical Note 8 (February 2009)

World Trade Organization, 'Communication from Armenia, Canada, the European Communities, the Kyrgyz Republic, Mongolia, New Zealand, Paraguay, and the Republic of Moldova', TN/TF/W/79 (15 February 2006)

World Trade Organization, 'Communication from Egypt and Turkey – Discussion Paper on the Inclusion of the Goods Moved via Fixed Infrastructure into the Definition of Traffic in Transit', Negotiating Group on Trade Facilitation, TN/TF/W/179 (4 June 2012)

World Trade Organization, 'Communication from Paraguay, Rwanda and Switzerland', TN/TF/W/39 (2 May 2005)

World Trade Organization, Council for Trade in Goods, 'WTO Trade Facilitation – Strengthening WTO Rules on GATT Article V on Freedom of Transit', Communication from the European Communities, G/C/W/422 (30 September 2002)

World Trade Organization, Ministerial Conference, 'Statement by Bolivia, Cuba, Nicaragua and Venezuela at the Ninth WTO Ministerial Conference', WT/MIN (13)/29 (7 December 2013)

World Trade Organization, Negotiating Group on Trade Facilitation, 'Draft Consolidated Negotiating Text', TN/TF/W/165 (14 December 2009)

World Trade Organization, Negotiating Group on Trade Facilitation, 'Draft Consolidated Negotiating Text', TN/TF/W/165/Rev.8 (21 April 2011)

World Trade Organization, Negotiating Group on Trade Facilitation, 'Draft Consolidated Negotiating Text', TN/TF/W/165/Rev.11 (7 October 2011)

World Trade Organization, Negotiating Group on Trade Facilitation, 'Article V of GATT 1994 – Scope and Application', TN/TF/W/2 (12 January 2005)

2.4.4 *Negotiations on Energy Services*

World Trade Organization, Council for Trade in Services Special Session, Communication from Norway, 'The Negotiations on Trade in Services', S/CSS/W/59, 21 March 2001

World Trade Organization, Council for Trade in Services Special Session, 'Energy Services', Communication from the United States, S/CSS/W/24 (18 December 2000)

World Trade Organization, Council for Trade in Services, 'Energy Services', Background Note by the Secretariat, S/C/W/52 (9 September 1998)

2.5 National Legal and Political Sources

Cuban Democracy Act of 1992 at http://www.treasury.gov/resource-center/sanctions/Documents/cda.pdf, accessed 20 December 2013

European Commission, 'Second Strategic Energy Review: An EU Energy Security and Solidarity Action Plan', COM 781 final (13 November 2008)

Directive 2009/73/EC of the European Parliament and of the Council of 13 July 2009 Concerning Common rules for the Internal Market in Natural Gas and Repealing Directive No. 2003/55/EC, OJ L 211, 14.8.2009, 94

Regulation No. 715/2009 of the European Parliament and the Council of 13 July 2009 on Conditions for Access to the Natural Gas Transmission Networks and Repealing Regulation, OJ 2009 L 211, 36

3. News, Newsletters and Miscellaneous Reference Materials and Electronic Sources

3.1 News and Newsletters

Agren, David, 'Mexico Opens up Petroleum Business to Private and Foreign Companies', Special to *USA Today*, at: http://www.usatoday.com/story/news/world/2013/12/20/mexico-petroleum-oil-foreign/4152521/ (20 December 2013), accessed 21 December 2013

'Armenia Intends to Join the Customs Union' ('Армения намерена вступить в Таможенный союз'), 3 September 2013, http://ria.ru/economy/20130903/960485143.html, accessed 10 September 2013

Clark, Pilita and Oliver, Christian, 'EU Energy Costs More than Twice Those of US', *Financial Times* (21 January 2014)

International Centre for Trade and Sustainable Development, 'Historic Bali Deal to Spring WTO, Global Economy Ahead' (Bridges, 7 December 2013)

International Centre for Trade and Sustainable Development, 'The 1996 Singapore Ministerial Declaration', Vol. 1 No. 6 (February 2003)

International Centre for Trade and Sustainable Development, 'Обзор Переговоров: Краткий Обзор к Переговорам на 9-й Министерской Конференции ВТО', Мосты/Bridges (специальный выпуск – декабрь 2013)

Jacobs, Justin, 'Bolivia's Fight for the Sea', *Petroleum Economist* (3 May 2013), http://www.petroleum-economist.com/Article/3201020/Bolivias-fight-for-the-sea.html, accessed 30 August 2013

Larry, Elliott, 'Bali summit invigorated World Trade Organisation, says Roberto Azevêdo', *The Guardian* (Wednesday, 18 December 2013), http://www.theguardian.com/business/2013/dec/18/roberto-azevedo-wto-bali-global-trading, accessed 15 January 2014

McElroy, Damien, 'Ukraine Receives Half Price Gas and $15 Billion to Stick with Russia', *The Telegraph* (17 December 2013), http://www.telegraph.co.uk/news

/worldnews/europe/ukraine/10523225/Ukraine-receives-half-price-gas-and-15-billion-to-stick-with-Russia.html, accessed 17 December 2013

WTO Press Release, Transcript of the video address of the WTO Director-General Pascal Lamy to the 2010 World Energy Congress, www.wto.org/english/news_e/sppl_e/sppl169_e.htm, accessed 30 March 2012

Yergin, Daniel, 'The Global Impact of US Shale' (8 January 2014), http://www.project-syndicate.org/commentary/daniel-yergin-traces-the-effects-of-america-s-shale-energy-revolution-on-the-balance-of-global-economic-and-political-power, accessed 21 January 2014

3.2 *Miscellaneous Reference Materials and Electronic Sources*

'Alaska Pipeline Project', http://thealaskapipelineproject.com/, accessed 5 September 2013

Association of International Petroleum Negotiators, https://www.aipn.org, accessed 28 July 2013

British Petroleum, 'Energy Outlook 2035' (January 2014), http://www.bp.com/content/dam/bp/pdf/Energy-economics/Energy-Outlook/Energy_Outlook_2035_booklet.pdf, accessed 2 April 2014

British Petroleum, 'Statistical Review of World Energy' (June 2009), www.bp.com/statisticalreview, accessed 30 March 2010

British Petroleum, 'Statistical Review of World Energy' (June 2011), www.bp.com/statisticalreview, accessed 10 March 2011

British Petroleum, 'Statistical Review of World Energy' (June 2013), www.bp.com/statisticalreview, accessed 21 August 2013

Energy Charter Secretariat, '1998 Trade Amendment', http://www.encharter.org/index.php?id=608&L=0 (last visited 1 December 2013)

Energy Charter Secretariat, http://www.encharter.org/, accessed 30 November 2011

Gazprom, http://www.gazprom.com/about/, accessed 10 January 2013

The New Encyclopaedia Britannica, vol. 10, 15th ed. (Chicago: Encyclopaedia Britannica, Inc., 1993)

Nord Stream, http://www.nord-stream.com/, accessed 30 September 2011

Organisation for Economic Co-operation and Development (OECD), 'Glossary of Statistical Terms', https://stats.oecd.org/glossary, accessed 15 September 2015

Oxford Dictionaries, http://oxforddictionaries.com, accessed 30 March 2012

The Soviet Encyclopaedia, 3rd ed. (Moscow, 1985)

Summary of Complaints and Resolution, http://www.worldtradelaw.net/databases/summary.php, accessed 22 August 2014

United Nations Office of the High Representative for the Least Developed Countries, 'Land-Locked Developing Countries and Small Island Developing States', http://unohrlls.org/, accessed 5 September 2014

United Nations Statistics Division, Provisional Central Product Classification, http://unstats.un.org/unsd/cr/registry/regcst.asp?cl=9&lg=1, accessed 30 September 2012

World Trade Organization, 'Energy Services', http://www.wto.org/english/tratop_e/serv_e/energy_e/energy_e.htm, accessed 30 November 2012

World Trade Organization, 'Members and Observers', www.wto.org/english/thewto_e/whatis_e/tif_e/org6_e.htm, accessed 30 November 2015

World Trade Organization, 'Nicaragua – Measures Affecting Imports from Honduras and Colombia', http://www.wto.org/english/tratop_e/dispu_e/cases_e/ds188_e.htm, accessed 1 February 2014

World Trade Organization, 'Trade Facilitation', http://www.wto.org/english/tratop_e/tradfa_e/tradfa_e.htm, accessed 30 November 2013

World Trade Organization, 'United States – The Cuban Liberty and Democratic Solidarity Act', http://www.wto.org/english/tratop_e/dispu_e/cases_e/ds38_e.htm, accessed 1 February 2014

World Trade Organization, Secretariat, 'Member/Sector Matrix Report: Transport Services', http://tsdb.wto.org, accessed, 10 January 2012

World Trade Organization, Secretariat, 'What Is the World Trade Organization?', http://www.wto.org/english/thewto_e/whatis_e/tif_e/fact1_e.htm, accessed 30 September 2011

INDEX

Lightning Source UK Ltd.
Milton Keynes UK
UKOW05n1337140517
301138UK00012B/258/P